Piezoelectric-Driven Flexible Robotic Microgrippers

Qingsong Xu

Piezoelectric-Driven Flexible Robotic Microgrippers

Springer

Qingsong Xu
Department of Electromechanical
Engineering
Faculty of Science and Technology
University of Macau
Avenida da Universidade
Macao, China

ISBN 978-981-95-4684-8 ISBN 978-981-95-4685-5 (eBook)
https://doi.org/10.1007/978-981-95-4685-5

This Springer imprint is published by the registered company Springer Nature Singapore Pte Ltd.
The registered company address is: 152 Beach Road, #21-01/04 Gateway East, Singapore 189721, Singapore

To my family

Preface

The relentless march of technology towards miniaturization has ushered in a new era of manufacturing, biomedical engineering, and scientific exploration, often operating at scales invisible to the naked eye. In this burgeoning domain of micro and nano-manipulation, the ability to precisely handle, assemble, and characterize miniature components—from biological cells and micro-optics to MEMS devices and optical fibers—is not just an advantage but a fundamental requirement. At the heart of these sophisticated tasks lies a critical device: the microgripper. This book, *Piezoelectric-Driven Flexible Robotic Microgrippers*, is dedicated to exploring the design, modeling, optimization, and control of these pivotal instruments, which serve as the dexterous fingertips of modern precision robotic systems.

The genesis of this work stems from a clear and pressing need in both academic research and industrial practice. Traditional rigid-link grippers, hampered by backlash, friction, and complex assembly processes, are ill-suited for the demands of high-precision, micro-scale applications. In their place, compliant mechanisms—monolithic structures that gain their motion from elastic deformation—have emerged as a superior alternative. They offer the compelling benefits of frictionless and wear-free motion, absence of backlash, and suitability for monolithic fabrication, which enhances precision and reduces maintenance. When paired with piezoelectric actuators, renowned for their nanometer-resolution positioning, high stiffness, and fast response, the resulting Piezoelectric-Actuated Compliant Microgrippers (PEACMs) represent a paradigm of precision motion generation.

This book is structured to provide a comprehensive journey from fundamental principles to advanced, application-specific designs. It begins with a foundational review of piezoelectric materials and compliant mechanisms, establishing the core physics and engineering concepts that underpin all subsequent designs. We then delve into the critical performance indices—such as displacement amplification, clamping range, force, and bandwidth—that define a microgripper's capability and guide its design.

The core of this volume presents a series of in-depth case studies, each chapter dedicated to a novel microgripper design that addresses a specific set of challenges. Readers will be guided through the complete development lifecycle of these devices: from initial conceptual mechanism design (leveraging principles like the Scott-Russell mechanism, parallelogram flexures, and multi-stage amplifiers) to rigorous analytical modeling for kinematics, statics, and dynamics. We place significant emphasis on modern computer-aided engineering techniques, detailing the use of Finite Element Analysis (FEA) for simulation and validation, and response surface methodology for multi-objective optimization to achieve desired performance characteristics.

Furthermore, we transition from theory to practice, documenting the prototype fabrication, experimental setup, and extensive testing procedures. These sections validate the models and demonstrate real-world performance through open-loop and closed-loop tests, displacement and force characterization, frequency response analyses, and successful gripping experiments on various micro-objects. Later chapters expand the scope to include more complex systems, such as biaxial grippers for planar manipulation, large-workspace XY compliant manipulators, and reconfigurable grippers with multiple working modes, highlighting the versatility of the PEACM technology. The concluding discussion on adaptive neural network-based control strategies underscores the importance of intelligent control in mitigating hysteresis and creep—inherent nonlinearities of piezoelectric materials—to achieve robust and precise motion tracking.

This book is intended for a broad audience, including senior undergraduate and graduate students in mechatronics, robotics, and mechanical engineering; academic researchers specializing in precision engineering and smart materials; and practicing engineers and R&D professionals in industries ranging from semiconductor manufacturing and biomedical device assembly to micro-robotics. It is our sincere hope that the insights, methodologies, and practical results contained within these pages will serve as a valuable resource, inspiring further innovation and contributing to the advancement of precision robotic systems that will shape the technologies of tomorrow.

The author would like to acknowledge the National Natural Science Foundation of China under Grant 52175556, the Macao Science and Technology Development Fund under Grant 0004/2022/AKP, 0102/2022/A2, and 0078/2023/RIB3, the University of Macau under Grant MYRG-GRG2024-00222-FST-UMDF, MYRG-CRG2022-00004-FST-ICI and MYRG-GRG2025-00121-FST and Guangdong Basic and Applied Basic Research Foundation under Grant 2023A1515011178 for co-funding the projects. The author is also grateful for the help provided by the editorial team from Springer Nature.

Macao, China Qingsong Xu
September 2025

Competing Interests The author has no competing interests to declare that are relevant to the content of this manuscript.

Contents

About the Author

Dr. Qingsong Xu is Professor at the Department of Electromechanical Engineering, Faculty of Science and Technology, University of Macau, and Director of the Smart and Micro/Nano Systems Laboratory. His research involves intelligent micro/nanosystems, precision robotics, and biomedical applications. He has published four books and over 450 papers in international journals and conferences, cited over 15000 times in Google Scholar with an H-index of 70. He was Associate Editor of IEEE T-RO, IEEE T-ASE, and IEEE RA-L and Technical Editor of IEEE/ASME T-MECH. Professor Xu has received more than a dozen best paper awards from international conferences and multiple times of Macao Science and Technology Awards from Macao SAR, China. He has been selected into the top 2% of the world's top scientists released by Stanford University since 2019. He is Fellow of IEEE, ASME, and AAIA.

Chapter 1
Review of Piezoelectric Compliant Precision Robotic Devices

Abstract Precision-manipulation robots play a vital role in small-scale object handling. Compliant manipulators utilizing piezoelectric actuation have found broad application in this field, offering exceptional resolution, precision, and repeatability. This chapter presents a review of state-of-the-art research on these piezoelectric-actuated compliant devices, particularly microgrippers, and concludes with a discussion on promising future research directions.

Keywords Microgripper · Piezoelectric transducer · Piezoelectric actuators · Compliant mechanism · Sensors and actuators

1.1 Introduction

Advances in small-scale research-including microscopy, microassembly, and cell surgery–have established precision manipulation as an essential capability for interacting with micro-scale objects [5, 6, 53]. As this field progresses, the required precision for manipulation continues to increase. This trend has spurred the growth of ultrahigh-precision engineering (UHPE), a field attracting significant attention due to its broad application prospects.

Conventional manipulators, which are based on rigid-body mechanisms employing rods and gears, exhibit inherent drawbacks that hinder high-precision performance. These include backlash, caused by gaps between connected parts, and friction, which induces wear and tear. These limitations significantly compromise motion accuracy and energy efficiency. Their detrimental effects are amplified at smaller scales, ultimately constraining the further development of rigid-body tools for UHPE applications. Consequently, there is a critical need to develop novel operational tools based on alternative principles that are suited to environments with ultrahigh-precision requirements.

Smart materials, which respond to external stimuli such as heat, electricity, and force, possess the ability to directly convert energy between other forms and mechanical energy [77]. This property makes them ideal for overcoming the drawbacks of rigid-body mechanisms. Over the past century, a variety of smart materials have

Q. Xu, *Piezoelectric-Driven Flexible Robotic Microgrippers*,
https://doi.org/10.1007/978-981-95-4685-5_1

been developed, including piezoelectric, thermoelectric, and magnetostrictive materials. These materials facilitate the creation of more efficient, reliable, and compact devices.

Among them, piezoelectric materials have been widely adopted due to their rapid response, high force output, substantial stiffness, and exceptional precision. They exhibit the piezoelectric effect: generating an internal electrical charge in response to an applied mechanical stress (the direct effect) and deforming mechanically in response to an applied electric field (the reverse effect) [57, 58]. These characteristics enable the development of high-performance piezoelectric transducers for various applications, such as actuators and sensors.

Conversely, compliant mechanisms transmit force and motion through the elastic deformation of their constituent material, thereby overcoming the inaccuracies induced by backlash and friction in rigid-body mechanisms [23, 26, 40, 74]. These mechanisms can be fabricated as monolithic structures, eliminating assembly errors entirely. Furthermore, their maintenance-free nature ensures higher reliability in practical applications. These advantages have led to the widespread adoption of compliant mechanisms in the field of micromanipulation.

When integrated with piezoelectric transducers, compliant mechanisms are frequently employed to provide mechanical motion or force amplification, addressing the limited displacement output of the piezoelectric material. The piezoelectric compliant robotic device, which embodies an integrated design of a piezoelectric transducer and a compliant mechanism, thus presents an effective and powerful solution for UHPE systems.

This chapter presents a review of recent advances in piezoelectric compliant robotic devices for ultra-high-precision engineering. As summarized in Fig. 1.1, the discussion focuses on state-of-the-art designs that capitalize on the distinct advantages of piezoelectric transducers and compliant mechanisms.

1.2 Piezoelectric Transducer

Over past decades, extensive research has led to the discovery and development of various piezoelectric materials, which are generally classified as crystalline, ceramic, or polymeric [55, 57]. Prominent examples include crystalline materials (e.g., quartz and topaz), ceramics (e.g., Barium Titanate, Lead Zirconate Titanate (PZT), and Macro Fiber Composites (MFC)), and polymers (e.g., Polyvinylidene Difluoride (PVDF)).

The defining characteristic of these materials is piezoelectricity–a phenomenon where an electrical potential is generated in response to an applied mechanical stress. As illustrated in Fig. 1.2a and b, piezoelectric transducers operate on the principle of reversible electromechanical energy conversion, known as the direct (mechanical to electrical) and reverse (electrical to mechanical) piezoelectric effects, respectively.

On the microscopic scale, the piezoelectric effect arises from a change in polarization within anisotropic crystalline materials. For certain ceramics and polymers,

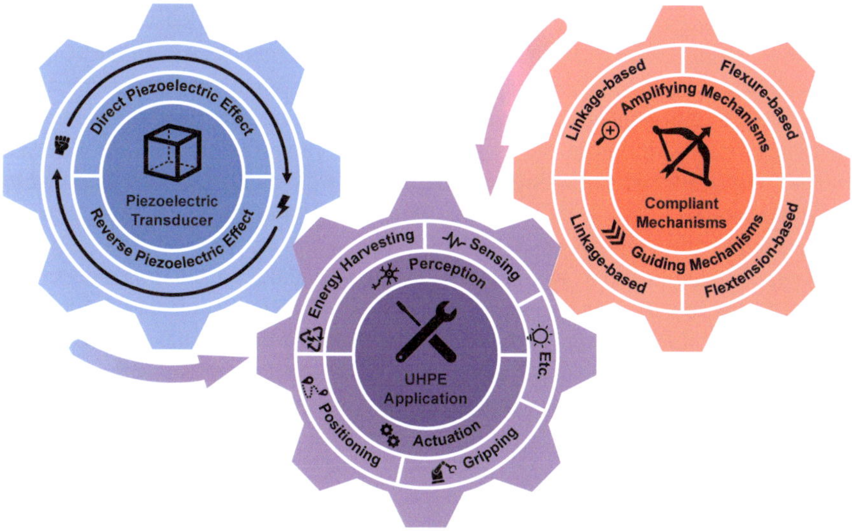

Fig. 1.1 Chapter overview, highlighting the integration of piezoelectric transducers with compliant mechanisms for ultra-high-precision engineering (UHPE) applications

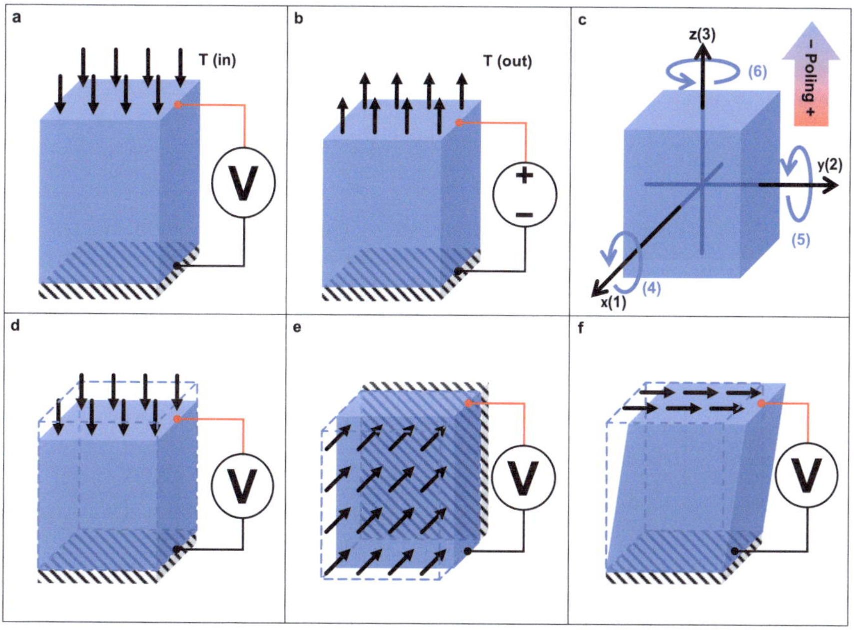

Fig. 1.2 Schematic diagrams of piezoelectric materials and transducers. **a** Direct and **b** reverse piezoelectric effects. **c** Standard coordinate system defined by the poling direction. Common working modes of piezoelectric transducers (PETs): **d** compression, **e** transverse, and **f** shear

this inherent property can be enhanced through a poling process, which involves applying a strong external electric field to align the internal electric dipoles.

As shown in Fig. 1.2c, the poling direction is conventionally defined as the z-axis to establish a right-handed Cartesian coordinate system. This standard framework facilitates the description of the material's anisotropic properties. Within this system, normal stress and strain directions along the x-, y-, and z-axes are denoted by the subscripts 1, 2, and 3, respectively, while shear motions about these axes are represented by the subscripts 4, 5, and 6.

In practical applications, two key constants are crucial for evaluating a material's suitability for actuation or sensing: the piezoelectric charge constant (d_{ij}) and the piezoelectric voltage constant (g_{ij}). Here, the subscript i denotes the direction of the electric field, while j indicates the direction of the applied stress or induced strain. These constants are intrinsically related by the expression:

$$d_{ij} = g_{ij}\varepsilon_{ij}^{T} \tag{1.1}$$

where ε_{jj}^{T} represents the absolute permittivity under a condition of constant stress.

The electromechanical coupling in piezoelectric materials is mathematically described by the following constitutive equations.

$$S_{ij} = \sum_{k,l} s_{ijkl}^{E} T_{kl} + \sum_{k} d_{kij} E_k, \tag{1.2}$$

$$D_i = \sum_{j,k} d_{ijk} T_{jk} + \sum_{j} \varepsilon_{ij}^{T} E_j, \tag{1.3}$$

where S, T, E, D, and s^E represent strain, stress, electric field strength, electric flux density, and elastic compliance under a constant electric field, respectively (with $i, j, k, l = 1, 2, 3$). Using Voigt notation–which condenses the tensor indices via the mapping ($11 \to 1, 22 \to 2, 33 \to 3, 23 \to 4, 13 \to 5, 12 \to 6$)–Eqs. (1.2) and (1.3) can be simplified into the following matrix form:

$$\begin{bmatrix} S_1 \\ S_2 \\ S_3 \\ S_4 \\ S_5 \\ S_6 \end{bmatrix} = \begin{bmatrix} s_{11}^E & s_{12}^E & s_{13}^E & 0 & 0 & 0 \\ s_{21}^E & s_{22}^E & s_{23}^E & 0 & 0 & 0 \\ s_{31}^E & s_{32}^E & s_{33}^E & 0 & 0 & 0 \\ 0 & 0 & 0 & s_{44}^E & 0 & 0 \\ 0 & 0 & 0 & 0 & s_{55}^E & 0 \\ 0 & 0 & 0 & 0 & 0 & s_{66}^E \end{bmatrix} \begin{bmatrix} T_1 \\ T_2 \\ T_3 \\ T_4 \\ T_5 \\ T_6 \end{bmatrix} + \begin{bmatrix} 0 & 0 & d_{31} \\ 0 & 0 & d_{32} \\ 0 & 0 & d_{33} \\ 0 & d_{24} & 0 \\ d_{15} & 0 & 0 \\ 0 & 0 & 0 \end{bmatrix} \begin{bmatrix} E_1 \\ E_2 \\ E_3 \end{bmatrix} \tag{1.4}$$

$$\begin{bmatrix} D_1 \\ D_2 \\ D_3 \end{bmatrix} = \begin{bmatrix} 0 & 0 & 0 & 0 & d_{15} & 0 \\ 0 & 0 & 0 & d_{24} & 0 & 0 \\ d_{31} & d_{32} & d_{33} & 0 & 0 & 0 \end{bmatrix} \begin{bmatrix} T_1 \\ T_2 \\ T_3 \\ T_4 \\ T_5 \\ T_6 \end{bmatrix} + \begin{bmatrix} \varepsilon_{11} & 0 & 0 \\ 0 & \varepsilon_{22} & 0 \\ 0 & 0 & \varepsilon_{33} \end{bmatrix} \begin{bmatrix} E_1 \\ E_2 \\ E_3 \end{bmatrix} \tag{1.5}$$

Figure 1.2d–f illustrate three fundamental operating modes of piezoelectric transducers: the compression mode, transverse mode, and shear mode. These modes are conventionally abbreviated as the d_{33}-mode, d_{31}-mode, and d_{15}-mode, respectively. In this notation, the two subscripts indicate the directions of the generated electrical polarization and the applied mechanical stress.

Depending on the specific application requirements, these operating modes can be implemented using various piezoelectric transducer architectures. For instance, unimorph and bimorph-type piezoelectric transducers exhibit low stiffness and can sustain large deflections, but they produce a relatively small output force. These characteristics make them suitable for force and strain sensing applications, as well as for actuating free loads (e.g., in micro-fans). In contrast, stack-type piezoelectric transducers possess high stiffness and generate large output forces, making them more appropriate for high-force actuation and large-force sensing applications [49].

1.2.1 Direct Piezoelectric Effect

According to the direct piezoelectric effect, piezoelectric sensors efficiently convert dynamic mechanical stress into electrical charge. This principle makes them suitable for applications in sensing and energy harvesting, where physical signals are transformed into electrical outputs. For a given application, the direction of the applied stress significantly influences the conversion performance.

The two most common operational modes for piezoelectric sensors are the d_{33}-mode and the d_{31}-mode. Typically, the d_{33}-mode excels in generating high voltage output, whereas the d_{31}-mode is superior for producing high current output. Consequently, the open-circuit voltage V_{oc} of a piezoelectric sensor can be expressed as follows:

$$V_{oc} = \frac{d_{ijk}}{\varepsilon_{ij}^{T}} T_{jk} g_e \tag{1.6}$$

where g_e denotes the distance between the electrodes. To achieve optimal output performance, the piezoelectric sensor should be positioned at the location of maximum strain, such as the terminal of a leaf flexure.

Furthermore, when designing a system that utilizes the direct piezoelectric effect, the stiffness and fracture strain threshold of the selected piezoelectric material are critical parameters. These properties primarily govern the device's sensitivity, working range, and maximum payload.

Therefore, both stiffness and fracture strain threshold must be carefully evaluated to select an appropriate piezoelectric sensor for specific sensing and energy harvesting applications.

1.2.2 Reverse Piezoelectric Effect

The reverse piezoelectric effect is typically employed for actuation, generating a mechanical output such as force or displacement [69]. For a given piezoelectric actuator, its elastic compliance s^E and piezoelectric charge constant d are fixed material properties determined by the composition, structure, and poling process. When a specific operating voltage is applied, the resulting electric field strength E can be precisely determined. Consequently, Eq. (1.2) reduces to a linear relationship between strain S and stress T.

From an actuation perspective, Fig. 1.3b illustrates the relationship between the output force and displacement under the maximum operating voltage. This curve reveals two key performance metrics of a piezoelectric actuator: the free displacement (ΔL_{nom}), achieved under zero external load, and the blocking force (F_{B}), generated under a fully constrained (blocked) condition. The ratio $F_{\mathrm{B}}/\Delta L_{\mathrm{nom}}$ defines the output stiffness (k_{A}) of the actuator. The linear force-displacement relationship can therefore be expressed as follows:

$$F = F_{\mathrm{B}} - k_{\mathrm{A}}\Delta L, \qquad \Delta L \in [0, \Delta L_{\mathrm{nom}}] \tag{1.7}$$

Referring to Fig. 1.3a, the actuation of a compliant mechanism with an input stiffness k_{CM} involves a force balance between the output force of the actuator and the reaction force from the mechanism. This equilibrium occurs at a specific operating point, defined by the effective force output F_{eff} and the effective displacement output ΔL_{eff}, as illustrated in Fig. 1.3b.

Using the force equilibrium condition, the expressions for F_{eff} and ΔL_{eff} can be derived as follows:

$$F_{\mathrm{eff}} = \frac{k_{\mathrm{CM}}}{k_{\mathrm{A}} + k_{\mathrm{CM}}} F_{\mathrm{B}} \tag{1.8}$$

$$\Delta L_{\mathrm{eff}} = \frac{k_{\mathrm{A}}}{k_{\mathrm{A}} + k_{\mathrm{CM}}} \Delta L_{\mathrm{nom}} \tag{1.9}$$

These equations provide a fundamental basis for designing piezoelectric-driven compliant mechanisms. For instance, they can be used to establish design objectives and constraints for the compliant mechanism once a piezoelectric actuator has been selected. Conversely, they can also guide the selection of a suitable piezoelectric actuator to drive a specific compliant mechanism in order to achieve a desired output displacement.

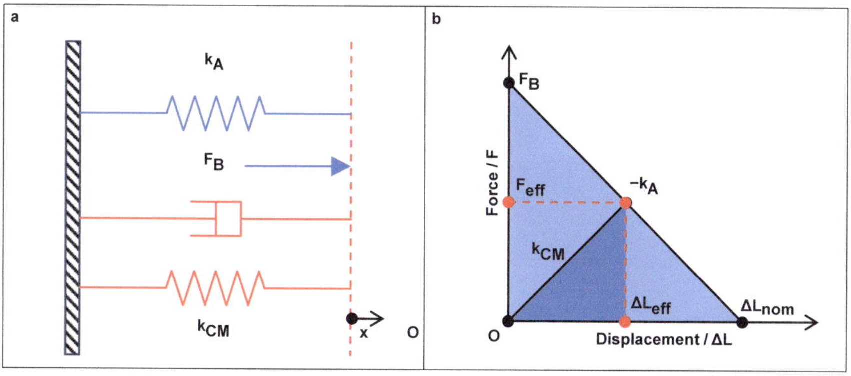

Fig. 1.3 Schematic of a piezo-driven compliant mechanism (**a**) and its corresponding actuator force-displacement relationship under maximum operating voltage (**b**)

1.3 Compliant Mechanisms

Compliant mechanisms utilize material flexibility to transmit force and motion through elastic deformation within the material's elastic region. This deformation process involves the storage and subsequent release of energy within the structure.

In contrast to rigid-body mechanisms, compliant systems offer significant advantages, including:

- Zero backlash, enabling precise motion control.
- Reduced friction, minimizing energy loss and wear.
- Maintenance-free operation, eliminating the need for lubrication.

Furthermore, compliant mechanisms can be designed and manufactured as monolithic structures, which eliminates assembly errors, simplifies production, and enhances reliability. These attributes also facilitate more complex, compact designs and improve overall cost-effectiveness. As a result, compliant mechanisms have been widely adopted in various fields, including precision manipulation, sensing in microelectromechanical systems (MEMS), and biomedical applications.

The structure of a compliant mechanism relies on compliant components as its essential functional elements. These components provide the required, on-demand deformation, meaning their characteristics directly determine the final performance of the mechanism.

Figure 1.4 illustrates several typical compliant elements. Their behavior varies significantly based on geometry and dimensions. For instance:

- Leaf flexures exhibit low stiffness along their working direction and can withstand large deflections due to their property of distributed compliance.
- Compliant hinges with different notch geometries–such as rectangular, semi-circular, and elliptical types–demonstrate distinct characteristics in terms of axial shift and stress concentration [34, 47].

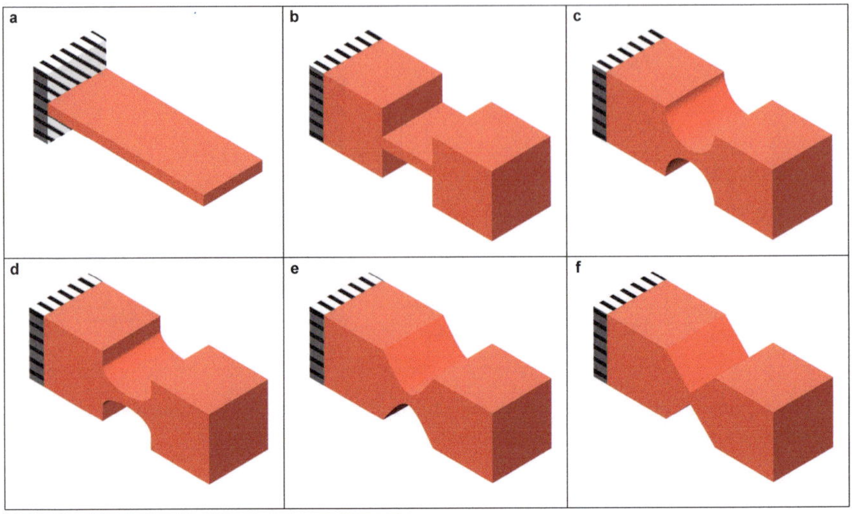

Fig. 1.4 Typical compliant elements: **a** a leaf flexure; and compliant hinges with various notch geometries, including **b** rectangular, **c** circular, **d** elliptical, **e** parabolic, and **f** triangular types

Since the working principle of compliant mechanisms is governed by elastic deformation, the relevant material properties directly determine their performance. Generally, material selection prioritizes a combination of:

- A low Young's modulus E, to minimize stress generation during deformation, and
- A high yield strength σ_y, to allow the material to withstand greater stress before the onset of plastic deformation.

This combination enables a larger range of elastic deformation or a higher payload capacity (see Fig. 1.5b). Consequently, metallic materials such as aluminum alloys, stainless steel, and titanium alloys are common choices. Among these, aluminum alloys are particularly widespread due to their favorable ratio of stiffness to density and their cost-effectiveness.

Due to their shared characteristics of high precision and high energy efficiency, compliant mechanisms are frequently integrated with piezoelectric materials to enhance specific performance metrics. Common enhancements include the amplification of force or displacement output and the guidance of output direction to minimize lateral disturbances. This section discusses the various implementation methods proposed in recent research for these two categories of performance enhancement.

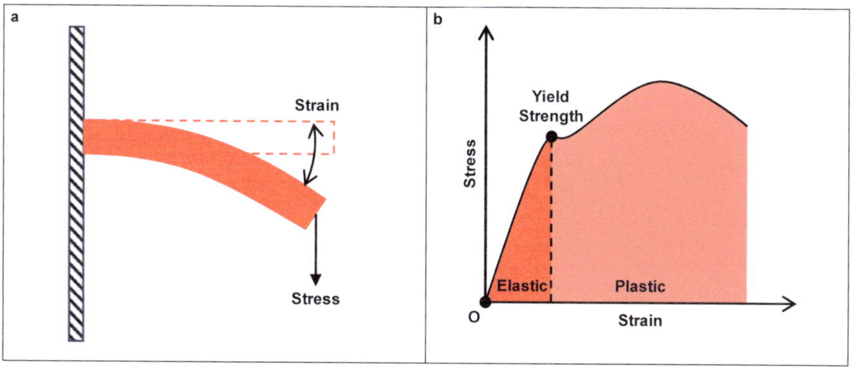

Fig. 1.5 **a** Stress and strain distribution in a bending flexure. **b** A typical material stress-strain curve, illustrating the elastic and plastic deformation regions

1.3.1 Amplifying Mechanisms

In practical applications, piezoelectric transducers often face inherent performance limitations. For instance, when employed as an actuator, the stroke of a stack-type piezoelectric transducer is typically only 0.1–0.2% of its length, making it difficult to meet specific stroke requirements [8, 24, 25, 28]. In sensing applications, the high stiffness of stack-type piezoelectric sensors results in minimal deformation, consequently weakening the output signal strength.

Rather than replacing these transducers with higher-performance alternatives, mechanical amplification presents a more cost-effective solution to mitigate these limitations [24, 49]. This approach can amplify the displacement output of piezoelectric actuators and enhance the force input sensitivity for sensors. This paper classifies the implementation methods for such amplifying mechanisms into two categories: linkage-based and flextension-based designs.

1.3.1.1 Linkage-Based Amplifier

The linkage-based amplifier is a type of mechanical amplifier that utilizes linkage mechanisms to achieve a specific mechanical advantage. Derived from its rigid-body counterpart, the compliant linkage amplifier employs compliant hinges to replace revolute joints and linear actuators (or guiders) to replace prismatic joints–a design strategy known as the rigid-body replacement method. Depending on the number of links and their arrangement, these amplifiers can be configured into various types, including lever-type, bridge-type, pantograph-type, and Scott-Russell-type mechanisms.

The compliant lever amplifier consists of a compliant hinge functioning as a pivot and a rigid link serving as the arm. It exhibits an amplification characteristic

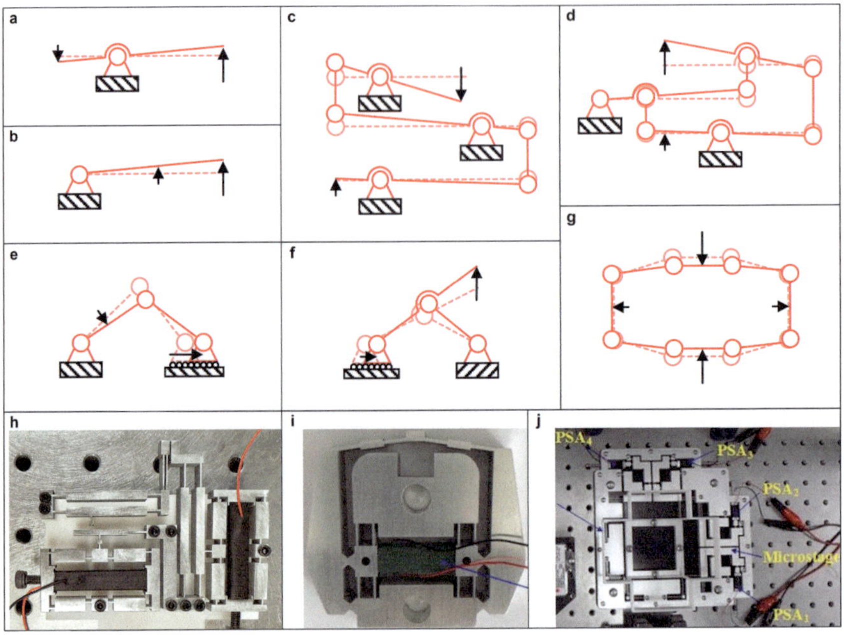

Fig. 1.6 Typical types of linkage-based amplifiers. Lever-type with configuration of **a** class 1 and **b** class 3, **c** multi-stage, and **d** differential; **e** triangular-type, **f** Scott-Russell-type, and **g** bridge-type, where the triangle, circle, line, and arrow represent the fixed support, compliant joint, rigid link, and motion, respectively. The same applies below unless specified otherwise. Examples of design: **h** Bridge-type and lever-type. **i** Lever-type and half bridge-type [35]. **j** Four-bar linkage and lever-type. Reproduced with permission from [76]

analogous to that of a rigid-body lever. A series of lever-type amplifiers are shown in Fig. 1.6a–d. The amplification ratio is determined by the lengths of the load and effort arms. This type of amplifier can also be designed to redirect the input direction through specific arm geometries and lever configurations.

Due to its straightforward operating principle, the lever-type amplifier is widely used to achieve high amplification ratios. However, its design often results in a bulky footprint and reduced efficiency, primarily due to the use of multiple compliant hinges. To improve compactness and efficiency, more complex linkages comprising three or more links can be employed. For instance, the triangular-type (Fig. 1.6e) and Scott-Russell-type (Fig. 1.6f) amplifiers are both derived from the crank-slider four-bar linkage mechanism.

In the literature, complementary designs often combine various linkage-based amplifiers to achieve a optimal compromise among competing performance requirements. For instance:

- Chen et al. [10] employed two compound bridge-type amplifiers aligned in the same output direction. These work in tandem to drive a parallelogram mechanism with symmetrical forces, thereby mitigating its inherent issue of parasitic motion.

- Lyu et al. [46] utilized a two-stage lever-type amplifier to increase the output stroke. They integrated the final stage lever into a parallelogram mechanism to effectively suppress parasitic displacement (see Fig. 1.6h).
- Ling et al. [35] presented a two-stage amplifier where the output from two symmetrical lever-type amplifiers serves as the input to a half bridge-type amplifier. This configuration was designed to enhance the dynamic bandwidth of the piezoelectric actuator (see Fig. 1.6i).
- Yang et al. [76] introduced a double-rocker mechanism–a four-bar linkage system with integrated lever-type amplifiers at both the input and output ends (see Fig. 1.6j).

1.3.1.2 Flextension-Based Amplifier

The flextension-based amplifier is another type of mechanical amplifier that utilizes the controlled flexibility of a monolithic structure [19]. As shown in Fig. 1.7a, when a horizontal input displacement is applied at the free end of an inclined flexure, it produces a vertical output displacement at the opposite end. The magnitude of this output depends on both the input displacement and the specific geometry of the flexure. This fundamental principle can be harnessed for the amplification of both force and displacement.

In comparison to linkage-based amplifiers, the distinguishing feature of flextension-based amplifiers is their use of compliant, hinge-free links that achieve motion through distributed compliance [8, 19]. A prominent example is the rhombus-type amplifier, which shares a topological similarity with bridge-type or triangular-type amplifiers but operates as a fully monolithic structure without discrete joints.

Owing to their compliant-link architecture, flextension-based amplifiers are characterized by distributed compliance, which significantly mitigates stress concentration during operation. Based on the geometry of these compliant links, flextension-based amplifiers can be categorized into various types, including rhombus-type (Fig. 1.7b) and ellipse-type (Fig. 1.7c) configurations.

For instance:

- Liu et al. [36] implemented a reverse ellipse-type amplifier in a piezo-driven middle ear implant to amplify the actuator's displacement. This design significantly reduced the power consumption required to achieve the same output, thereby extending the device's battery life (Fig. 1.7d).
- Li et al. [30] developed a fully compliant stroke amplifier that achieved a high amplification ratio exceeding 13. Their design incorporated a half rhombus-type amplifier as the final stage (Fig. 1.7e).
- Wu and Xu [71] introduced a spatial reverse rhombus-type amplifier for vertical positioning. This sandwich-like structure separates the amplifiers and piezoelectric actuator across three different planes, reducing the overall stage height and resulting in a compact form factor (Fig. 1.7f).

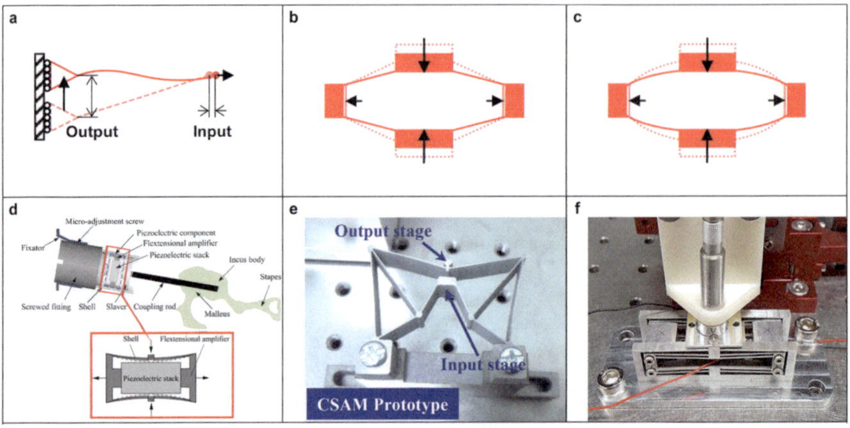

Fig. 1.7 **a** Operating principle of flextension-based amplification. **b–c** Typical flextension-based amplifier configurations: **b** rhombus-type and **c** ellipse-type. **d–f** Application examples: **d** Piezo-driven middle ear implant [36], **e** fully compliant stroke amplifier [30], and **f** sandwich-style vertical positioning stage

1.3.2 Guiding Mechanisms

The compliant hinge, functioning as a revolute joint with a limited range of motion, is a fundamental component of compliant mechanisms. However, the rotation axis of a compliant hinge exhibits a certain degree of shift during deflection, the extent of which depends on the hinge's specific geometry [18, 59]. This axis shift can introduce deviations in the intended motion path. Furthermore, practical manufacturing and assembly tolerances can exacerbate these deviations. Additionally, interactions with external objects or environments may impose lateral forces on the mechanism, further misdirecting its motion.

To maintain precise motion direction and isolate the system from undesired disturbances, it is essential to integrate a guider into the compliant mechanism's design. In multi-degree-of-freedom (multi-DOF) systems, compliant guiders can also function as decouplers to isolate the primary motion along one axis from parasitic motions in others. This chapter classifies compliant guiders into two main categories: linkage-based and flexure-based guiders.

1.3.2.1 Linkage-Based Guider

Similar to the previously discussed linkage-based amplifiers, linkage-based guiders can be designed by adapting perfect or approximate straight-line linkages, such as the Scott-Russell linkage.

Consider a compliant mechanism where a rigid-body beam is connected to a fixed base via a compliant hinge, as illustrated in Fig. 1.8a. Due to the relatively

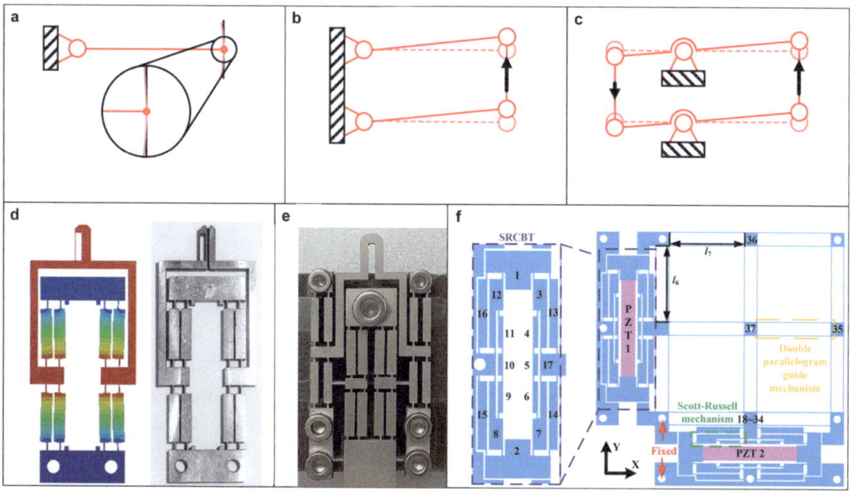

Fig. 1.8 Linkage-based guiding mechanisms. **a** Approximate linear motion generated by a hinged-free beam. **b** Parallelogram-type guider. **c** Compound lever-type guider. Design examples: **d** Compound bridge-type mechanism. Reproduced with permission from [16]. **e** Hybrid design combining compound bridge-type, compound lever-type, and parallelogram-type guiders. Reproduced with permission from [11]. **f** Bridge-type mechanism integrated with a Scott-Russell linkage. Reproduced with permission from [66]

small allowable angular deflection of compliant hinges, the free end of the rigid link traces a short arc trajectory, which can be approximated as a linear motion over a limited range. Consequently, linkages that generate near-linear or specific motion paths from circular inputs are also highly relevant for compliant mechanism design.

For example, Fig. 1.8b shows a parallelogram-type linkage-based guider. This configuration offers superior directionality due to its inherent constraint against yaw motion, effectively maintaining a precise translation path.

Certain linkage-based amplifiers, by virtue of providing a directed output, can also function as a specialized form of linkage-based guider. In practical applications, the directionality of these guiders is often enhanced through compound configurations, such as the compound lever-type linkage-based guider illustrated in Fig. 1.8c. These integrated designs simultaneously fulfill both amplification and guidance functions within a compact structure, making them widely adopted in the literature.

For instance:

- Das et al. [15, 16] utilized a compound bridge-type linkage-based guider to achieve precise gripping with minimal parasitic motion (Fig. 1.8d).
- Zhao et al. [83] and Lyu et al. [42, 44, 46] both employed a compound lever-type linkage-based guider with a single input to actuate gripping jaws while suppressing parasitic angular motion.

- Chen et al. [11] proposed a piezoelectric microgripper incorporating both parallelogram-type and compound bridge-type linkage-based guiders, which further reduces parasitic linear motion (Fig. 1.8e).
- Wu et al. [66] employed a Scott-Russell mechanism to replace one arm of a compound bridge-type linkage-based guider, enhancing its lateral stiffness and natural frequency in the working direction (Fig. 1.8f).

1.3.2.2 Flexure-Based Guider

A single-degree-of-freedom (single-DOF) compliant structure, such as the fixed-guided leaf flexure illustrated in Fig. 1.9a, exhibits direction-dependent stiffness when a load is applied at its guided end. This anisotropic stiffness property can be effectively utilized for motion guiding.

In practice, a pair of mirror-symmetric leaf flexures constitutes the simplest form of a flexure-based guider. As a fundamental building block, leaf flexures can be arranged and superimposed in multiple configurations to achieve specific performance characteristics. For example, Fig. 1.9b shows a two-stage compound parallelogram flexure guider. This configuration enhances the restriction of yaw and pitch motions while maintaining the desired input stiffness.

Compared to linkage-based guiders, flexure-based guiders (FBGs) are characterized by distributed compliance, which enables a relatively larger range of motion. This property grants FBGs greater agility across various applications.

In the literature:

- Xun et al. [75] designed a spatial screw compliant mechanism utilizing multiple helically arranged flexures to convert linear motion into rotary motion.
- Shao et al. [54] connected two leaf flexures in series–oriented orthogonally in different planes–to create a spatial 2-DOF decoupler. This design was employed to guide and decouple tilt motions in two directions for a piezo-driven orientation stage (see Fig. 1.9e).

Building upon linkage-based guiders, constant-force mechanisms (CFMs) have been developed to exhibit a nearly constant reaction force and quasi-zero stiffness over a specific range of motion [33]. To achieve high guiding performance–characterized by near-zero input stiffness while maintaining high lateral stiffness–CFMs can be realized through two primary design approaches: a direct zero-stiffness configuration, or by superimposing positive-stiffness and negative-stiffness mechanisms. The required negative stiffness is typically generated through structural buckling or specialized flexure designs.

Figure 1.9c illustrates the schematic of an inclined flexure that provides negative stiffness within a specific motion range. The solid curves in Fig. 1.9d show the respective force-displacement relationships of the negative-stiffness and positive-stiffness mechanisms, along with their superimposed constant-force result.

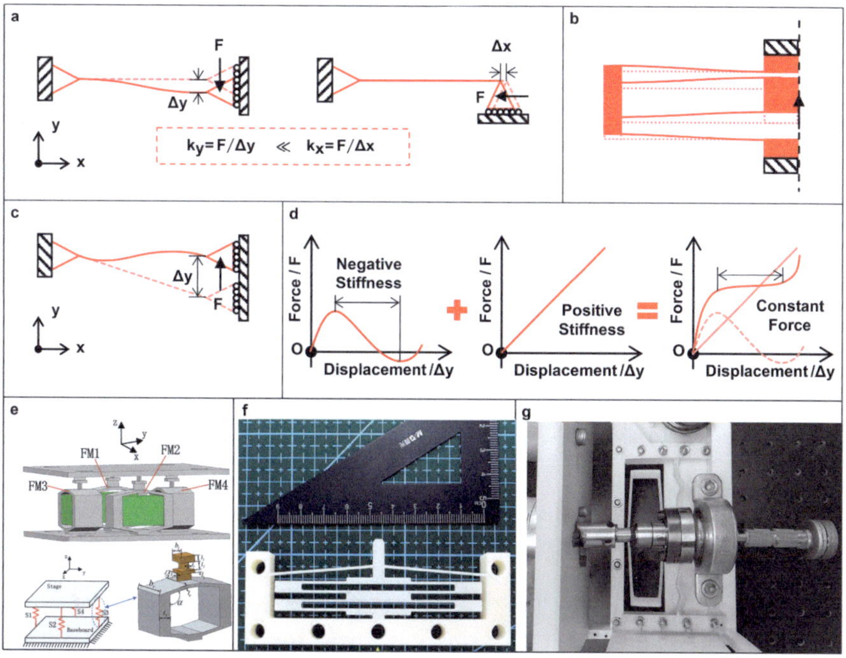

Fig. 1.9 **a** A fixed-guided leaf flexure under vertical and horizontal loading. **b** A two-stage compound parallelogram flexure guider. **c** Schematic of an inclined flexure exhibiting negative stiffness. **d** Force-displacement relationships for negative-stiffness and positive-stiffness mechanisms, and their superposition yielding a constant-force region. Design examples: **e** A spatial 2-DOF decoupler for tilt motion guidance and decoupling. Reproduced with permission from [54]. **f** Constant-force mechanism (CFM) for salmon egg manipulation. Reproduced with permission from [45]. **g** A CFM-based tool head for robotic polishing

Within the constant-force region, these mechanisms offer several advantages beyond guidance: they reduce the actuation force required to achieve a specific strain and limit the maximum force applied to a target object. Leveraging these characteristics, researchers have developed various CFM designs for specialized applications:

- Ding et al. [17] presented a quasi-zero stiffness vibration isolator comprising three sets of constant-force mechanisms for low-frequency vibration isolation.
- Lyu and Xu [45] designed a CFM and demonstrated its feasibility for the delicate manipulation of salmon eggs (Fig. 1.9f).
- Wu et al. [68] and Wei and Xu [63] incorporated constant-force mechanisms into tool head designs to implement constant-force robotic polishing (Fig. 1.9g).

1.4 Integrated Applications

Leveraging the reversible piezoelectric effect of piezoelectric transducers and the continuous force and motion transmission capabilities of compliant mechanisms, their integration into piezoelectric compliant robotic devices has been widely adopted in UHPE systems. These integrated systems enable a diverse range of applications, including high-resolution sensing, high-efficiency energy harvesting, and high-precision actuation.

This section introduces recent advances in these applications across various fields, highlighting key design distinctions that arise from differing performance objectives.

1.4.1 Perception and Sensing

Leveraging the direct piezoelectric effect, piezoelectric sensors can detect dynamic deformation in themselves or in attached structures, enabling tactile perception capabilities. Complementing visual perception, tactile sensing has garnered significant research interest [2, 13, 41, 50, 80], particularly for dexterous manipulation tasks where accurate and timely haptic feedback is crucial for achieving flexible and safe interactions with target objects [4, 9, 27, 51].

For instance, to minimize cell injury during microinjection, Wang and Xu [61] developed a piezo-driven microinjection system that integrated a PVDF film into the injection micropipette for real-time force sensing (see Fig. 1.10a). They employed a store-and-hold charge amplifier to convert the generated piezoelectric charge into a stable voltage signal, enabling clear measurement of the force during the cell piercing process. However, the cantilever-based configuration of the PVDF film was susceptible to external disturbances, compromising measurement accuracy.

To overcome this limitation, Wei and Xu [62] introduced an improved design incorporating a two-stage compound parallelogram flexure guider. The flexures were constructed from PVDF and Macro Fiber Composite (MFC) films (Fig. 1.10b). This setup ensures controlled deformation of the piezoelectric films, allowing them to accurately reflect forces applied along a specific direction.

Due to inherent charge leakage, the direct measurement approach–often termed the non-resonant method–is primarily suitable for dynamic force sensing, as exemplified in the piercing process. For static or quasi-static force measurement, the resonant method provides a more effective solution. This technique operates on the principle that changes in a structure's boundary conditions, such as an applied force, alter its resonant characteristics. These shifts in resonance frequency can be precisely measured and mapped to quantify the applied stimulus.

For instance, Gehin et al. [21] developed a piezoelectric sensor comprising two piezoelectric disks bonded to a steel diaphragm (Fig. 1.10c). One piezoelectric disk excited the diaphragm at its first resonant frequency, while the other detected the resultant vibrations. When an external force was applied to the diaphragm, the struc-

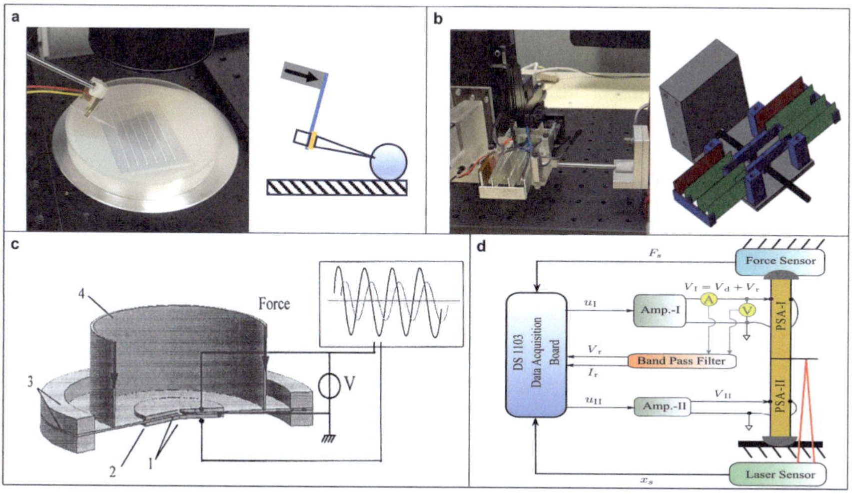

Fig. 1.10 **a** PVDF film integrated onto an injection micropipette for cell piercing force sensing. **b** Two-stage compound parallelogram flexure guider constructed from PVDF and MFC films for directional force measurement. Static force piezoelectric sensors based on: **c** structural resonance method, reproduced with permission from [21] and **d** capacitance measurement principle, reproduced with permission from [48]

ture's resonant frequency shifted, and this change was calibrated to determine the magnitude of the applied force.

Beyond analyzing the electrical signals generated by piezoelectric sensors, static force measurement can also leverage changes in the physical properties of the sensor itself. Several methods have been developed based on monitoring variations in electrical admittance [32], decay time constant [37], and capacitance [1, 48] (see Fig. 1.10d).

Notably, the electrical admittance and decay time constant methods are analogous to the resonant method, as they similarly require time-series data analysis and are suited for comparable application scenarios. Although the capacitance method can operate in real time, its applicability is constrained by the inherent characteristics of different piezoelectric form factors: the small capacitance of film-type sensors and the high stiffness of stack-type sensors. Consequently, this method is predominantly suitable for stack-type sensors where the applied force is either large in magnitude or mechanically amplified [64, 65, 70].

1.4.2 Actuation

Piezoelectric compliant robotic devices for actuation harness the reverse piezoelectric effect to provide motive force, using compliant mechanisms as functional components to achieve active manipulation of target objects. In general, the core requirements for manipulation are focused on two fundamental operations: moving and holding (gripping). These capabilities fulfill essential needs across diverse fields, including microassembly, biomedical applications, and microscopy.

1.4.2.1 Positioning

The positioning operation involves translating end-effectors or target objects to a desired location. The performance of a positioning stage is typically evaluated using two key indicators: workspace and operating speed. For multi-degree-of-freedom (multi-DOF) positioning stages, the number of DOFs directly determines the flexibility and dexterity available for micromanipulation tasks. Each additional DOF significantly expands the range of potential application scenarios. Common multi-DOF stage configurations include XY, XYθ (planar motion with rotation), and XYZ (three-dimensional translational) stages. These multi-DOF systems are implemented through two primary kinematic configurations: serial kinematics and parallel kinematics, each offering distinct advantages in precision, stiffness, and workspace.

The serial-kinematic configuration implements multi-DOF output by connecting the fixed base of a subsequent substage to the output platform of a preceding one. This configuration maintains relatively independent substages, facilitating the integration of distinct functionalities–such as long stroke and high resolution–within a single system. In the literature, serial kinematics are commonly employed in dual-stage systems, where two substages, each with the same number of DOFs, are combined to achieve complementary performance characteristics. For instance:

- Xu [73] designed a dual-stage positioning system with a 10 mm stroke and 500 nm resolution, comprising two serially connected stages driven by a voice coil motor and a piezoelectric actuator, respectively.
- Ekbatani et al. [78] developed a fully piezo-driven dual-stage system. The first substage used a compound bridge-type amplifier to achieve an amplified stroke of 648 μm, while the subsequent high-stiffness, direct-drive substage provided a resolution of 1 μm.
- Zhang et al. [79] designed a dual-range rotational positioning stage using two sets of double slider-crank mechanisms (Fig. 1.11a). These mechanisms utilized compliant hinges for high-resolution displacement output and leaf flexures for large-stroke motion, respectively. Simulation studies identified two distinct rotational ranges: 0.88 and 17.40 mrad.

However, in a serial-kinematic configuration, each subsequent substage must actuate the cumulative inertial mass of all preceding stages, leading to varied dynamic performance across substages and ultimately restricting the maximum number of integrable degrees of freedom. For example, Lin et al. [31] developed a 6-DOF positioning stage by serially stacking three substages: the first provides linear motion along the X- and Y-axes, the second enables rolling, pitching, and linear Z-axis motion, and the third implements yaw motion (see Fig. 1.11b). To mitigate the adverse

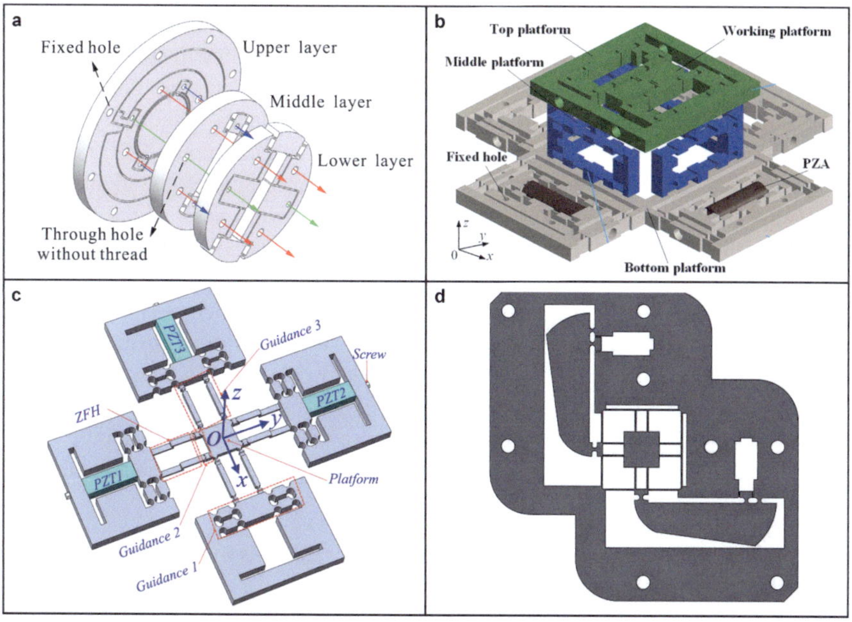

Fig. 1.11 **a** Dual-range rotational positioning stage employing two double slider-crank mechanisms. Reproduced with permission from [79]. **b** 6-DOF serial-kinematic positioning stage comprising three serially stacked substages. Reproduced with permission from [31]. **c** Compact planar XYZ parallel-kinematic positioning stage. Reproduced with permission from [72]. **d** Widely-adopted decoupling scheme for XY positioning stages using four symmetric planar 2-DOF decouplers. Reproduced with permission from [52]

effects of cumulative inertia, design principles emphasizing structural compactness and mass reduction have been proposed. For instance, Wu and Xu [71] designed a compact vertical positioning stage specifically to minimize additional inertial mass when connected in series with an XY stage, thereby preserving dynamic performance in the XY plane.

In contrast to serial-kinematic configurations, the parallel-kinematic configuration connects all substages in parallel, with each substage driving a common output platform directly. This architecture enables consistent dynamic performance through symmetrical design, tailored to the required degrees of freedom. Parallel-kinematic configurations can be further categorized into coupled and decoupled types, distinguished by whether the motion generated by each substage is independent. Specifically, a coupled-type stage functions as a multi-input multi-output (MIMO) system. Although it requires sophisticated kinematic modeling and advanced control algorithms to manage cross-coupling between axes, it offers significant advantages, including high stiffness, large load capacity, and rapid dynamic response. For instance, Xie et al. [72] developed an XYZ parallel positioning stage with a compact, planar profile. This stage achieves 3-DOF motion by employing Z-shaped flexure hinges to connect the actuating components to the output platform (see Fig. 1.11c).

In contrast, the decoupled-type stage employs dedicated decoupling mechanisms or decouplers to isolate the output motion between substages, effectively dividing the complex system into multiple simpler single-input single-output (SISO) systems. By minimizing cross-talk between axes, the decoupler simplifies the identification and compensation of axis-specific errors and nonlinearities, such as the hysteresis inherent in piezoelectric actuators [14].

Consequently, the overall performance of a decoupled-type stage is predominantly determined by the effectiveness of its motion decoupler, to which most performance improvements can be attributed. For instance, Fig. 1.11d illustrates a widely adopted decoupling scheme for an XY positioning stage [3, 52]. This design utilizes four identical planar 2-DOF decouplers arranged symmetrically, with each unit serving a dual function: guiding the primary motion and providing lateral decoupling.

Further advancements in decoupling performance have been demonstrated through various design innovations:

- Zhang et al. [81] enhanced lateral decoupling in each axis by implementing compound parallelogram mechanisms.
- Wang et al. [60] and Wu et al. [67] replaced all leaf flexures with double parallelogram mechanisms, significantly improving the decoupler's guiding precision and load-bearing capacity.

1.4.2.2 Gripping

The gripping operation involves temporarily securing target objects to enable subsequent manipulation tasks. In principle, gripping can be viewed as the relative positioning of the gripper's jaws. Consequently, it shares similar performance metrics with positioning operations, such as jaw stroke and speed, which determine the gripping range and operational efficiency, respectively [43]. For example:

- Chen et al. [7] developed a gripper featuring a high amplification ratio of 52, achieved through a two-stage amplification mechanism, resulting in a maximum jaw displacement of 900 μm.
- Lyu and Xu [42] employed two independent actuation modules to drive each jaw in the same direction (Fig. 1.12a). This design enables two working modes (normally open and normally closed) and achieves a large gripping range from 54.44 to 862.2 μm.

During operation, grippers must provide sufficient force to generate adequate friction between the jaws and the target object for secure manipulation. However, excessive force must be avoided to prevent damage to delicate objects such as biological cells or optical fibers. Therefore, maintaining a precise and appropriate gripping force is critical for successful micromanipulation.

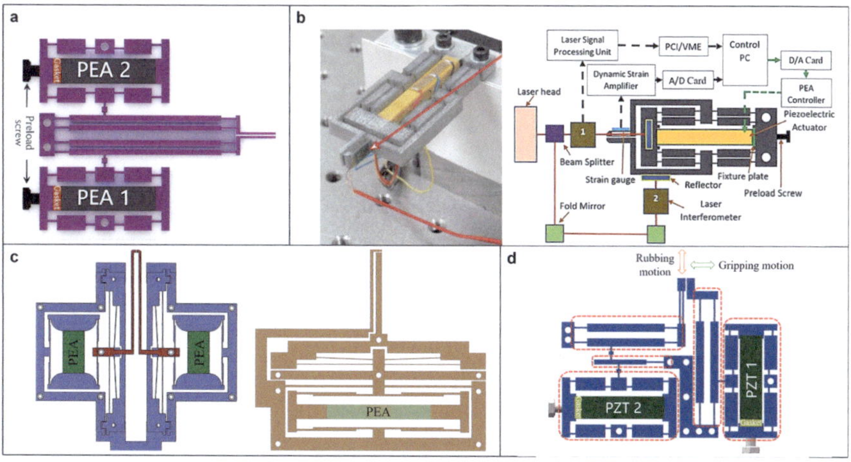

Fig. 1.12 The microgrippers with: **a** Normally open and normally close jaws. **b** Active-type force control. Reproduced with permission from [16]. **c** Passive-type force control. **d** Gripping and rubbing functions

A primary solution involves integrating force sensors into the gripper system and implementing feedback control to achieve the desired force output. This active force-control method, as discussed in the previous section, has been widely adopted. For instance:

- Zhao et al. [83] and Das et al. [16] both integrated strain gauges onto flexible gripping jaws to function as force sensors, enabling precise closed-loop force control (see Fig. 1.12b).

An alternative solution employs a passive approach by integrating a constant-force mechanism (CFM) directly into the gripper's mechanical design. Within its specified constant-force region, the CFM isolates the gripping force from undesired factors such as control overshoot and external disturbances, maintaining a near-constant contact force to protect delicate target objects. Related research primarily focuses on optimizing the size of the constant-force motion region. For instance:

- Liu et al. [38, 39] developed two different grippers providing constant forces of 1.1 N over a 200 μm range and 530 mN over a 220 μm range, respectively (see Fig. 1.12c).

As different tasks require specific contact forces, developing grippers with an adjustable constant-force capability can significantly expand their application scope. For example:

- Gan et al. [20] designed a gripper with an adjustable preloading mechanism for its CFM, enabling a variable constant force from 0.110 to 1.735 N.
- He et al. [22] utilized a bridge-type amplified piezoelectric actuator as an adjustment module to achieve online tuning of the constant-force magnitude.

- Zhang and Yan [82] incorporated an adjustable bolt on one gripper jaw to modify the gap distance, thereby extending the range of object sizes that can be manipulated with a constant force.

Furthermore, in scenarios requiring the rotation of a single target object, a conventional single-DOF gripper must either rotate itself or the entire carrying platform after gripping. This approach is often inefficient and complex to implement.

A more practical and realistic solution is a 2-DOF gripper. By incorporating an additional translational degree of freedom into the jaws, the gripper can perform a rubbing operation to facilitate rotation. For example, Lyu et al. [46] proposed a 2-DOF gripper with independently actuated jaws, achieving gripping and rubbing strokes of 251.2 μm and 225.0 μm, respectively (see Fig. 1.12d).

1.5 Further Discussion

Through the literature review, a prominent research trend can be identified: the majority of piezoelectric compliant robotic designs emphasize minimizing the physical footprint of the devices. Performance comparisons are generally based on a specific output metric relative to the occupied area or volume. This tendency can be attributed to the following reasons.

- *Improved Flexibility and Integration*: Devices with a reduced footprint offer significant advantages for integration with existing equipment, particularly in fields such as MEMS, biomedicine, and aerospace. For instance, a micromanipulator operating under a microscope must fit within the limited available space without obstructing optical paths or interfering with adjacent instruments. Furthermore, enhanced compactness improves modularity and portability. This facilitates the replacement or reconfiguration of specific modules for tailored applications, thereby significantly increasing the overall versatility of the robotic system.
- *Increased Cost-Effectiveness*: From a manufacturing standpoint, smaller devices consume less material and require reduced processing time, leading to significantly lower production costs. Furthermore, their compact design enhances modularity and portability. This allows for the easier replacement or reconfiguration of specific components to suit various applications, thereby reducing long-term operational and maintenance costs.

Consequently, the miniaturization of piezoelectric compliant robotic devices is poised to become a major focus of future research, which will facilitate their wider adoption in UHPE systems. Achieving this objective will require concerted investigation into the two core components of these devices: piezoelectric transducers and compliant mechanisms.

Since the structural design and spatial arrangement of a piezoelectric compliant robotic device are fundamentally constrained by its overall size, the adoption of new materials with superior piezoelectric constants offers a promising path forward. These advanced materials can enable the construction of smaller piezoelectric transducers without compromising performance. Furthermore, by reducing the physical size of the transducer core, they minimize the need for extensive design adjustments to accommodate larger actuators, thereby allowing for more flexible, compact-oriented, and efficient overall structural designs.

Conversely, the static and dynamic characteristics, along with the safety factor of compliant components, are intrinsically linked to their physical dimensions. Consequently, miniaturization profoundly influences the overall performance of piezoelectric compliant robotic devices, necessitating careful compensatory design. Specifically regarding static performance, manufacturing limitations and material properties constrain the minimal achievable thickness of compliant components, often resulting in increased structural stiffness. In addition to employing piezoelectric transducers with enhanced performance, integrating constant-force mechanisms (CFMs) presents an effective alternative for mitigating high stiffness. Certain CFM designs can even reduce the reaction force to near zero [29, 56]. Regarding dynamic performance, passive damping techniques [12] can be employed to enhance system response. These methods typically involve attaching viscoelastic damping layers to compliant components to improve linear dynamic behavior and suppress vibrations.

1.6 Conclusion

In summary, this chapter has reviewed recent advances in the integrated design of piezoelectric compliant robotic devices for Ultra-High Precision Engineering (UHPE) applications. The fundamental principles of piezoelectric materials—namely, the direct and reverse piezoelectric effects—were discussed, highlighting how they enable diverse functionalities as sensors and actuators. Key material properties crucial for selection were also examined. Compliant mechanisms were categorized based on their primary functions and structural configurations, with a focus on amplification and guidance mechanisms. The applications of integrated piezoelectric compliant robotic devices were subsequently summarized in two key areas: high-resolution sensing and high-precision manipulation. The chapter concludes by identifying device miniaturization as a predominant future trend, underscoring the need for continued research into advanced materials and innovative design strategies to achieve more compact, efficient, and high-performance systems.

References

1. Ahmadi, S., Seethaler, R.: On the effect of force, displacement and voltage on the capacitance of piezoelectric stack actuators for self-sensing applications. In: Proceedings of 2021 4th International Conference on Mechatronics, Robotics and Automation (ICMRA), pp. 50–54 (2021). https://doi.org/10.1109/icmra53481.2021.9675783

2. Asaga, E., Takemura, K., Maeno, T., Ban, A., Toriumi, M.: Tactile evaluation based on human tactile perception mechanism. Sens. Actuators, A **203**, 69–75 (2013). https://doi.org/10.1016/j.sna.2013.08.013
3. Bazaei, A., Boudaoud, M., Hemmasian Ettefagh, M., Chen, Z., Regnier, S.: Displacement sensing by piezoelectric transducers in high-speed lateral nanopositioning. IEEE Sens. J. **19**(20), 9156–9165 (2019). https://doi.org/10.1109/jsen.2019.2923428
4. Beltran-Hernandez, C.C., Petit, D., Ramirez-Alpizar, I.G., Nishi, T., Kikuchi, S., Matsubara, T., Harada, K.: Learning force control for contact-rich manipulation tasks with rigid position-controlled robots. IEEE Robot. Autom. Lett. **5**(4), 5709–5716 (2020)
5. Bolopion, A., Regnier, S.: A review of haptic feedback teleoperation systems for micromanipulation and microassembly. IEEE Trans. Autom. Sci. Eng. **10**(3), 496–502 (2013). https://doi.org/10.1109/tase.2013.2245122
6. Cecil, J., Powell, D., Vasquez, D.: Assembly and manipulation of micro devices-a state of the art survey. Robot. Comput.-Integr. Manuf. **23**(5), 580–588 (2007). https://doi.org/10.1016/j.rcim.2006.05.010
7. Chen, F., Gao, Y., Dong, W., Du, Z.: Design and control of a passive compliant piezo-actuated micro-gripper with hybrid flexure hinges. IEEE Trans. Industr. Electron. **68**(11), 11168–11177 (2021). https://doi.org/10.1109/tie.2020.3032921
8. Chen, F., Zhang, Q., Gao, Y., Dong, W.: A review on the flexure-based displacement amplification mechanisms. IEEE Access **8**, 205919–205937 (2020). https://doi.org/10.1109/access.2020.3037827
9. Chen, F., Zhao, H., Li, D., Chen, L., Tan, C., Ding, H.: Contact force control and vibration suppression in robotic polishing with a smart end effector. Robot. Comput.-Integr. Manuf. **57**, 391–403 (2019)
10. Chen, X., Tai, K., Tan, H., Xie, Z.: Design of a two-stage compliant asymmetric piezoelectrically actuated microgripper with parasitic motion compensation. Mech. Syst. Signal Process. **208**, 110950 (2024). https://doi.org/10.1016/j.ymssp.2023.110950
11. Chen, X., Xie, Z., Tai, K., Tan, H.: Design of low parasitic motion microgripper based on symmetrical parallelogram mechanism. Sens. Actuators, A **367**, 115072 (2024). https://doi.org/10.1016/j.sna.2024.115072
12. Chen, Z., Zhong, X., Shi, J., Zhang, X.: Damping-enabling technologies for broadband control of piezo-stages: a survey. Annu. Rev. Control. **52**, 120–134 (2021). https://doi.org/10.1016/j.arcontrol.2021.10.007
13. Cheng, X., Gong, Y., Liu, Y., Wu, Z., Hu, X.: Flexible tactile sensors for dynamic triaxial force measurement based on piezoelectric elastomer. Smart Mater. Struct. **29**(7), 075007 (2020)
14. Chi, Z., Xu, Q.: Recent advances in the control of piezoelectric actuators. Int. J. Adv. Rob. Syst. **11**, 11 (2014). https://doi.org/10.5772/59099
15. Das, T.K., Shirinzadeh, B., Ghafarian, M., Al-Jodah, A.: Design, analysis, and experimental investigation of a single-stage and low parasitic motion piezoelectric actuated microgripper. Smart Mater. Struct. **29**(4), 045028 (2020). https://doi.org/10.1088/1361-665X/ab79b6
16. Das, T.K., Shirinzadeh, B., Ghafarian, M., Al-Jodah, A., Pinskier, J.: Characterization of a compact piezoelectric actuated microgripper based on double-stair bridge-type mechanism. J. Micro-Bio Robot. **16**(1), 79–92 (2020). https://doi.org/10.1007/s12213-020-00132-5
17. Ding, B., Li, X., Chen, S.C., Li, Y.: Modular quasi-zero-stiffness isolator based on compliant constant-force mechanisms for low-frequency vibration isolation. J. Vib. Control **30**(13–14), 3006–3020 (2024). https://doi.org/10.1177/10775463231188160
18. Ding, B., Li, X., Li, C., Li, Y., Chen, S.C.: A survey on the mechanical design for piezo-actuated compliant micro-positioning stages. Rev. Sci. Instrum. **94**(10), 101502 (2023). https://doi.org/10.1063/5.0162246
19. Galeti, J.H., Higuti, R.T., Silva, E.C.N., Kitano, C.: Nanodisplacement measurements of piezoelectric flextensional actuators using a new interferometry homodyne method. IEEE Trans. Instrum. Meas. **64**(5), 1256–1265 (2015). https://doi.org/10.1109/tim.2015.2395551
20. Gan, J., Xu, H., Zhang, X., Ding, H.: Design of a compliant adjustable constant-force gripper based on circular beams. Mech. Mach. Theory **173**, 104843 (2022). https://doi.org/10.1016/j.mechmachtheory.2022.104843

21. Gehin, C., Barthod, C., Teisseyre, Y.: Design and characterisation of a new force resonant sensor. Sens. Actuators, A **84**(1–2), 65–69 (2000)
22. He, J., Liu, Y., Yang, C., Tong, Z., Wang, G.: Design and evaluation of an adjustable compliant constant-force microgripper. Micromachines **15**(1), 52 (2023). https://doi.org/10.3390/mi15010052
23. Howell, L.L.: Compliant Mechanisms, pp. 189–216. Springer London, New York (2013)
24. Hricko, J., Havlík, Š.: Compliant Mechanisms for Motion/Force Amplifiers for Robotics, pp. 26–33. Springer International Publishing (2019)
25. Iqbal, S., Malik, A.: A review on mems based micro displacement amplification mechanisms. Sens. Actuators, A **300**, 111666 (2019). https://doi.org/10.1016/j.sna.2019.111666
26. Jagtap, S.P., Deshmukh, B.B., Pardeshi, S.: Applications of compliant mechanism in today's world - a review. J. Phys: Conf. Ser. **1969**(1), 012013 (2021). https://doi.org/10.1088/1742-6596/1969/1/012013
27. Kim, H., Torres, F., Wu, Y., Villagran, D., Lin, Y., Tseng, T.L.B.: Integrated 3d printing and corona poling process of pvdf piezoelectric films for pressure sensor application. Smart Mater. Struct. **26**(8), 085027 (2017)
28. Kiziroglou, M.E., Temelkuran, B., Yeatman, E.M., Yang, G.Z.: Micro motion amplification-a review. IEEE Access **8**, 64037–64055 (2020). https://doi.org/10.1109/access.2020.2984606
29. Kuppens, P.R., Bessa, M.A., Herder, J.L., Hopkins, J.B.: Compliant mechanisms that use static balancing to achieve dramatically different states of stiffness. J. Mech. Robot. **13**(2) (2021). https://doi.org/10.1115/1.4049438
30. Li, H., Guo, F., Wang, Y., Wang, Z., Li, C., Ling, M., Hao, G.: Design and modeling of a compact compliant stroke amplification mechanism with completely distributed compliance for ground-mounted actuators. Mech. Mach. Theory **167**, 104566 (2022). https://doi.org/10.1016/j.mechmachtheory.2021.104566
31. Lin, C., Shen, Z., Wu, Z., Yu, J.: Kinematic characteristic analysis of a micro-/nano positioning stage based on bridge-type amplifier. Sens. Actuators, A **271**, 230–242 (2018). https://doi.org/10.1016/j.sna.2017.12.030
32. Lin, C.H., Tsai, M.C., Hsiao, S.W.: Static force measurement for automation assembly systems. Sens. Actuators, A **187**, 147–153 (2012)
33. Ling, J., Ye, T., Feng, Z., Zhu, Y., Li, Y., Xiao, X.: A survey on synthesis of compliant constant force/torque mechanisms. Mech. Mach. Theory **176**, 104970 (2022). https://doi.org/10.1016/j.mechmachtheory.2022.104970
34. Ling, M., Yuan, L., Lai, J., Wei, H., Zhang, X.: Compliance and precision modeling of general notch flexure hinges using a discrete-beam transfer matrix. Precis. Eng. **82**, 233–250 (2023). https://doi.org/10.1016/j.precisioneng.2023.03.014
35. Ling, M., Yuan, L., Luo, Z., Huang, T., Zhang, X.: Enhancing dynamic bandwidth of amplified piezoelectric actuators by a hybrid lever and bridge-type compliant mechanism. Actuators **11**(5), 134 (2022). https://doi.org/10.3390/act11050134
36. Liu, H., Wang, M., Huang, X., Yang, J., Zhou, L.: Design and analysis of a flextensional piezoelectric actuator for incus-body driving type middle ear implant. J. Vibroeng. **19**(5), 3842–3854 (2017). https://doi.org/10.21595/jve.2017.18422
37. Liu, J., Luo, X., Liu, J., Li, M., Qin, L.: Development of a commercially viable piezoelectric force sensor system for static force measurement. Meas. Sci. Technol. **28**(9), 095103 (2017)
38. Liu, Y., Xu, Q.: Design of a compliant constant force gripper mechanism based on buckled fixed-guided beam. In: Proceedings of 2016 International Conference on Manipulation, Automation and Robotics at Small Scales (MARSS), pp. 1–6 (2016). https://doi.org/10.1109/marss.2016.7561731
39. Liu, Y., Zhang, Y., Xu, Q.: Design and control of a novel compliant constant-force gripper based on buckled fixed-guided beams. IEEE/ASME Trans. Mechatron. **22**(1), 476–486 (2017). https://doi.org/10.1109/tmech.2016.2614966
40. Lobontiu, N.: Compliant Mechanisms: design of Flexure Hinges. CRC Press (2002). https://doi.org/10.1201/9781420040272

41. Luo, S., Bimbo, J., Dahiya, R., Liu, H.: Robotic tactile perception of object properties: a review. Mechatronics **48**, 54–67 (2017)
42. Lyu, Z., Xu, Q.: Design and analysis of a new compliant microgripper with two working modes. In: Proceedings of 2021 IEEE International Conference on Manipulation, Manufacturing and Measurement on the Nanoscale (3M-NANO), pp. 65–70 (2021). https://doi.org/10.1109/3m-nano49087.2021.9599736
43. Lyu, Z., Xu, Q.: Recent design and development of piezoelectric-actuated compliant microgrippers: a review. Sens. Actuators, A **331**, 113002 (2021). https://doi.org/10.1016/j.sna.2021.113002
44. Lyu, Z., Xu, Q.: Design and testing of a new piezoelectric-actuated symmetric compliant microgripper. Actuators **11**(3), 77 (2022). https://doi.org/10.3390/act11030077
45. Lyu, Z., Xu, Q.: A compliant constant-force mechanism with sub-newton force and millimeter stroke output. Sens. Actuators, A **374**, 115465 (2024). https://doi.org/10.1016/j.sna.2024.115465
46. Lyu, Z., Xu, Q., Zhu, L.: Design and development of a new piezoelectric-actuated biaxial compliant microgripper with long strokes. IEEE Trans. Autom. Sci. Eng. **20**(1), 206–217 (2023). https://doi.org/10.1109/tase.2022.3145670
47. Ma, W., Wang, R., Zhou, X., Meng, G.: The performance comparison of typical notched flexure hinges. Proc. Inst. Mech. Eng. C J. Mech. Eng. Sci. **234**(9), 1859–1867 (2019). https://doi.org/10.1177/0954406219897941
48. Mansour, S.Z., Seethaler, R.: Simultaneous quasi-static displacement and force self-sensing of piezoelectric actuators by detecting impedance. Sens. Actuators, A **274**, 272–277 (2018)
49. Mohith, S., Upadhya, A.R., Navin, K.P., Kulkarni, S.M., Rao, M.: Recent trends in piezoelectric actuators for precision motion and their applications: a review. Smart Mater. Struct. **30**(1), 013002 (2020). https://doi.org/10.1088/1361-665x/abc6b9
50. Morales, A.R.R., Zaghloul, M.E.: Highly sensitive wearable piezoelectric force sensor with quasi-static load testing. IEEE Sens. J. **18**(24), 9910–9918 (2018)
51. Nan, Z., Xu, Q., Zhang, Y., Ge, W.: Force-sensing robotic microinjection system for automated multi-cell injection with consistent quality. IEEE Access **7**, 55543–55553 (2019)
52. Pankaj, Chandra, V., Murat, R., Kumar, V.: Design and analysis of piezo actuated flexure guided nanopositioning stage. Mater. Today: Proc. **80**, 327–332 (2023). https://doi.org/10.1016/j.matpr.2023.01.405
53. Savia, M., Koivo, H.: Contact micromanipulation-survey of strategies. IEEE/ASME Trans. Mechatron. **14**(4), 504–514 (2009). https://doi.org/10.1109/tmech.2008.2011986
54. Shao, S., Shao, Y., Song, S., Xu, M., Ma, X.: Structure and controller design of a piezo-driven orientation stage for space antenna pointing. Mech. Syst. Signal Process. **138**, 106525 (2020). https://doi.org/10.1016/j.ymssp.2019.106525
55. Smith, M., Kar-Narayan, S.: Piezoelectric polymers: theory, challenges and opportunities. Int. Mater. Rev. **67**(1), 65–88 (2021). https://doi.org/10.1080/09506608.2021.1915935
56. Tolman, K.A., Merriam, E.G., Howell, L.L.: Compliant constant-force linear-motion mechanism. Mech. Mach. Theory **106**, 68–79 (2016). https://doi.org/10.1016/j.mechmachtheory.2016.08.009
57. Uchino, K.: The development of piezoelectric materials and the new perspective, pp. 1–85. Elsevier (2010). https://doi.org/10.1533/9781845699758.1
58. Uchino, K.: The Development of Piezoelectric Materials and the New Perspective, pp. 1–92. Elsevier (2017). https://doi.org/10.1016/b978-0-08-102135-4.00001-1
59. Verotti, M.: Analysis of the center of rotation in primitive flexures: uniform cantilever beams with constant curvature. Mech. Mach. Theory **97**, 29–50 (2016). https://doi.org/10.1016/j.mechmachtheory.2015.10.007
60. Wang, F., Huo, Z., Liang, C., Shi, B., Tian, Y., Zhao, X., Zhang, D.: A novel actuator-internal micro/nano positioning stage with an arch-shape bridge-type amplifier. IEEE Trans. Industr. Electron. **66**(12), 9161–9172 (2019). https://doi.org/10.1109/tie.2018.2885716
61. Wang, G., Xu, Q.: Design and precision position/force control of a piezo-driven microinjection system. IEEE/ASME Trans. Mechatron. **22**(4), 1744–1754 (2017). https://doi.org/10.1109/tmech.2017.2698139

62. Wei, Y., Xu, Q.: Design of a pvdf-mfc force sensor for robot-assisted single cell microinjection. IEEE Sens. J. **17**(13), 3975–3982 (2017). https://doi.org/10.1109/jsen.2017.2701835
63. Wei, Y., Xu, Q.: Design of a new passive end-effector based on constant-force mechanism for robotic polishing. Robot. Comput.-Integr. Manuf. **74**, 102278 (2022). https://doi.org/10.1016/j.rcim.2021.102278
64. Wen, S., Xu, Q.: Design of a novel piezoelectric energy harvester based on integrated multistage force amplification frame. IEEE/ASME Trans. Mechatron. **24**(3), 1228–1237 (2019)
65. Wen, S., Xu, Q., Zi, B.: Design of a new piezoelectric energy harvester based on compound two-stage force amplification frame. IEEE Sens. J. **18**(10), 3989–4000 (2018)
66. Wu, H., Lai, L., Zhang, L., Zhu, L.: A novel compliant xy micro-positioning stage using bridge-type displacement amplifier embedded with scott-russell mechanism. Precis. Eng. **73**, 284–295 (2022). https://doi.org/10.1016/j.precisioneng.2021.09.014
67. Wu, H., Tang, H., Qin, Y.: Design and test of a 2-dof compliant positioning stage with antagonistic piezoelectric actuation. Machines **12**(6), 420 (2024). https://doi.org/10.3390/machines12060420
68. Wu, Z., Wang, X., Lyu, Z., Xu, Q.: Design of a novel passive polishing end-effector with adjustable constant force and wide operating angle. IEEE/ASME Trans. Mechatron. 1–11 (2024). https://doi.org/10.1109/tmech.2024.3375516
69. Wu, Z., Xu, Q.: Survey on recent designs of compliant micro-/nano-positioning stages. Actuators **7**(1), 5 (2018). https://doi.org/10.3390/act7010005
70. Wu, Z., Xu, Q.: Design and testing of a novel bidirectional energy harvester with single piezoelectric stack. Mech. Syst. Signal Process. **122**, 139–151 (2019)
71. Wu, Z., Xu, Q.: Design, fabrication, and testing of a new compact piezo-driven flexure stage for vertical micro/nanopositioning. IEEE Trans. Autom. Sci. Eng. **16**(2), 908–918 (2019). https://doi.org/10.1109/tase.2018.2875711
72. Xie, Y., Li, Y., Cheung, C.F., Zhu, Z., Chen, X.: Design and analysis of a novel compact xyz parallel precision positioning stage. Microsyst. Technol. **27**(5), 1925–1932 (2020). https://doi.org/10.1007/s00542-020-04968-6
73. Xu, Q.: Design and development of a flexure-based dual-stage nanopositioning system with minimum interference behavior. IEEE Trans. Autom. Sci. Eng. **9**(3), 554–563 (2012). https://doi.org/10.1109/tase.2012.2198918
74. Xu, Q.: Design and Implementation of Large-Range Compliant Micropositioning Systems. Wiley (2016)
75. Xun, M., Yu, H., Liu, Y., Deng, J., Zhang, S., Li, K.: A precise rotary piezoelectric actuator based on the spatial screw compliant mechanism. IEEE/ASME Trans. Mechatron. **28**(1), 223–232 (2023). https://doi.org/10.1109/tmech.2022.3197454
76. Yang, Y., Wu, G., Wei, Y.: Design, modeling, and control of a monolithic compliant x-y-θ microstage using a double-rocker mechanism. Precis. Eng. **71**, 209–231 (2021). https://doi.org/10.1016/j.precisioneng.2021.03.014
77. Yildirim, M., Candan, Z.: Smart materials: the next generation in science and engineering. Mater. Today: Proc. (2023). https://doi.org/10.1016/j.matpr.2023.10.116
78. Zarrabi Ekbatani, R., Zheng, J., Chen, X., Nikzad, M., Man, Z.: Design and control of a flexure-based dual stage piezoelectric micropositioner. Int. J. Precis. Eng. Manuf. **25**, 1793–1811 (2024). https://doi.org/10.1007/s12541-024-00990-0
79. Zhang, J., Gan, J., Ding, H., Li, H.: Design of a pure rotation micropositioning stage with dual-range. Mech. Mach. Theory **169**, 104631 (2022). https://doi.org/10.1016/j.mechmachtheory.2021.104631
80. Zhang, J., Yao, H., Mo, J., Chen, S., Xie, Y., Ma, S., Chen, R., Luo, T., Ling, W., Qin, L., et al.: Finger-inspired rigid-soft hybrid tactile sensor with superior sensitivity at high frequency. Nat. Commun. **13**(1), 5076 (2022)
81. Zhang, Q., Zhao, J., Shen, X., Xiao, Q., Huang, J., Wang, Y.: Design, modeling, and testing of a novel xy piezo-actuated compliant micro-positioning stage. Micromachines **10**(9), 581 (2019). https://doi.org/10.3390/mi10090581

82. Zhang, Y., Yan, P.: A large range constant force microgripper with a three-stage compliant amplification mechanism. IEEE Access **10**, 58225–58232 (2022). https://doi.org/10.1109/access.2022.3179588
83. Zhao, Y., Huang, X., Liu, Y., Wang, G., Hong, K.: Design and control of a piezoelectric-driven microgripper perceiving displacement and gripping force. Micromachines **11**(2), 121 (2020). https://doi.org/10.3390/mi11020121

Chapter 2
Recent Development of Piezoelectric-Actuated Compliant Microgrippers

Abstract This chapter presents a comprehensive review of recent advances in the design, modeling, and control of piezoelectric-actuated compliant microgrippers (PEACM). First, standard performance indices and classification schemes are elaborated to provide a foundation for defining design objectives. Subsequently, the compliant mechanisms fundamental to microgripper design are discussed, focusing on flexible hinges, displacement amplifiers, and guiding mechanisms. Furthermore, prevailing optimization and modeling methodologies for these systems are presented. The review also surveys popular types of integrated position/force sensors and assesses various displacement and force control strategies. Finally, future development trends for PEACMs are predicted and discussed.

Keywords Microgripper · Gripper performance · Gripper classification · Compliant mechanism · Optimization and modeling · Sensors and controllers

2.1 Introduction

The microgripper is a crucial tool in the field of micro/nano-manipulation [64, 89]. Functioning as an end-effector for executing microassembly and micromanipulation tasks, it directly interfaces with manipulated objects. Microgrippers have found broad applications across diverse domains, including biomedical engineering, advanced optics, and precision instrumentation [8, 38, 73, 90]. For instance, in the construction of micro-optoelectromechanical systems (MOEMS), such as Fourier transform microscopic spectrometers, the application of a four-degree-of-freedom (4-DOF) microgripper can significantly enhance robotic dexterity, achieving alignment tolerances with arc-second precision [75]. Furthermore, the assembly of ignition targets for the US National Ignition Facility's Inertial Confinement Fusion program relies on high-precision microgripping technology due to the components' extreme requirements for precision, compactness, and fragility [3, 51]. In biomedical research, affinity microgrippers are essential for capturing and grasping individual cells, thereby facilitating morphological observation and precise substance injection [122].

Q. Xu, *Piezoelectric-Driven Flexible Robotic Microgrippers*,
https://doi.org/10.1007/978-981-95-4685-5_2

A microgripper typically comprises one or more actuators, a motion transmission mechanism, and gripper jaws [67]. It functions by converting electrical voltage or other forms of energy into precise jaw displacement and force. Operations such as picking, transporting, and inserting micro-objects are reliably performed using the generated friction force. The selection of an appropriate actuator is a critical design task in microgripper development [2]. As the system's power source, the actuator directly provides the force and displacement necessary for successful operation. Consequently, the microgripper's motion stroke and resolution are fundamentally determined by the actuator's performance. Furthermore, an optimal actuation method can significantly enhance the overall compactness of the gripper, for instance, by being fully integrated into the gripper's structure or by operating as a discrete, standalone component. Based on their driving principles, actuators suitable for microgrippers are primarily categorized into electrostatic, electrothermal, electromagnetic, shape memory alloy (SMA), and piezoelectric types [4, 47, 53, 71, 77, 129], as illustrated in Fig. 2.1. A comprehensive review [113] has summarized the comparative advantages and disadvantages of these five actuator types, along with their compatible materials and fabrication technologies.

Among these, the piezoelectric-actuated compliant microgripper (PEACM) has been widely designed and adopted due to its compelling advantages: a small footprint, extremely high displacement resolution, rapid response speed, low operating voltage, and large output force. In a PEACM, a piezoelectric actuator (PZT) works in concert with a compliant mechanism to efficiently convert the ceramic's linear input motion into the jaw's clamping movement. Research into PZT-driven compli-

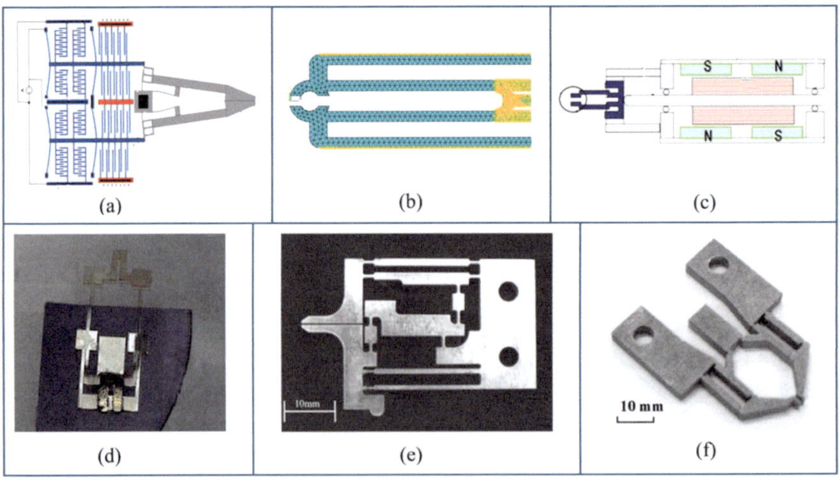

Fig. 2.1 Microgrippers driven by different actuation principles: **a** electrostatic, reproduced with permission from [4], **b** electrothermal, reproduced with permission from [47], **c** electromagnetic, reproduced with permission from [53], **d** shape memory alloy (SMA), reproduced with permission from [77], **e** normally-open piezoelectric-actuated compliant microgripper (PEACM), reproduced with permission from [129], and **f** normally-closed PEACM, reproduced with permission from [71]

ant microgrippers has been an intensely active topic in recent decades. This review aims to summarize the state-of-the-art developments in this field to facilitate the design of more precise, efficient, and advanced microgripping systems. Certain crystalline materials generate electrical charges when subjected to mechanical stress, a phenomenon known as the direct piezoelectric effect. Conversely, the same materials undergo controlled deformation when exposed to an electric field, based on the inverse piezoelectric effect. As an electromechanical transducer, a PZT utilizes the inverse piezoelectric effect to convert electrical signals into precise mechanical displacements.

However, PZT actuators exhibit inherent limitations that must be addressed for precision applications. Specifically, they suffer from hysteresis and creep effects, which degrade positioning accuracy. Fortunately, these nonlinearities can be mitigated through suitable control strategies and parameter compensation [40, 87]. A further limitation is the inherently small output displacement of PZT materials, typically on the order of 0.1% of the actuator's length. A widely adopted solution is the integration of compliant displacement amplification mechanisms, such as lever-type and bridge-type amplifiers. Compliant mechanisms are particularly suited for this role due to their significant advantages: they operate without friction, require no assembly or lubrication, exhibit no backlash, and are compatible with vacuum environments. In a PEACM, motion is transmitted entirely through the material's elastic deformation. This monolithic design simplifies the system and provides exceptional convenience for precise position and force control.

In addition to the actuation method, the selection of materials and fabrication processes plays a vital role in the practical application of microgrippers. Currently, most PEACMs are manufactured from aluminum alloys using wire electrical discharge machining (wire-EDM) technology. The fundamental principle of wire-EDM involves using a continuously moving fine metal wire to erode the workpiece through controlled pulsed spark discharges. This manufacturing technique offers high machining accuracy, high production efficiency, low power consumption, and relatively low cost. Aluminum alloy 7075 (Al 7075) is widely adopted for fabricating grippers due to its favorable combination of high yield strength and low elastic modulus. With ongoing advancements in manufacturing technology, an increasing number of micromachining processes are being applied to gripper fabrication, including LIGA (Lithographie, Galvanoformung, Abformung), deep reactive-ion etching (DRIE), and photolithography (PL) [31].

Furthermore, materials such as nickel (Ni), nickel-titanium (Ni-Ti) alloys, silicon, and various polymers (e.g., polyurethane, SU-8) are excellent choices for microgrippers dedicated to biological applications, owing to their specific biocompatibility and mechanical properties. It is foreseeable that continued progress in manufacturing technologies will further facilitate the development of high-performance microgrippers [33]. Moreover, the integration of displacement/force sensors and sophisticated control systems is essential to enable the precision micro/nano-manipulation capabilities that these microgrippers are designed to perform.

This chapter reviews recent advances in the design and implementation of PEACMs. A generalized design flowchart for a PEACM is presented in Fig. 2.2.

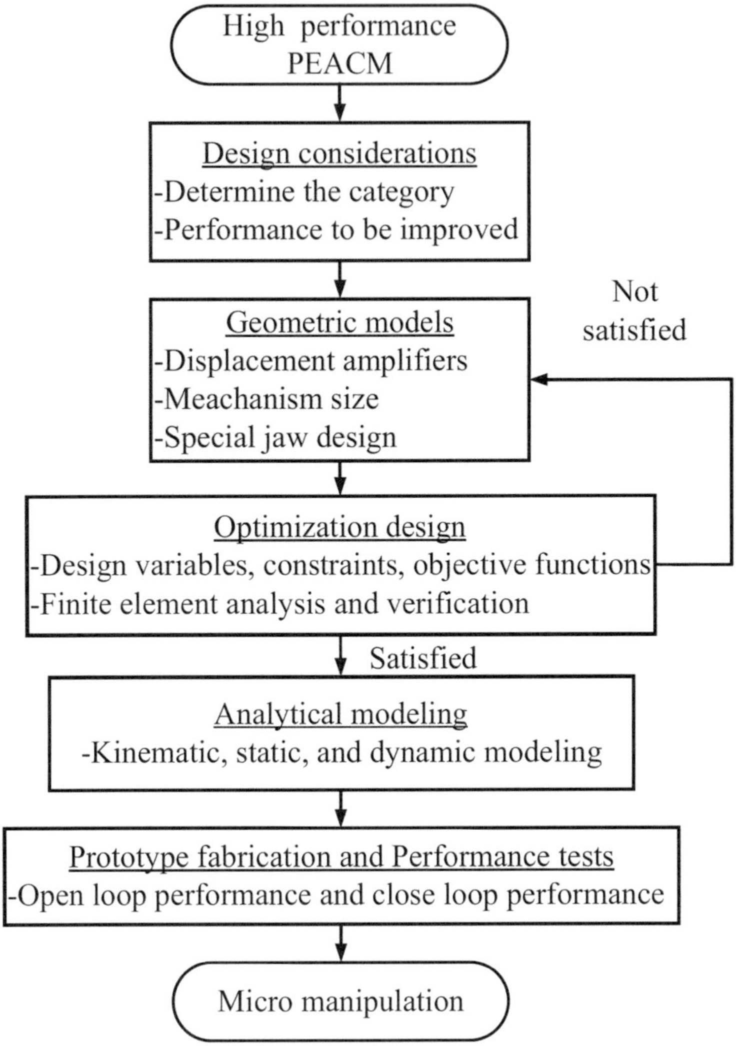

Fig. 2.2 Design methodology flowchart for a piezoelectric-actuated compliant microgripper (PEACM)

It should be noted that the PEACMs discussed herein are primarily driven by stack-type PZTs, although alternative constructions using piezoelectric bimorphs are also possible [33, 49]. The design process for a high-performance PEACM begins with defining the gripper's category and target performance specifications. Subsequently, during the geometric modeling phase, key components such as the displacement amplification mechanism, structural dimensions, and specialized jaw geometry must be designed. Optimization and finite element analysis (FEA) are then employed to refine the design and enhance performance. Following this, comprehensive kinematic, static, and dynamic modeling and analyses are conducted. Upon finalizing the design, a prototype is fabricated and subjected to rigorous performance testing. This testing phase involves the selection of appropriate sensors and control strategies to characterize both open-loop and closed-loop behavior. Finally, the validated gripper is deployed to execute targeted micromanipulation tasks. Furthermore, the chapter concludes by predicting and discussing future development trends in PEACM technology, providing valuable insights to guide designers toward creating more versatile and high-performing microgrippers.

2.2 Performance Indices of Microgrippers

The working performance of a PEACM is evaluated through a comprehensive set of qualitative and quantitative indices, encompassing both static and dynamic characteristics. Prior to designing a new gripper, a thorough understanding of existing evaluation metrics and their current state of development is indispensable [31]. Nikoobin et al. surveyed several key parameters influencing microgripper performance based on available designs [72]. However, a unified and complete set of indicators specifically for assessing PEACM performance has not yet been fully established. To address this gap, the following section presents ten fundamental indices for evaluating the universal capability of PEACMs.

2.2.1 Clamping Stroke and Clamping Range

The clamping stroke (S) is defined as the maximum achievable displacement of the gripper jaw. For symmetrical grippers, this stroke is the sum of the output displacements from both jaws, as illustrated in Fig. 2.3a. A larger clamping stroke enhances the gripper's adaptability in micro-operations by accommodating a wider range of object sizes.

For a given initial tip gap (D) between the jaws, a larger clamping stroke allows for the manipulation of smaller objects. The range of object sizes a gripper can handle is defined as its clamping range (R). As shown in Fig. 2.3b for cylindrical objects, this range is represented by the difference in diameter between the largest (blue circle) and smallest (red circle) objects that can be securely gripped.

A fundamental design trade-off exists between clamping stroke and resolution. Generally, achieving a larger stroke compromises the gripper's resolution, and vice versa. Consequently, designs prioritizing high-resolution clamping often feature a minimized stroke [10].

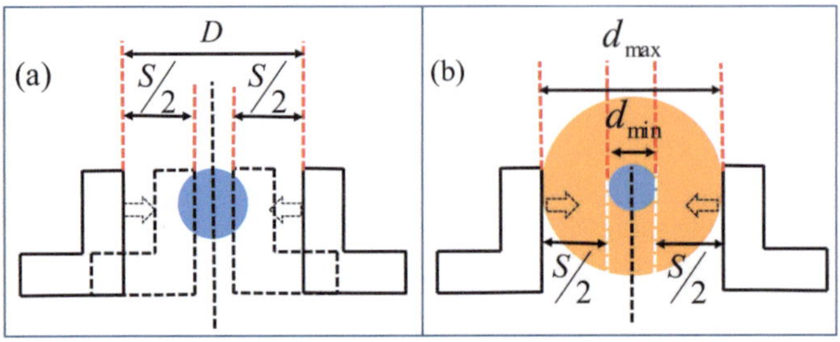

Fig. 2.3 Schematic diagram of the **a** clamping stroke and **b** clamping range

2.2.2 *Displacement Magnification and Area Efficiency*

The displacement magnification ratio (R_m) of a PEACM is defined as the ratio of the gripper's output jaw displacement to the input displacement from the piezoelectric actuator (PZT), as illustrated in Fig. 2.4a. Given the inherently small displacement of PZTs, a primary design goal is to achieve a high magnification ratio to ensure a sufficient clamping stroke.

However, a high magnification ratio does not directly guarantee a large clamping stroke, as the final output is also constrained by the PZT's nominal displacement. While increasing the length of amplification levers or optimizing flexible hinge parameters can enhance the magnification ratio, an excessively long lever compromises the system's compactness. Consequently, the compliant mechanism responsible for displacement amplification must be carefully designed and precisely optimized.

A significant design challenge is that the additional mechanical structure required for amplification can often negate the benefits gained. Therefore, the overall footprint or area (A) of the gripper structure must be considered. To evaluate this trade-off, the metric of area efficiency (AE) is introduced, as shown in Fig. 2.4b. Area efficiency is defined as the ratio of the clamping stroke to the mechanical area of the device,

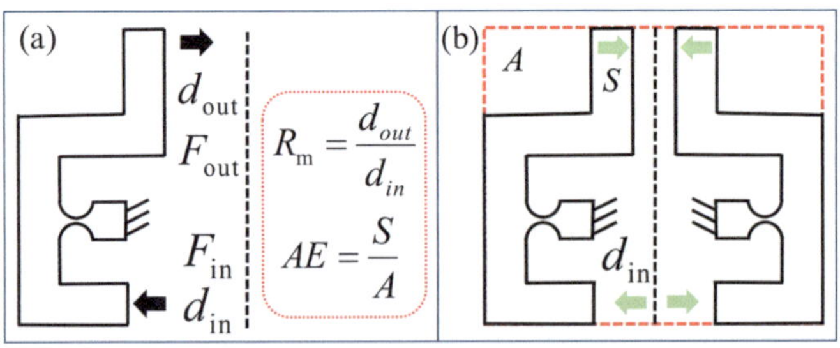

Fig. 2.4 Schematic diagram illustrating **a** displacement magnification ratio and **b** area efficiency for piezoelectric-actuated compliant microgrippers

representing the stroke output per unit area [87]. A higher area efficiency indicates a more compact design capable of achieving the same clamping stroke.

2.2.3 Normally-Open Gripper and Normally-Closed Gripper

A normally-open gripper is characterized by a maximum initial gap between its jaws in the resting state, as shown in Fig. 2.1e [7, 52, 70, 98]. Its gripping process is illustrated in Fig. 2.5a. The target object must be smaller than this initial jaw gap. To clamp, a voltage is applied to actuate the jaws, causing them to close until they contact and secure the target. The gripping force can be monitored and controlled in real-time during this process. When the voltage is removed, the jaws return to their open position, releasing the object.

Conversely, a normally-closed gripper features a minimal initial gap at rest, as sketched in Fig. 2.1f [21, 22, 67, 71, 76]. Its operating sequence is shown in Fig. 2.5b. Here, the target object is typically larger than the initial gap. To grip, a voltage is applied to open the jaws to a sufficient width. Once the target is positioned, the driving voltage is gradually reduced or disconnected. The elastic restoring force of the pre-deformed flexible hinges then drives the jaws to close, clamping the object. Compared to the normally-open type, the normally-closed gripper consumes less energy, as it only requires power to open, not to close. However, the sustained elastic deformation in the hinges can accelerate fatigue and increase the mechanism's failure rate.

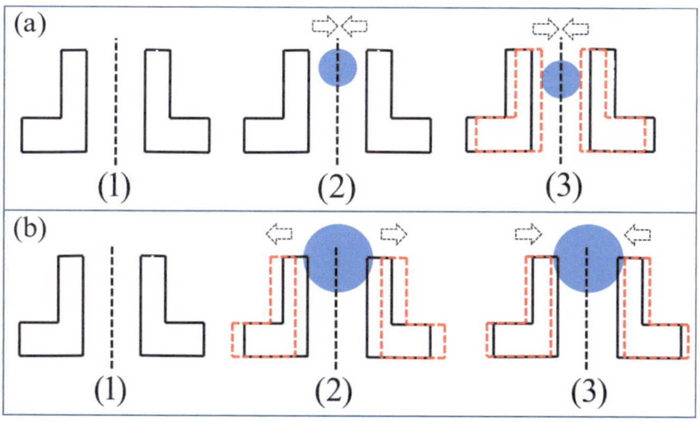

Fig. 2.5 Schematic diagram of the **a** normally-open gripper and **b** normally-closed gripper

2.2.4 Degree of Freedom and Parasitic Motion Rate

Conventional microgrippers typically possess a single degree of freedom (DOF) in Cartesian space. By controlling reciprocating motion along one axis, they perform basic operations such as approaching, clamping, and releasing. However, for advanced high-efficiency micromanipulation, a single DOF is often insufficient to meet complex precision requirements. This has led to the development of 2-DOF microgrippers for specialized applications like optical fiber alignment [15].

As shown in Fig. 2.6a, a 2-DOF gripper can achieve independent clamping in the x-direction and rubbing or positioning in the y-direction, offering significantly enhanced dexterity and agility. This functionality typically requires two independent piezoelectric actuators (PZTs), one for each axis. Such designs provide greater operational freedom, provided that motion coupling between the axes is effectively minimized.

Even in single-DOF grippers, motion coupling is a prevalent issue. As exhibited in Fig. 2.6b, grippers based on rotational principles often exhibit parasitic motion in the y-direction while performing the primary clamping motion in the x-direction. This parasitic movement subjects the target object to unintended lateral forces, resulting in unstable clamping–a critical concern when handling spherical or cylindrical objects. Therefore, ensuring motion parallelity is a fundamental design consideration.

To quantify this effect, researchers have introduced the parasitic motion ratio (R_p), which characterizes the gripper's ability to maintain parallel motion. It is defined as the ratio of the parasitic displacement (d_y) to the primary clamping displacement (d_x). A smaller R_p value indicates higher clamping reliability and precision. In practice, integrating a flexible parallelogram guide mechanism is an effective strategy for suppressing parasitic motion and achieving near-pure translational movement.

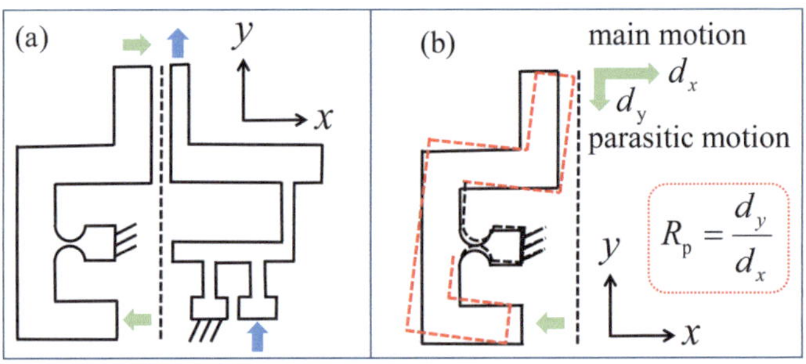

Fig. 2.6 Schematic diagram of **a** a 2-DOF gripper and **b** parasitic motion rate calculation

2.2.5 *Clamping Force and Natural Frequency*

The closing motion of the jaws achieves gripping through both form closure and friction. For frictional gripping, a substantial clamping force must be applied to generate sufficient positive pressure on the target object. Consequently, the maximum achievable clamping force is a critical performance index for evaluating a microgripper. For fragile targets, such as biological cells or micro-optical components, precise control of this force is essential for non-destructive operation. Therefore, clamping force capability and control must be central considerations in the gripper's design process.

Beyond macroscopic clamping forces, microscopic surface forces–such as Van der Waals, electrostatic, and capillary forces–become significant at small scales and can cause severe object release difficulties. A prominent strategy to overcome these adhesion forces is to employ high-frequency mechanical vibration to detach the object from the jaw surfaces [35]. This vibration-based release method utilizes the inertial effects of the end-effector and the micro-object to break adhesion, a process that requires a high control bandwidth from the actuation system [34].

In engineering practice, the achievable control bandwidth is typically limited to approximately one-third of the system's first natural frequency. This relationship makes the natural frequency of the gripper structure a crucial dynamic performance index. A high natural frequency is doubly beneficial: it directly enables a higher control bandwidth for effective vibration-based release, and it generally enhances the performance of the entire control system. Conversely, a low natural frequency severely limits the maximum operating speed, degrading performance in both open-loop and closed-loop control schemes.

2.3 Classification of PEACM

The categories of PEACMs have become increasingly abundant due to advancements over recent decades. These devices can be classified into various types based on different criteria, such as motion parallelity, structural symmetry, and degrees of freedom. Understanding these classifications is essential for researchers to select the appropriate gripper type for a specific application. The following sections will introduce these categories in detail.

2.3.1 *Parallel Gripper*

Parallel grippers are characterized by a high degree of parallelism in the motion of their tips, as illustrated in Fig. 2.7. The parasitic motion ratio is the standard metric for quantifying their translational performance.

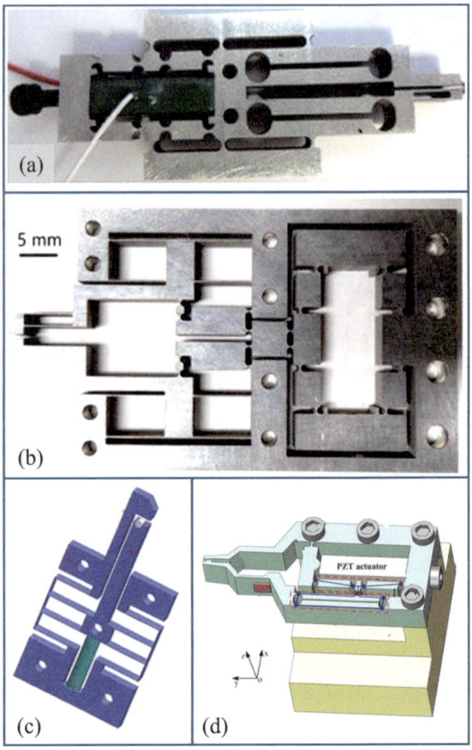

Fig. 2.7 Parallel microgripper designs: **a** employing a standard parallelogram mechanism, reproduced with permission from [29] and **b** utilizing over-constrained parallelogram mechanisms for enhanced motion linearity, reproduced with permission from [26]. Asymmetric microgripper configurations: **c** longitudinal clamping implementation, reproduced with permission from [24] and **d** transverse clamping approach, reproduced with permission from [56]

In contrast, the movement of an ordinary, scissor-like gripper can cause spherical targets (e.g., microlenses or microspheres) to be pushed out of the jaws during clamping. To prevent this, achieving parallel jaw movement is a critical design objective [5, 9, 44, 45, 101, 115]. Parallel clamping is particularly advantageous for irregular spherical or cylindrical parts, enabling stable gripping and precise assembly. Furthermore, it minimizes stress concentration on brittle objects, reducing the risk of damage.

A highly effective method for minimizing parasitic motion is the integration of a parallelogram flexible guide mechanism. For instance:

- Das et al. employed a parallelogram mechanism as the final amplification stage to linearize the gripping motion, as shown in Fig. 2.7a [29].
- In a subsequent design, Das et al. achieved a total output displacement of 184.04 μm with a remarkably low parasitic motion ratio of 0.59% by utilizing over-constrained parallelogram mechanisms [26]. While this approach increases mechanical complexity, the significant improvement in performance is often justified.

It is worth noting that some research suggests that a slight rotational error at the clamping end may be negligible within a typical micro clamping range (e.g., 200 μm) [106].

2.3.2 Asymmetric Gripper

The asymmetrical PEACM features a single moving jaw, while the opposing tip remains fixed, as illustrated in Fig. 2.7c and d [17, 20, 23, 24, 56, 103, 104, 120].

Although most current grippers employ a fully symmetrical mechanism to output clamping force and displacement, this theoretically ideal configuration suffers from two inherent limitations that compromise accuracy. First, inevitable manufacturing and assembly errors introduce asymmetry in the output of the two jaws. Second, symmetrical designs often exhibit dense modal clusters, which can degrade dynamic performance.

In contrast, research demonstrates that asymmetrical grippers can overcome these issues. Simulation results indicate that the first natural frequency of an asymmetric design can be its dominant working frequency, effectively mitigating the problem of dense modes [56]. Furthermore, controlling a single jaw enables more precise and stable motion, which helps prevent damage to fragile objects [120].

A primary trade-off is that the output displacement of an asymmetric PEACM is typically half that of an equivalent symmetric gripper. To compensate for this and achieve a larger amplification ratio, designs often incorporate additional actuators or amplification stages. For example, Chen et al. developed an asymmetric microgripper utilizing two PZTs to achieve a substantial gripping stroke of 632 μm [17].

2.3.3 2-DOF Gripper

A 2-DOF microgripper possesses the capability to manipulate objects along two independent axes, typically utilizing two actuators [14, 15, 30, 32, 68, 83, 95, 105, 121]. This functionality is particularly suited for precision alignment tasks, such as those required for polarization-maintaining optical fibers [95].

Fiber alignment necessitates both a gripping motion in the horizontal direction and a rubbing/frictional motion in the vertical direction to induce rotation. Employing two separate, dedicated grippers for this task increases system complexity. A 2-DOF microgripper elegantly solves this problem by integrating both functions into a single, compact end-effector with multifunctional jaws.

Notable implementations from the literature include:

- Chen et al., who designed a 2-DOF gripper that mimics the dexterous manipulation of human fingers for optical assembly processes, achieving strokes of 190 μm and 85 μm for the left and right tips, respectively [15].

- Zhang et al., who introduced a monolithic 2-DOF microgripper capable of simultaneous grasping and rubbing for the rotational alignment of polarization-maintaining optical fibers. Experimental results demonstrated its ability to clamp a 250 μm diameter fiber and rotate it through an angle exceeding 90° [121].

Two primary challenges must be addressed in the current design of 2-DOF micro-grippers. First, the displacement magnification ratio of both actuation arms requires further improvement to achieve larger strokes. Second, the achievable angular range for frictional rotation of optical fibers must exceed 180° to enable full rotational manipulation [68]. Overcoming these challenges is critical for replacing manual assembly with fully automated systems in optical fiber alignment applications.

Beyond these core issues, other advanced 2-DOF designs have been demonstrated:

- One design enhances functionality by adding independent longitudinal motion capability to the primary x-direction clamping [30].
- Furthermore, Xu et al. have developed a 2-DOF gripper capable of parallel translational motion based on a modular design concept [105].

2.3.4 Constant-Force Gripper

Non-destructive operation is a critical requirement in fields such as medical equipment, semiconductor manufacturing, and precision instrumentation. When a micro-gripper's jaws contact a fragile object–such as a microlens, biological cell, or micro-electronic component–the applied clamping force must remain below a damage threshold to preserve the target's structural integrity and functionality.

Integrating position or force sensors with closed-loop control can achieve precise and safe manipulation. However, such measurement and control systems add complexity, reduce reliability, and increase the cost, thereby heightening the risks associated with commercial deployment.

To address this challenge, constant-force microgrippers have been developed. These devices utilize specialized compliant mechanisms to maintain an nearly constant output force across a range of displacements when manipulating small targets [60, 61, 63, 110]. This output force can be either fixed at a predetermined safe level or made adjustable to accommodate a variety of micro-objects with different fragility limits, providing a robust and elegant solution for non-destructive micro-manipulation.

Various methodologies exist for generating a constant output force in compliant mechanisms [96]. A common approach involves creating a zero-stiffness mechanism by combining positive and negative stiffness elements in series. For instance, connecting an oblique beam (negative stiffness) with a straight beam (positive stiffness) can effectively provide overload protection at the micro-scale. Notable implementations include:

- Liu et al. demonstrated a gripper utilizing this negative stiffness effect, achieving a constant force output of 1.1 N over a 200 μm range, as shown in Fig. 1.12c. Motion control remains essential for achieving precise jaw positioning during the clamping operation.
- Xu et al. reported a clamp based on buckled fixed-guided beams, providing a constant force of 530 mN across a 220 μm displacement, as displayed in Fig. 1.12c. They employed a discrete-time variable structure control strategy to achieve precise motion output, thereby simplifying the overall control system architecture.

For advanced applications like the micro-assembly of diverse small targets, future research must focus on developing grippers with multiple constant-force levels or tunable force output. Additional constant-force clamp designs, driven by various actuation principles, continue to be explored [42, 43, 118, 123].

2.3.5 Standard Gripper

It is appropriate to classify the remaining, most common type of PEACM as the standard clamp, which represents the majority of designs in the literature [1, 16, 79, 81, 85, 128, 129]. While lacking the specialized functions of the previously discussed categories, the standard clamp remains highly effective for a wide range of fundamental gripping tasks.

A standard clamp typically comprises three core components: gripping jaws, a PZT, and a displacement amplification mechanism. For instance, Shi et al. developed a compliant microgripper featuring a PZT, jaws, and a symmetric lever mechanism that achieved a displacement amplification ratio of 8.26 [81].

The ongoing development direction for the standard clamp focuses on enhancing its capabilities, particularly through the implementation of controllable clamping force to increase its precision and versatility for delicate operations.

2.4 Compliant Mechanism Used in PEACM

During the geometric model design phase, the selection and integration of compliant mechanisms is a critical consideration. These mechanisms form the core structural foundation of any PEACM. Compliant mechanisms in PEACMs can be broadly categorized into three functional types:

- *Basic Compliant Hinges*: Providing localized rotation through elastic deformation.
- *Displacement Amplification Mechanisms*: Magnifying the small stroke of the piezoelectric actuator.
- *Flexible Guide Mechanisms*: Constraining motion to a desired path and minimizing parasitic movements.

This section will detail the design and application of these compliant mechanisms within PEACMs.

2.4.1 Compliant Hinge

Compliant mechanisms play a crucial role in transmitting and transforming displacement within piezoelectric-actuated microgrippers. The fundamental building block of any compliant mechanism is the flexible hinge, which utilizes elastic material deformation to generate controlled relative motion between adjacent rigid linkages [13, 119]. The notch-type flexible hinge is the most commonly employed rotational joint. Its primary advantages are a simple structure and ease of manufacturing. However, a significant limitation is its restricted range of motion; the rotational angle typically does not exceed 5° before exceeding the material's elastic limits or inducing unacceptable stress concentrations.

Notch-type flexible hinges are classified based on their contour geometry, with common types including circular, rectangular, elliptical, and parabolic (chamfered) profiles. Each type exhibits distinct mechanical characteristics:

- Circular hinges are characterized by a small, precise rotation angle, a relatively fixed instantaneous center of rotation, high rotational accuracy, but significant stress concentration.
- Rectangular hinges offer a larger range of rotation and are easier to manufacture; however, they suffer from a shifting center of rotation and lower rotational accuracy.
- The performance of other hinge types (e.g., elliptical, parabolic) generally falls between these two extremes.

The selection of hinge type and quantity is a critical design factor, as it directly governs the overall stiffness and performance of the compliant mechanism. While increasing the number of hinges can provide greater flexibility and motion range, it also reduces the mechanism's overall rigidity. Furthermore, flexible hinges are particularly susceptible to lateral forces; therefore, their design must ensure that load paths are aligned to avoid transmitting significant lateral loads, which can lead to premature failure. Representative implementations include:

- Liang et al., who developed a microgripper utilizing double-notch circular flexure hinges, achieving a displacement amplification ratio of 19.4 [57].
- Lin et al., who designed a three-stage amplification gripper incorporating multiple double-notch rectangular compliant hinges [58].

2.4.2 *Displacement Amplification Mechanism*

The displacement amplification mechanism is a critical component that significantly enhances the operational performance of PEACMs. It magnifies the small native displacement of the piezoelectric actuator to achieve a practical and effective gripping range. Common types of displacement amplification mechanisms include:

- Lever mechanisms, which provide amplification through a rigid arm rotating about a flexure pivot.
- Bridge-type mechanisms, known for their compact size and symmetrical force output.
- Scott-Russell (SR) mechanisms, which convert input displacement into a perpendicular, amplified output.

2.4.2.1 Lever Mechanism

The lever mechanism is a fundamental and effective method for displacement amplification in compliant systems [48]. A basic lever mechanism and its application in microgrippers are illustrated in Fig. 2.8. While simple in design, a single-stage lever typically provides limited amplification, usually less than a factor of 10.

To achieve higher magnification, compound lever mechanisms are employed. These can be categorized into two configurations:

- *Serial Levers*: The output of one lever acts as the input for the subsequent lever, cascading the amplification effect [17, 29, 92]. This is the more common and versatile type.
- *Parallel Levers*: Multiple levers act on a common output stage, often to increase force or improve stability.

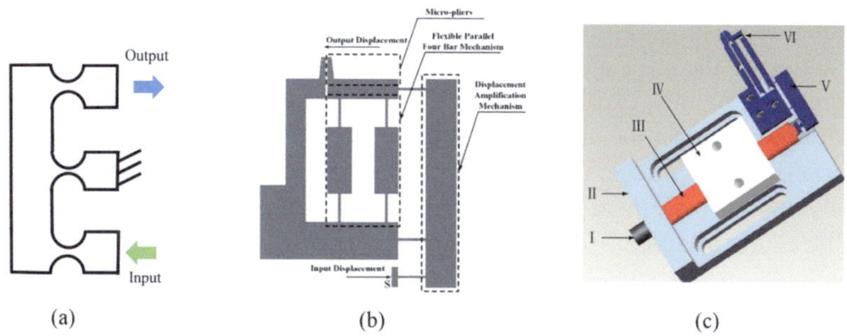

Fig. 2.8 Lever-based amplification mechanisms and their microgripper applications: **a** Fundamental compliant lever principle, **b** single-lever implementation [104], **c** series-lever configuration, reproduced with permission from [120]

Notable implementations from the literature include:

- Xing et al. utilized a single lever mechanism combined with a flexible parallel four-bar mechanism to design an asymmetric gripper, achieving a displacement amplification ratio of 4.16, as shown in Fig. 2.8b [104].
- Zhang et al. constructed a monolithic compliant gripper using a two-lever mechanism, attaining an amplification ratio of 6.0 [120].
- Wang et al. adopted a serial lever architecture to build a three-stage PEACM, achieving a significantly higher total amplification ratio of 22.8 [92].

Beyond compound levers, the differential lever mechanism can achieve a larger magnification ratio within the same spatial footprint compared to traditional designs. For instance, Zhang et al. implemented a differential amplifying lever mechanism to extend the PZT's output in their gripper design, achieving a theoretical amplification ratio of 18.12 [68]. A significant challenge for these mechanisms, however, is the presence of longitudinal coupling displacement, which compromises motion accuracy.

Employing a symmetrical structure is an effective strategy to mitigate longitudinal coupling errors. A symmetrical lever mechanism can thus simultaneously provide a high magnification ratio and eliminate parasitic displacements. Shi et al. demonstrated this by integrating a symmetrically arranged dual-stage lever amplifier into a clamp structure, achieving an amplification ratio exceeding 20 [82]. The gripper's performance was successfully validated through picking and placing experiments. A noted trade-off of this complex symmetrical design is its requirement for a larger driving force. Alternatively, adding a guide mechanism, such as a parallelogram structure, at the output end of the lever can also enhance the directionality of the output motion. Dsouza et al. utilized this approach, employing a parallelogram mechanism to linearize the output motion of the leverage system in their clamp design [32].

Furthermore, numerous variations of lever mechanisms exist, including levered parallelograms and L-shaped levers. The L-shaped lever is particularly notable for enabling an orthogonal change in displacement direction, with its input and output ends typically perpendicular to each other [26, 36, 37, 81, 91, 95]. Wang et al. successfully utilized an L-shaped lever mechanism in a gripper to achieve precise manipulation [91].

In summary, the lever mechanism offers several advantages, including a simple structure, high rigidity, excellent energy efficiency, and linear amplification. Its primary drawback is that the amplification effect is constrained by spatial limitations. Furthermore, in accordance with the principle of conservation of energy, increasing the output displacement through a longer lever arm correspondingly reduces the available clamping force.

2.4.2.2 Bridge Mechanism

The bridge-type mechanism is another widely used displacement amplification method in compliant mechanisms [48]. Its principle and various applications are

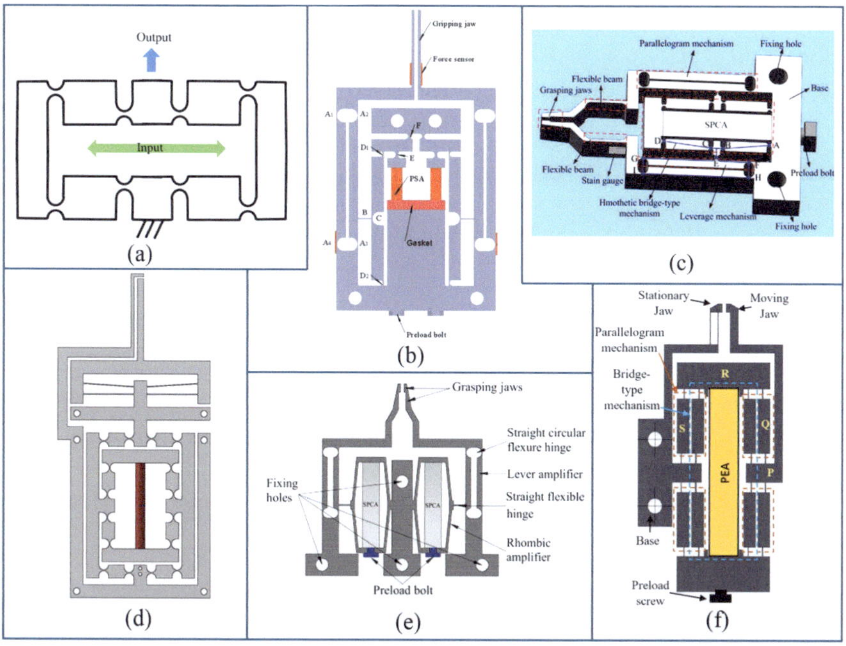

Fig. 2.9 Bridge-type amplification mechanisms and their microgripper applications: **a** Fundamental compliant bridge mechanism; **b** implementation using a half-bridge mechanism [125]; **c** application of a full bridge mechanism, reproduced with permission from [57]; **d** series configuration of bridge mechanisms, reproduced with permission from [20]; **e** parallel arrangement of bridge mechanisms, reproduced with permission from [18]; **f** implementation using a double-stair bridge mechanism, reproduced with permission from [27]

illustrated in Fig. 2.9. The mechanism operates based on the mechanical principle of compressive linkage instability. When an input force is applied, the deformation of the flexible hinges produces a significantly amplified output displacement in a direction perpendicular to the input.

Bridge amplifiers typically employ rectangular or circular flexible hinges to achieve large magnification ratios. Compared to lever-type amplifiers, the bridge mechanism offers a more compact design for an equivalent magnification ratio, albeit with a more complex displacement analysis model [100]. Significant analytical work has been dedicated to modeling these mechanisms, with researchers like Lobontiu, Ma, Qi, Ye, and Lin deriving amplified displacement models based on various kinematic and mechanical principles [59, 65, 69, 74, 117].

Design variations and configurations are shown as follows.

- *Half-Bridge Mechanisms*: To save space, some designs utilize only half of the full bridge structure [56, 125].

- *Actuator Integration*: The PZT is typically embedded at the mechanism's center. A longitudinal placement generally facilitates double-sided symmetric output, while a horizontal placement is used for single-sided output.
  - Longitudinal/Double-Sided: Liang et al. symmetrically placed the bridge mechanism and PZT to prevent shear forces and bending moments from acting on the actuator [57].
  - Horizontal/Single-Sided: This configuration is used for applications requiring output from one side [26, 66].
- *Series and Parallel Connections*: Bridge mechanisms can be cascaded to enhance performance.
  - Series Connection: Achieves a larger overall magnification ratio within a limited space [20]. For instance, Chen et al. connected two bridge mechanisms in series to maximize end-effector stroke.
  - Parallel Connection: Features a fixed common end, often used to increase load capacity or stability [18, 115].
- *Composite and Multi-Dimensional Bridges*: Advanced designs push the limits of performance.
  - Xu et al. developed a flexible composite bridge mechanism renowned for its large magnification ratio, compact size, and high lateral stiffness [106, 107, 111].
  - Das et al. designed a single-stage amplifier based on a double-stair bridge mechanism, achieving a magnification of 19.3 for reliable grasp/release tasks [27, 28].
  - Multi-dimensional bridge mechanisms combine two or three bridges to achieve output in vertical space and an even larger magnification ratio, leading to higher spatial utilization efficiency [12, 54].
- *Semi-Rhombic Amplifiers*: Chen et al. designed a novel gripper using a semi-rhombic amplifier, which operates on a principle similar to the bridge mechanism [19].

2.4.2.3 Scott-Russell Mechanism

The Scott-Russell (SR) mechanism is a displacement amplification mechanism that provides orthogonal displacement conversion [55], as illustrated in Fig. 2.10a. A key feature of the SR mechanism is its ability to produce a highly linear output displacement in response to an input displacement. Two or more SR mechanisms can be connected in series to achieve greater amplification.

Research into the characteristics of SR mechanisms has been extensive:

- Ha et al. (2006) investigated an SR mechanism with two flexure hinges, finding that the hinge's orientation and offset significantly impact both the magnification ratio and output linearity [41].

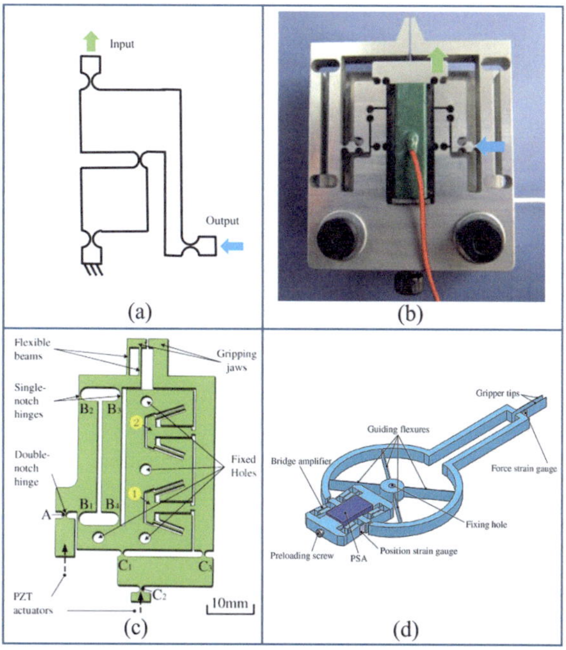

Fig. 2.10 **a** Schematic of a Scott-Russell (SR) mechanism. **b** Microgripper implementing an SR mechanism, reproduced with permission from [86]. **c** Microgripper utilizing a Roberts parallel guiding mechanism, reproduced with permission from [15]. **d** Microgripper employing a rotary bearing mechanism

- Chen et al. (2012) performed system identification on a biased SR amplifier driven by a piezoelectric actuator [11].
- Tian et al. (2009) developed a design methodology and dynamic model for a piezoelectrically driven SR nanomanipulation mechanism [88].

The SR mechanism is most powerfully employed in combination with other amplification mechanisms:

- Sun et al. (2013) arranged an SR mechanism in series with a lever mechanism, constructing a gripper with a maximum clamping displacement of 134 μm (Fig. 2.10b) [86].
- Ai et al. (2014) combined an SR mechanism with a parallelogram mechanism to achieve pure translational motion at the gripper's output [1].
- Liu et al. (2016) connected an improved SR mechanism with a parallel mechanism, achieving a magnification ratio 3.56 times greater than that of a conventional design [62].
- Zhu et al. (2020) designed a novel gripper utilizing a deformed SR mechanism for micromanipulation tasks [127].
- Wang et al. (2019) implemented an SR mechanism as a guiding stage in their micromanipulator design [95].
- Wu et al. (2021) developed a micro-gripper using two SR mechanisms, achieving a high natural frequency of 815.77 Hz [99].

These developments indicate a clear trend towards increasingly sophisticated and hybridized applications of the SR mechanism to enhance microgripper performance.

2.4.3 Compliant Guiding Mechanism

The displacement guide mechanism is an indispensable component for ensuring precise motion transmission and amplification in compliant microgrippers. The flexible notched parallel four-bar mechanism is commonly employed to achieve parallel clamping [25, 91, 130]. A primary limitation of this design is its inherently small motion stroke. To increase the range of motion, multiple leaf springs are often combined to form a flexible prismatic joint, replacing the notched hinge. However, such arrangements can introduce significant parasitic motion in non-actuated directions.

A promising solution to mitigate parasitic error is the implementation of overconstrained structures, which enhance guidance precision by reducing unwanted degrees of freedom [26]. Numerous innovative guiding mechanisms have been demonstrated in the literature:

- Xu et al. constructed a 2-DOF gripper based on a compound parallelogram flexure and a modular design approach [105].
- Dao et al. utilized a composite multi-stage leaf spring mechanism to design a single-sided clamping device [23, 24].
- Beyond guidance, double-leaf moving pairs can also provide an optimal surface for mounting strain gauges for force sensing [26, 95].
- Yang et al. designed a microgripper incorporating a double-rocker and parallelogram mechanism, achieving a large output displacement of 213.9 μm [114].
- Chen et al. applied two compliant Roberts mechanisms to enhance parallel motion performance, surpassing that of traditional four-bar linkages [15] (Fig. 2.10c).
- Xu et al. integrated a rotary bearing mechanism into a microgripper to achieve precise displacement orientation [106] (Fig. 2.10d).
- Huang et al. innovatively leveraged the parasitic motion of a flexible structure itself to design a functional compliant gripper [46].

Consequently, the critical role of advanced flexible guiding mechanisms is receiving increasingly significant attention from researchers in the field.

2.5 Structure Optimization and Modeling

Once a clamp's geometric configuration is finalized, the subsequent critical steps involve structure optimization and mathematical modeling.

Structure optimization techniques are essential for determining the optimal parameters of the flexible mechanism, ensuring peak performance. Simultaneously, establishing an accurate mathematical model of the clamp is fundamental for deepening our understanding of its performance characteristics, including kinematics, statics, and dynamics.

This section reviews prevalent methodologies employed for the structure optimization and mathematical modeling of PEACMs.

2.5.1 Structure Optimization

Advancements in mathematics, computational mechanics, and computer technology have made structural optimization an essential tool for efficiently and economically achieving target performance within specified constraints. Consequently, adopting an optimal design process is crucial for maximizing the performance of microgripper models.

Optimization of the flexible continuum in compliant mechanisms occurs at three hierarchical levels: topology optimization, shape optimization, and size (scale) optimization. In microgripper design, topology and size optimization are the most widely applied methods.

1. Topology Optimization: This highest-level optimization determines the optimal material layout and connectivity within a given design space. It allows designers to specify support and load locations, enabling software algorithms to generate the most efficient material distribution [126]. For instance:

   - Grossard et al. developed the FlexIn topological optimization method specifically for monolithic piezoelectric microgrippers, validating its effectiveness experimentally [39].
   - Numerous other microgripper designs have been successfully realized through topology optimization [6, 78].

2. Size Optimization: Once the topology and shape of the mechanism are fixed, size optimization focuses on determining the optimal dimensions of the constituent elements. Key design variables often include the length and thickness of flexure hinges. The process involves iterating these variables within a defined domain to extremize an objective function, such as maximizing the displacement amplification ratio.

A variety of optimization algorithms have been developed to address these complex design challenges, including sorting algorithms, Particle Swarm Optimization (PSO), and Genetic Algorithms (GA). The significant benefits of this approach are demonstrated by specific cases:

- Xiao et al. employed a multi-objective Genetic Algorithm based on a Radial Basis Function Network (RBFN) to design a gripper with perfectly parallel jaw motion [102].
- Das et al. initially designed a gripper with a displacement amplification ratio of 12.76 [29]. After applying Mixed-Integer Sequential Quadratic Programming

(MISQP) to optimize the structural parameters, they achieved a significantly improved ratio of 23.96 [25].

These examples underscore that optimizing a microgripper's structural parameters prior to fabrication is a highly worthwhile endeavor, leading to substantial performance enhancements.

2.5.2 Gripper Modeling

A comprehensive performance analysis of a designed microgripper necessitates the establishment of three fundamental mathematical models: kinematic, static, and dynamic. These models are essential for predicting and quantifying critical performance parameters, including the displacement amplification ratio, input stiffness, and natural frequency. Through rigorous modeling, designers can accurately evaluate a gripper's behavior before physical prototyping, enabling optimization and ensuring it meets specific application requirements.

The most commonly used analytical models for microgrippers are the Pseudo-Rigid-Body Model (PRBM) and the Finite Element Model (FEM). The PRBM simplifies the analysis of compliant mechanisms by making two key assumptions:

- Elastic deformation is localized only at the flexure hinges, while all other components are treated as rigid bodies.
- The deformation at each flexure hinge is pure bending, and the resulting rotation angles are small.

Within this framework, the motion characteristics of a flexible segment are simulated by a rigid link connected by a pin joint, while the segment's stiffness is represented by a torsional spring at the joint. This approach allows the well-established theory of rigid-body mechanisms to be effectively applied to the analysis of compliant systems. Wang et al., among others, have successfully performed kinematic and dynamic analyses of their gripper designs using the PRBM [57, 81, 92]. The method is also frequently employed for static analysis to determine input stiffness using matrix-based compliance modeling [81] and is widely adopted to simplify model analysis in the literature [29, 82, 91, 129].

In contrast, the Finite Element Method (FEM) uses mathematical approximations to simulate the behavior of a real physical system. A Finite Element Model is the specific computational model generated for this analysis. Leveraging mature commercial software like ANSYS and COMSOL, FEM enables accurate and reliable simulation of microgripper performance. This approach offers significant advantages for design and analysis, including low computational cost, high efficiency, and powerful visualization capabilities. Consequently, FEM has become an indispensable tool, and nearly all modern papers on gripper design utilize it to assist in performance validation and optimization [1, 71, 115].

2.6 Sensors and Controllers

Following fabrication, a comprehensive testing system must be established to experimentally characterize the performance of a PEACM. Sensors and controllers are the essential components of such a system.

The most commonly employed sensors include:

- Capacitive sensors and laser displacement sensors for high-resolution position measurement of the gripper's output end.
- Strain gauges and vision-based sensors, which can also be utilized to infer gripping force based on a pre-calibrated force/position relationship or through direct image analysis.

For control, various advanced strategies have been implemented to achieve precise manipulation:

- Proportional-Integral-Derivative (PID) control
- Sliding Mode Control (SMC)
- Fuzzy Logic Control (FLC).

A central challenge in this field is the development of robust and smooth position/force switching control algorithms to enable versatile and stable manipulation across different tasks.

This section provides a survey of the sensors and controllers typically employed in microgripper testing systems.

2.6.1 Position and Force Sensors

2.6.1.1 Capacitive Sensors

Capacitive displacement sensors operate on the fundamental principle of parallel plate capacitance. These sensors utilize capacitors as sensitive elements to convert mechanical displacement into a measurable change in capacitance.

In a typical configuration, the two electrodes of the capacitor function as the sensor and the target object being measured. When a stable alternating current is applied to the sensor, the resulting output voltage is proportional to the distance between the sensor electrode and the target object. Consequently, the displacement or distance information is derived by precisely measuring this voltage change.

In the literature, capacitive sensors are widely employed to measure the output displacement of microgrippers with high precision [15, 17, 18, 121, 125]. For instance, Zhang et al. utilized a high-precision commercial capacitive displacement sensor, achieving a resolution of approximately 1 nm in their gripper's testing setup, as illustrated in Fig. 2.11 [121]. Capacitive displacement sensors are precision non-contact measuring instruments characterized by their simple structure, high resolution, operational reliability, and excellent dynamic response.

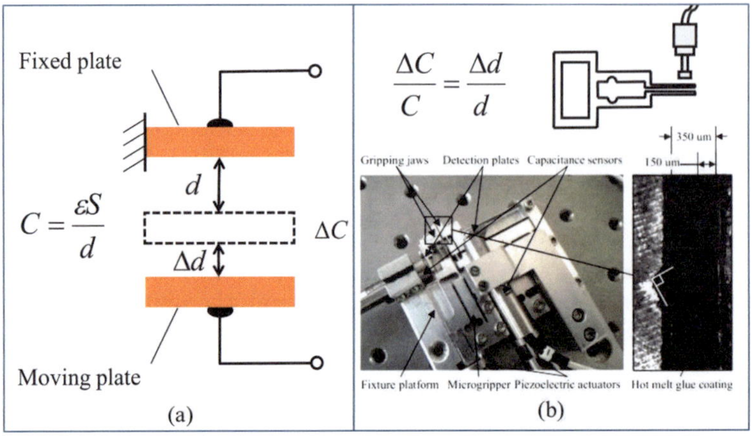

Fig. 2.11 Operating principle and application of a capacitive displacement sensor. **a** Fundamental working principle of capacitive sensing. **b** Practical implementation for measuring microgripper jaw displacement, reproduced with permission from [121]

2.6.1.2 Strain Sensors

Strain sensors operate based on the piezoresistive effect (or strain effect), wherein the electrical resistance of a conductor or semiconductor material changes when it undergoes mechanical deformation due to an external force, as illustrated in Fig. 2.12a.

Resistance strain gauges are categorized into two types: metal strain gauges and semiconductor strain gauges. Metal resistance strain gauges are predominantly used

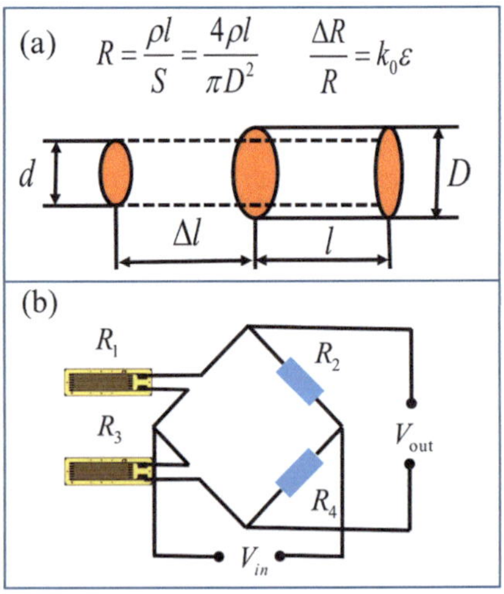

Fig. 2.12 Strain sensing methodology. **a** Fundamental operating principle of strain gauges; **b** Wheatstone bridge circuit configuration for signal conditioning

in force sensing applications. A typical strain gauge consists of a sensitive grid (the sensing element), lead wires, a substrate, and a protective covering layer.

In practice, the gauge is firmly bonded to a measurement point on a component. When the component is stressed, the sensitive grid deforms accordingly, causing a change in its electrical resistance. This resistance change is detected by a measurement circuit–typically a Wheatstone bridge, as shown in Fig. 2.12b–and converted into a corresponding strain value. Subsequently, the surface stress state of the component is determined using the appropriate stress-strain relationship.

Within the literature, strain gauges are commonly employed to characterize microgripper performance:

- They are used to measure motion displacement [91].
- They are also widely adhered to flexure hinges to measure gripping force directly [56, 91, 95, 106, 115]. For example, Xu et al. integrated a gripping sensor system into their prototype by mounting semiconductor strain gauges on compliant hinges [106].

Strain gauges effectively leverage the physical and geometric properties of conductive materials. Their advantages include small size, rapid dynamic response, high measurement accuracy, and ease of use. However, the accuracy of detection is highly sensitive to the precise placement of the gauges. Prior to testing, it is essential to use simulation software to identify the regions of the mechanism that experience significant strain during operation.

Compared to other contact measurement methods, strain sensors offer the distinct benefits of low cost and ease of integration into microsystems.

2.6.1.3 Laser Interferometry Sensors

The laser interferometer is a precision instrument that utilizes the wavelength of light as a fundamental standard for length measurement. It is globally recognized as a benchmark method for detecting and calibrating machine tools requiring extremely high geometric accuracy.

Figure 2.13a shows a schematic of a Michelson interferometer, which employs an amplitude-splitting method to generate dual beams for optical interference. The core components consist of a light source, a beam splitter, a reference mirror, and a detection plane. Based on this fundamental principle, numerous specialized interferometer variants have been developed.

Laser interferometric displacement measurement systems are employed in select laboratories due to their exceptional precision, though their use is limited by high cost [25–29]. For instance, Das et al. utilized high-stability plane mirror interferometers to measure gripper motion displacement, achieving nanoscale resolution, as shown in Fig. 2.13b.

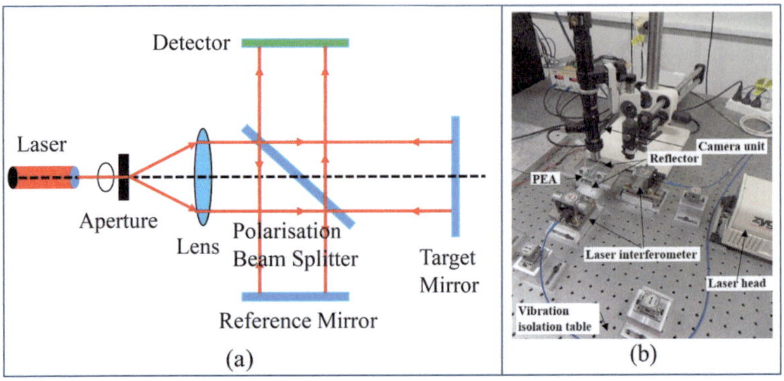

Fig. 2.13 **a** Operating principle of a laser interferometry displacement sensor. **b** Practical application in microgripper displacement measurement, reproduced with permission from [29]

Laser interferometers offer significant advantages, including:

- Exceptionally high precision and accuracy
- Non-contact measurement
- Large measurement range
- High measurement speed
- Excellent sensitivity.

However, these systems also present several disadvantages:

- High acquisition cost
- Complex optical alignment and adjustment
- Significant technical expertise required (high barrier to entry)
- Requirement for an uninterrupted laser beam path during measurement.

2.6.1.4 Laser Triangulation Sensors

The laser triangulation method is particularly suitable for high-precision, short-distance displacement measurement. This technique operates based on the optical law of reflection and the geometric principle of similar triangles formed during light propagation in space, as illustrated in Fig. 2.14a.

In this configuration, the object space and the image space of the receiving lens form a similar triangular relationship. The displacement of the target is then calculated using angular relationships within these triangles. To ensure optimal focus across the entire measurement field, the optical path must be designed to satisfy the Scheimpflug condition.

Laser triangulation displacement sensors are widely adopted in microgripper testing due to their excellent cost-performance ratio [1, 44, 56, 62, 85, 93, 95, 106, 115].

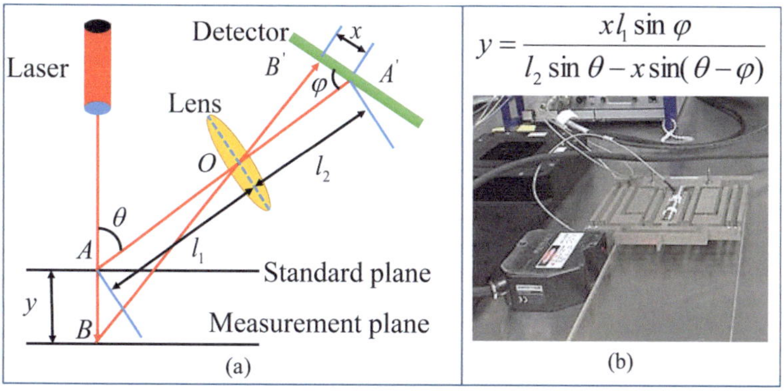

Fig. 2.14 **a** Operating principle of a laser triangulation displacement sensor. **b** Practical application for microgripper jaw displacement measurement [62]

For instance, Liu et al. employed a commercial laser sensor to verify the actual gripping displacement of a microgripper, as shown in Fig. 2.14b [62].

The primary advantages of this method include non-contact measurement, high precision, and high speed. However, its accuracy is influenced by several error sources, including laser source jitter, spot detection resolution, and the limited depth of field of the laser system. A significant practical advantage is its relatively low requirement for surface finish on the target object, unlike interferometric methods which typically require very smooth surfaces. Furthermore, laser triangulation systems offer a more compact structure and greater cost-effectiveness compared to interferometers.

2.6.1.5 Vision-Based Sensors

Micromanipulation tasks performed by PEACMs typically require visual assistance from an optical microscope [120]. Consequently, vision-based feedback has become a significant method for information perception, integrating principles from mechanics, image processing, and computer vision.

The process involves capturing and processing microscopic images of the gripper's compliant structure before and after deformation. The clamping force is then estimated using a theoretical model or a pre-calibrated deformation coefficient derived from the observed displacement. This approach has been widely implemented:

- Nah et al. measured gripper displacement using a microscope integrated with a CCD camera [71].
- Jayaram et al. equipped a gripper system with a vision system for integrated force measurement and control [50].

- Dsouza et al. developed an optical image processing system using LabVIEW to comprehensively measure both clamp displacement and gripping force [32].
- Shi et al. developed an online calibration method for displacement sensing, eliminating the need for superfluous external equipment [82].
- Yao et al. proposed a high-precision motion measurement system (0.06 μm) based on computer microscopic vision and optical flow methods [116].

A key advantage of vision-based sensing is its ability to perform non-contact measurement of both displacement and force using the system's existing optical components, thereby minimizing any impact on the microgripper's inherent performance.

However, this method also presents several limitations. Its effectiveness relies heavily on the accuracy of feature detection and the processing speed of the image analysis algorithms. Furthermore, other system components can obstruct the camera's view during complex manipulation tasks, causing the system to fail. Most critically, the resolution of this force sensing method is fundamentally limited by the magnification and quality of the microscope's optical components, which inhibits its application in scenarios requiring high-bandwidth, direct force feedback [97].

Therefore, vision-based force and position detection is most suitable for the performance verification stage of a gripper, where the operating environment is simple and the required operational speed is low.

2.6.2 Control Strategy

Micro/nano motion control is the fundamental core of precision equipment. In microgripper systems, this control is inherently nonlinear due to the piezoelectric nonlinear hysteresis and creep effects inherent to the actuators.

A common approach to compensate for this nonlinearity is feedforward control based on an inverse hysteresis model and an inverse creep model [40]. However, developing accurate models for these phenomena is often prohibitively time-consuming. Furthermore, the performance of this method is highly sensitive to the precision of the model, and even small inaccuracies can lead to significant control errors.

Consequently, an alternative strategy is to treat the combined nonlinearities as a lumped perturbation within the system dynamics, deliberately avoiding explicit modeling. This approach makes easily implementable robust controllers particularly attractive for practical applications.

A variety of advanced control strategies have been successfully implemented in PEACMs to achieve high-performance precise positioning and real-time motion control. A typical testing system integrating both sensors and a controller for a microgripper is displayed in Fig. 2.15. The following section provides an overview of the development of control strategies specifically for PEACMs.

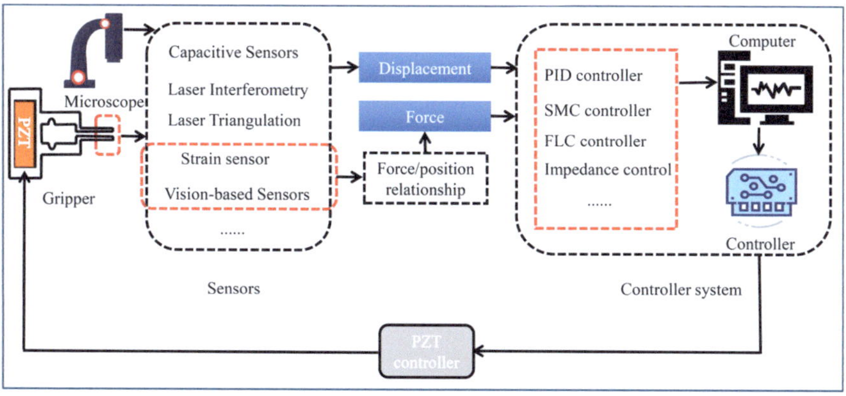

Fig. 2.15 Sensor and control system configuration for microgripper performance characterization

2.6.2.1 PID Control

Proportional-Integral-Derivative (PID) control remains widely used in practical applications due to its model-free nature and general effectiveness.

Carrozza et al. adopted a Proportional-Integral (PI) strategy to achieve stable and accurate grasping force control for assembling biomedical microdevices [7]. Their results demonstrated that a PI controller can ensure closed-loop system stability within a suitable bandwidth. Zhang et al. employed an incremental PID algorithm to control gripping motion, enabling precise regulation and maintaining a continuously stable gripping force [120]. This algorithm responds rapidly to feedback signals to prevent error accumulation. Their work also indicated that the derivative (D) term can often be neglected in microgripper control due to the relatively slow operating speeds involved.

Das et al. utilized a PI control strategy to perform closed-loop experiments on their designed microgripper, validating its dynamic performance [25, 26]. The gripper's tracking performance was verified using sinusoidal wave signals of varying frequencies and amplitudes, while its displacement resolution was characterized through multi-step response tests.

Further advancements include a combined controller proposed by Wang et al., which integrates a PI feedforward model with PID feedback control to enhance output linearity [94]. Similarly, Zhao et al. implemented a PID controller for closed-loop feedback to achieve precise control over both displacement and clamping force [125]. Experimental results consistently show that closed-loop control effectively suppresses hysteresis nonlinearity compared to open-loop operation.

2.6.2.2 SMC Control

However, traditional PID control often proves inadequate for micromanipulation tasks due to high-frequency dynamic vibrations that can occur during the clamping process. In contrast, Sliding Mode Control (SMC) typically demonstrates superior force regulation performance compared to incremental PID force controllers.

To address this challenge, researchers have developed sophisticated hybrid control schemes:

- Wang et al. combined an incremental PID controller with discrete SMC to create a position/force switching control strategy [56]. This scheme employs PID for precise position tracking prior to contact [93] and switches to robust SMC for force regulation during high-speed contact, implemented physically using MATLAB and dSPACE DS1103. Experimental results confirmed this approach achieves high-speed, accurate, and robust micro-operations.
- Similarly, Liang et al. utilized a PID controller for motion positioning–leveraging its simple feedback loop and straightforward parameter tuning–and adopted SMC for the clamping phase to suppress vibrations, capitalizing on its strong robustness against uncertainties and disturbances [57].
- To specifically counteract hysteresis effects, Wu et al. developed an adaptive backstepping SMC algorithm, which demonstrates superior stability, robustness, and tracking performance compared to both PID and conventional SMC [99].
- Zhang et al. integrated online estimation of unknown parameters, Kalman filtering, and Adaptive Sliding Mode Control (ASMC) to achieve precise control. Their controller significantly reduced overshoot and steady-state error compared to traditional PID and non-predictive ASMC. The Kalman filter-based position prediction enhanced tracking accuracy and effectively compensated for hysteresis [124].
- Yang et al. proposed a hybrid position/force control scheme combining nonlinear Fuzzy Logic Control (FLC) with an incremental PID controller, validating its effectiveness through clamping experiments [115].

2.6.2.3 Position/Force Control

For micromanipulation tasks, combined position and force control is essential to manage both the approach phase (position control) and the contact phase (force control) between the gripper jaw and target objects.

Xu et al. conducted extensive research on precise microgripper control [108]. Initially, they employed a position/force switching control strategy based on PID feedback for a microassembly task, demonstrating good feasibility and effectiveness [106]. A fundamental challenge in such switching control is maintaining stability during the transition between control modes [84].

To overcome this, Xu et al. developed a new robust control method integrating both position and force sensors [107]. By adopting a sliding mode-based robust position strategy, they effectively converted the position/force control problem into a position control problem, thereby eliminating the stability issues associated with mode switching. However, maintaining a constant position or force remains challenging due to the inherent creep effect in piezoelectric materials.

Impedance control presents an alternative strategy to address these issues by regulating the dynamic relationship between position and contact force, rather than controlling them separately. A rapid response speed is critical for high-throughput micro-assembly and non-destructive biomedical testing [80, 112]. To this end, Xu et al. presented a digital sliding mode impedance control strategy that operates without a state observer. This approach significantly improves calculation speed and enables high-speed operation by utilizing a single, efficient controller [109]. Furthermore, implementing this control scheme on an FPGA (Field-Programmable Gate Array) platform further accelerates gripper operation compared to previous implementations. A noted limitation is that impedance controllers are generally complex and require meticulous design to achieve the desired accuracy.

While the aforementioned strategies enable accurate clamping, the development of more efficient and practical controllers remains necessary for complex micromanipulation environments. A primary focus for future research is achieving smooth and stable transitions between force and position control modes.

2.7 Future Research Trend

The overall performance of a PEACM is characterized by both its static and dynamic properties. A key static performance metric is the clamping stroke, which determines the size range of micro-objects that can be manipulated. Conversely, a critical dynamic performance metric is the natural frequency, which dictates the achievable control bandwidth and response speed. An optimal gripper design must therefore strike a balance between achieving a large stroke and maintaining a high natural frequency.

Figure 2.16 plots the current performance distribution of grippers from the literature, with clamping stroke on the abscissa and natural frequency on the ordinate. The data, represented by blue hollow diamonds, are predominantly clustered in the lower-left quadrant, indicating a common trade-off where most designs achieve either a large stroke or a high frequency, but not both simultaneously. The red solid square represents the mean of the dataset, indicating the average performance level of current technology.

By establishing this mean point as the origin of a new orthogonal coordinate system (indicated by the red arrows), a large area is revealed in the first quadrant. This space represents the untapped design potential for future high-performance grippers. We can therefore predict that future development will progress along the direction of the black arrow in Fig. 2.16, toward designs that simultaneously offer both a larger clamping stroke and a higher natural frequency.

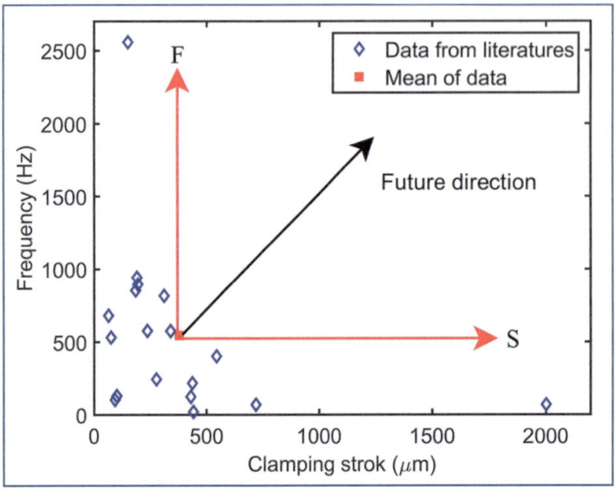

Fig. 2.16 Performance diagram of available microgrippers

Beyond these core performance metrics, the geometry of the gripper jaws is another critical area for future development. A prior survey has summarized the application of various tip shapes [72]. For instance:

- Flat-surface tips are versatile for manipulating both uniform and irregularly shaped objects.
- Pyramidal-shaped tips facilitate easier object release.
- Tips combining multiple geometries (e.g., flat and cylindrical surfaces) have also been developed.

Lofroth et al. demonstrated the value of this approach with a modular microgripper featuring interchangeable tips fabricated using Deep Reactive-Ion Etching (DRIE). This design enabled the manipulation of diverse targets, from single biological cells to microparticles [66]. This suggests a future trend toward modular, application-specific jaws to meet the demands of complex assembly tasks and increasingly stringent requirements.

Finally, while most current grippers operate in air, future research will likely expand into challenging environments. Grippers designed for operation in liquid media or under high-humidity conditions are anticipated to become a significant focus of upcoming research.

2.8 Conclusion

The piezoelectric-actuated compliant microgripper (PEACM) has gained significant popularity in the field of micromanipulation due to its rapid response speed and high displacement accuracy. The development and application of high-performance PEACMs continue to be an area of intense research investigation. This survey has first presented a comprehensive overview of the complete design and implementation procedure for these microgrippers. Subsequently, ten key performance indicators for characterizing gripper capability have been presented and discussed. Next, typical gripper architectures and their respective advantages were introduced. The review also detailed the critical roles of flexure hinges, displacement amplifiers, and guiding mechanisms based on compliant structures, along with their practical applications. Furthermore, the importance of structure optimization and analytical modeling in enhancing PEACM performance has been addressed. Finally, recent advancements in position/force sensors and precision control strategies have been thoroughly reviewed. Although PEACM design has been extensively investigated in the literature, several challenging issues remain to be solved to further enhance gripper performance. As these devices continue to improve, they are poised to play an increasingly crucial role in advanced micromanipulation and microassembly tasks in the future.

References

1. Ai, W., Xu, Q.: New structure design of a flexure-based compliant microgripper. In: 2014 IEEE International Conference on Robotics and Biomimetics (ROBIO 2014), pp. 2588–2593 (2014)
2. Ai, W., Xu, Q.: Overview of flexure-based compliant microgrippers. Adv. Robot. Res. **1**(1), 1 (2014)
3. Alger, E.T., Dzenitis, E.G., Mapoles, E.R., Klingmann, J.L., Bhandarkar, S.D., Reynolds, J.G., Florio, J.W., Lord, D.M., Castro, C., Segraves, K.: Experimental DT ice-layering target assembly. Fusion Sci. Technol. **55**(3), 269–275 (2009)
4. Bazaz, S.A., Khan, F., Shakoor, R.I.: Design, simulation and testing of electrostatic SOI MUMPs based microgripper integrated with capacitive contact sensor. Sens. Actuators, A **167**(1), 44–53 (2011)
5. Beroz, J., Awtar, S., Bedewy, M., Tawfick, S., Hart, A.J.: Compliant microgripper with parallel straight-line jaw trajectory for nanostructure manipulation. In: Proceedings of 26th American Society of precision engineering annual meeting, Denver, USA (2011)
6. Bharanidaran, R., Ramesh, T.: A modified post-processing technique to design a compliant based microgripper with a plunger using topological optimization. Int. J. Adv. Manuf. Technol. **93**(1), 103–112 (2017)
7. Carrozza, M.C., Eisinberg, A., Menciassi, A., Campolo, D., Micera, S., Dario, P.: Towards a force-controlled microgripper for assembling biomedical microdevices. J. Micromech. Microeng. **10**(2), 271 (2000)
8. Cecchi, R., Verotti, M., Capata, R., Dochshanov, A., Broggiato, G.B., Crescenzi, R., Balucani, M., Natali, S., Razzano, G., Lucchese, F., et al.: Development of micro-grippers for tissue and cell manipulation with direct morphological comparison. Micromachines **6**(11), 1710–1728 (2015)

9. Chang, P.L., Chi, I.T., Tran, N.D.K., Wang, D.A.: Design and modeling of a compliant gripper with parallel movement of jaws. Mech. Mach. Theory **152**, 103942 (2020)
10. Chen, B.K., Zhang, Y., Sun, Y., et al.: Active release of microobjects using a MEMS microgripper to overcome adhesion forces. J. Microelectromech. Syst. **18**(3), 652–659 (2009)
11. Chen, C.M., Hsu, Y.C., Fung, R.F.: System identification of a Scott-Russell amplifying mechanism with offset driven by a piezoelectric actuator. Appl. Math. Model. **36**(6), 2788–2802 (2012)
12. Chen, F., Du, Z.j., Yang, M., Gao, F., Dong, W., Zhang, D.: Design and analysis of a three-dimensional bridge-type mechanism based on the stiffness distribution. Precis. Eng. **51**, 48–58 (2018)
13. Chen, G., Liu, X., Du, Y.: Elliptical-arc-fillet flexure hinges: toward a generalized model for commonly used flexure hinges. J. Mech. Design **133**(8) (2011)
14. Chen, W., Qu, J., Chen, W., Zhang, J.: A compliant dual-axis gripper with integrated position and force sensing. Mechatronics **47**, 105–115 (2017)
15. Chen, W., Shi, X., Chen, W., Zhang, J.: A two degree of freedom micro-gripper with grasping and rotating functions for optical fibers assembling. Rev. Sci. Instrum. **84**(11), 161–3440 (2013)
16. Chen, W., Zhang, X., Li, H., Wei, J., Fatikow, S.: Nonlinear analysis and optimal design of a novel piezoelectric-driven compliant microgripper. Mech. Mach. Theory **118**, 32–52 (2017)
17. Chen, X., Deng, Z., Hu, S., Gao, J., Gao, X.: Design of a compliant mechanism based four-stage amplification piezoelectric-driven asymmetric microgripper. Micromachines **11**(1), 25 (2020)
18. Chen, X., Deng, Z., Hu, S., Gao, J., Gao, X.: Designing a novel model of 2-DOF large displacement with a stepwise piezoelectric-actuated microgripper. Microsyst. Technol. **26**(9), 2809–2816 (2020)
19. Chen, X., Deng, Z., Hu, S., Gao, X., Gao, J.: Research on three-stage amplified compliant mechanism-based piezo-driven microgripper. Adv. Mech. Eng. **12**(3), 1687814020911470 (2020)
20. Chen, X., Li, Y.: A new structure to achieve large-scale damage-avoiding capture based on compliant mechanism. Microsyst. Technol. 1–8 (2020)
21. Choi, H., Shin, D., Ryuh, Y., Han, C.: Development of a micro manipulator using a microgripper and PZT actuator for microscopic operations. In: 2011 IEEE International Conference on Robotics and Biomimetics, pp. 744–749 (2011)
22. Choi, H.S., Lee, D.C., Kim, S.S., Han, C.S.: The development of a microgripper with a perturbation-based configuration design method. J. Micromech. Microeng. **15**(6), 1327 (2005)
23. Dao, T.P., Ho, N.L., Nguyen, T.T., Le, H.G., Thang, P.T., Pham, H.T., Do, H.T., Tran, M.D., Nguyen, T.T.: Analysis and optimization of a micro-displacement sensor for compliant microgripper. Microsyst. Technol. **23**(12), 5375–5395 (2017)
24. Dao, T.P., Huang, S.C., Le Chau, N.: Robust parameter design for a compliant microgripper based on hybrid taguchi-differential evolution algorithm. Microsyst. Technol. **24**(3), 1461–1477 (2018)
25. Das, T.K., Shirinzadeh, B., Al-Jodah, A., Ghafarian, M., Pinskier, J.: Computational parametric analysis and experimental investigations of a compact flexure-based microgripper. Precis. Eng. **66**, 363–373 (2020)
26. Das, T.K., Shirinzadeh, B., Al-Jodah, A., Ghafarian, M., Pinskier, J.: A novel compliant piezoelectric actuated symmetric microgripper for the parasitic motion compensation. Mech. Mach. Theory **155**, 104069 (2021)
27. Das, T.K., Shirinzadeh, B., Ghafarian, M., Al-Jodah, A.: Design, analysis, and experimental investigation of a single-stage and low parasitic motion piezoelectric actuated microgripper. Smart Mater. Struct. **29**(4), 045028 (2020)
28. Das, T.K., Shirinzadeh, B., Ghafarian, M., Al-Jodah, A., Pinskier, J.: Characterization of a compact piezoelectric actuated microgripper based on double-stair bridge-type mechanism. J. Micro-Bio Rob. **16**(1), 79–92 (2020)
29. Das, T.K., Shirinzadeh, B., Ghafarian, M., Al-Jodah, A., Zhong, Y., Smith, J.: Design, analysis and experimental investigations of a high precision flexure-based microgripper for micro/nano manipulation. Mechatronics **69**, 102396 (2020)

30. Das, T.K., Shirinzadeh, B., Ghafarian, M., Pinskier, J.: A flexure-based 2-DOF microgripper for handling micro-objects. In: 2018 International Conference on Manipulation, Automation and Robotics at Small Scales (MARSS), pp. 1–6 (2018)
31. Dochshanov, A., Verotti, M., Belfiore, N.P.: A comprehensive survey on microgrippers design: operational strategy. J. Mech. Design **139**(7) (2017)
32. Dsouza, R.D., Navin, K.P., Theodoridis, T., Sharma, P.: Design, fabrication and testing of a 2 DOF compliant flexural microgripper. Microsyst. Technol. **24**(9), 3867–3883 (2018)
33. El-Sayed, A.M., Abo-Ismail, A., El-Melegy, M.T., Hamzaid, N.A., Osman, N.A.A.: Development of a micro-gripper using piezoelectric bimorphs. Sensors **13**(5), 5826–5840 (2013)
34. Fang, Y., Tan, X.: A dynamic JKR model with application to vibrational release in micromanipulation. In: 2006 IEEE/RSJ International Conference on Intelligent Robots and Systems, pp. 1341–1346 (2006)
35. Fantoni, G., Porta, M.: A critical review of releasing strategies in microparts handling. In: International Precision Assembly Seminar, pp. 223–234 (2008)
36. Feng, F.Y., Cui, Y.G., Xue, F., Wu, L.E.: Design of a new piezo-electric micro-gripper based on flexible magnifying mechanism. Appl. Mech. Mater. **201**, 907–911 (2012)
37. Gao, Q., Zhang, D., Xu, D., Zhang, Z.: A kinematics modeling and stress analysis method for flexible micro-gripper. In: 2012 IEEE International Conference on Mechatronics and Automation, pp. 825–830 (2012)
38. Garcés-Schröder, M., Leester-Schädel, M., Schulz, M., Böl, M., Dietzel, A.: Micro-gripper: a new concept for a monolithic single-cell manipulation device. Sens. Actuators A: Phys. **236**, 130–139 (2015)
39. Grossard, M., Rotinat-Libersa, C., Chaillet, N., Boukallel, M.: Mechanical and control-oriented design of a monolithic piezoelectric microgripper using a new topological optimization method. IEEE/ASME Trans. Mechatron. **14**(1), 32–45 (2009)
40. Gu, G.Y., Zhu, L.M., Su, C.Y., Ding, H.: Motion control of piezoelectric positioning stages: modeling, controller design, and experimental evaluation. IEEE/ASME Trans. Mechatron. **18**(5), 1459–1471 (2012)
41. Ha, J.L., Kung, Y.S., Hu, S.C., Fung, R.F.: Optimal design of a micro-positioning Scott-Russell mechanism by taguchi method. Sens. Actuators, A **125**(2), 565–572 (2006)
42. Hajhashemi, M., Barazandeh, F., Nejad, S.N., Nadafi DB, R.: Design and microfabrication of a constant-force microgripper. Proc. Inst. Mech. Eng. Part C: J. Mech. Eng. Sci. **225**(11), 2739–2748 (2011)
43. Hao, G., Mullins, J., Cronin, K.: Simplified modelling and development of a bi-directionally adjustable constant-force compliant gripper. Proc. Inst. Mech. Eng. C J. Mech. Eng. Sci. **231**(11), 2110–2123 (2017)
44. Hao, G., Zhu, J.: Design of a monolithic double-slider based compliant gripper with large displacement and anti-buckling ability. Micromachines **10**(10), 665 (2019)
45. Ho, N.L., Dang, M.P., Dao, T.P.: Design and analysis of a displacement sensor-integrated compliant microgripper based on parallel structure. Vietnam J. Mech. **42**(4), 363–374 (2020)
46. Huang, H., Zhao, H., Yang, Z., Mi, J., Fan, Z., Wan, S., Shi, C., Ma, Z.: A novel driving principle by means of the parasitic motion of the microgripper and its preliminary application in the design of the linear actuator. Rev. Sci. Instrum. **83**(5), 055002 (2012)
47. Iamoni, S., Somà, A.: Design of an electro-thermally actuated cell microgripper. Microsyst. Technol. **20**(4–5), 869–877 (2014)
48. Iqbal, S., Malik, A.: A review on MEMS based micro displacement amplification mechanisms. Sens. Actuators, A **300**, 111666 (2019)
49. Jain, R.K., Majumder, S., Ghosh, B., Saha, S.: Design and manufacturing of mobile micro manipulation system with a compliant piezoelectric actuator based micro gripper. J. Manuf. Syst. **35**, 76–91 (2015)
50. Jayaram, K., Joshi, S.S.: Development of a flexure-based, force-sensing microgripper for micro-object manipulation. J. Micromech. Microeng. **20**(1), 015001 (2009)
51. Johal, Z.Z., Crippen, J.W., Forsman, A.C., Lundgren, E.H., Moreno, K.A., Nikroo, A.: Robust capsule and fill tube assemblies for the national ignition campaign. Fusion Sci. Technol. **55**(3), 331–336 (2009)

52. Kim, B., Park, J.S., Moon, C., Jeong, G.M., Ahn, H.S.: A precision robot system with modular actuators and MEMS micro gripper for micro system assembly. J. Mech. Sci. Technol. **22**(1), 70–76 (2008)
53. Kim, D.H., Lee, M.G., Kim, B., Sun, Y.: A superelastic alloy microgripper with embedded electromagnetic actuators and piezoelectric force sensors: a numerical and experimental study. Smart Mater. Struct. **14**(6), 1265 (2005)
54. Kim, J.H., Kim, S.H., Kwak, Y.K.: Development of a piezoelectric actuator using a three-dimensional bridge-type hinge mechanism. Rev. Sci. Instrum. **74**(5), 2918–2924 (2003)
55. Kiziroglou, M.E., Temelkuran, B., Yeatman, E.M., Yang, G.Z.: Micro motion amplification-a review. IEEE Access **8**, 64037–64055 (2020)
56. Liang, C., Wang, F., Shi, B., Huo, Z., Zhou, K., Tian, Y., Zhang, D.: Design and control of a novel asymmetrical piezoelectric actuated microgripper for micromanipulation. Sens. Actuators, A **269**, 227–237 (2018)
57. Liang, C., Wang, F., Tian, Y., Zhao, X., Zhang, D.: Development of a high speed and precision wire clamp with both position and force regulations. Rob. Comp.-Integr. Manuf. **44**, 208–217 (2017)
58. Lin, C., Ren, Y.h., Ji, J.x., Cai, L.z., Shao, J.m.: The bond graph method for analysis of the micro-motion characteristics of a micro gripper. Strojniški vestnik-J. Mech. Eng. **62**(9), 494–502 (2016)
59. Lin, C., Shen, Z., Wu, Z., Yu, J.: Kinematic characteristic analysis of a micro-/nano positioning stage based on bridge-type amplifier. Sens. Actuators, A **271**, 230–242 (2018)
60. Liu, Y., Xu, Q.: Design and analysis of a micro-gripper with constant force mechanism. In: 2016 12th World Congress on Intelligent Control and Automation (WCICA), pp. 2142–2147 (2016)
61. Liu, Y., Xu, Q.: Design of a compliant constant force gripper mechanism based on buckled fixed-guided beam. In: 2016 International Conference on Manipulation, Automation and Robotics at Small Scales (MARSS), pp. 1–6 (2016)
62. Liu, Y., Xu, Q.: Mechanical design, analysis and testing of a large-range compliant microgripper. Mech. Sci. **7**(1), 119–126 (2016)
63. Liu, Y., Zhang, Y., Xu, Q.: Design and control of a novel compliant constant-force gripper based on buckled fixed-guided beams. IEEE/ASME Trans. Mechatron. **22**(1), 476–486 (2016)
64. Llewellyn-Evans, H., Griffiths, C., Fahmy, A.: Microgripper design and evaluation for automated μ-wire assembly: a survey. Microsyst. Technol. **26**(6), 1745–1768 (2020)
65. Lobontiu, N., Garcia, E.: Analytical model of displacement amplification and stiffness optimization for a class of flexure-based compliant mechanisms. Comput. & Struct. **81**(32), 2797–2810 (2003)
66. Lofroth, M., Avci, E.: Development of a novel modular compliant gripper for manipulation of micro objects. Micromachines **10**(5), 313 (2019)
67. Long, Z., Zhang, J., Liu, Y., Han, C., Li, Y., Li, Z.: Dynamics modeling and residual vibration control of a piezoelectric gripper during wire bonding. IEEE Trans. Compon. Pack. Manuf. Technol. **7**(12), 2045–2056 (2017)
68. Lu, K., Zhang, J., Chen, W., Jiang, J., Chen, W.: A monolithic microgripper with high efficiency and high accuracy for optical fiber assembly. In: 2014 9th IEEE Conference on Industrial Electronics and Applications, pp. 1942–1947 (2014)
69. Ma, H.W., Yao, S.M., Wang, L.Q., Zhong, Z.: Analysis of the displacement amplification ratio of bridge-type flexure hinge. Sens. Actuators, A **132**(2), 730–736 (2006)
70. Menciassi, A., Eisinberg, A., Carrozza, M.C., Dario, P.: Force sensing microinstrument for measuring tissue properties and pulse in microsurgery. IEEE/ASME Trans. Mechatron. **8**(1), 10–17 (2003)
71. Nah, S.K., Zhong, Z.W.: A microgripper using piezoelectric actuation for micro-object manipulation. Sens. Actuators, A **133**(1), 218–224 (2007)
72. Nikoobin, A., Niaki, M.H.: Deriving and analyzing the effective parameters in microgrippers performance. Scientia Iranica **19**(6), 1554–1563 (2012)

73. Panepucci, R.R., Martinez, J.A.: Novel SU-8 optical waveguide microgripper for simultaneous micromanipulation and optical detection. J. Vacuum Sci. & Technol. B: Microelectr. Nanometer Struct. Proc. Measur. Phenomena **26**(6), 2624–2627 (2008)
74. Qi, K.q., Xiang, Y., Fang, C., Zhang, Y., Yu, C.s.: Analysis of the displacement amplification ratio of bridge-type mechanism. Mech. Mach. Theory **87**, 45–56 (2015)
75. Rabenorosoa, K., Clevy, C., Lutz, P., Das, A.N., Murthy, R., Popa, D.: Precise motion control of a piezoelectric microgripper for microspectrometer assembly. In: International Design Engineering Technical Conferences and Computers and Information in Engineering Conference, vol. 49033, pp. 769–776 (2009)
76. Raghavendra, M.R.A., Kumar, A.S., Jagdish, B.N.: Design and analysis of flexure-hinge parameter in microgripper. Int. J. Adv. Manuf. Technol. **49**(9–12), 1185–1193 (2010)
77. Roch, I., Bidaud, P., Collard, D., Buchaillot, L.: Fabrication and characterization of an SU-8 gripper actuated by a shape memory alloy thin film. J. Micromech. Microeng. **13**(2), 330 (2003)
78. Ruiz, D., Díaz-Molina, A., Sigmund, O., Donoso, A., Bellido, J.C., Sánchez-Rojas, J.L.: Optimal design of robust piezoelectric unimorph microgrippers. Appl. Math. Model. **55**, 1–12 (2018)
79. Salim, R., Wurmus, H., Harnisch, A., Hülsenberg, D.: Microgrippers created in microstructurable glass. Microsyst. Technol. **4**(1), 32–34 (1997)
80. Sariola, V., Jääskeläinen, M., Zhou, Q.: Hybrid microassembly combining robotics and water droplet self-alignment. IEEE Trans. Robot. **26**(6), 965–977 (2010)
81. Shi, B., Wang, F., Huo, Z., Tian, Y., Zhao, X., Zhang, D.: Design and characteristics of a novel compliant symmetric microgripper mechanism. In: 2018 IEEE International Conference on Manipulation, Manufacturing and Measurement on the Nanoscale (3M-NANO), pp. 65–69 (2018)
82. Shi, Q., Yu, Z., Wang, H., Sun, T., Huang, Q., Fukuda, T.: Development of a highly compact microgripper capable of online calibration for multisized microobject manipulation. IEEE Trans. Nanotechnol. **17**(4), 657–661 (2018)
83. Shi, X., Zhang, J., Chen, W., Chen, W.: Design, modeling, and simulation of a 2-DOF microgripper for grasping and rotating of optical fibers. In: Proceedings of IEEE/ASME International Conference on Advanced Intelligent Mechatronics, pp. 1597–1602 (2013)
84. Shimada, N., Yoshioka, T., Ohishi, K., Miyazaki, M.: Smooth touch control between position control and force control for industrial robots. In: The 36th Annual Conference on IEEE Industrial Electronics Society (IECON 2010), pp. 1949–1954 (2010)
85. Sun, X., Chen, W., Fatikow, S., Tian, Y., Zhou, R., Zhang, J., Mikczinski, M.: A novel piezo-driven microgripper with a large jaw displacement. Microsyst. Technol. **21**(4), 931–942 (2015)
86. Sun, X., Chen, W., Tian, Y., Fatikow, S., Zhou, R., Zhang, J., Mikczinski, M.: A novel flexure-based microgripper with double amplification mechanisms for micro/nano manipulation. Rev. Sci. Instrum. **84**(8), 085002 (2013)
87. Tang, H., Li, Y.: A new flexure-based y-θ nanomanipulator with nanometer-scale resolution and millimeter-scale workspace. IEEE/ASME Trans. Mechatron. **20**(3), 1320–1330 (2014)
88. Tian, Y., Shirinzadeh, B., Zhang, D., Alici, G.: Development and dynamic modelling of a flexure-based Scott-Russell mechanism for nano-manipulation. Mech. Syst. Signal Process. **23**(3), 957–978 (2009)
89. Verotti, M., Dochshanov, A., Belfiore, N.P.: A comprehensive survey on microgrippers design: mechanical structure. J. Mech. Design **139**(6) (2017)
90. Vijayasai, A.P., Sivakumar, G., Mulsow, M., Lacouture, S., Holness, A., Dallas, T.E., Member, I.: Haptic controlled three degree-of-freedom microgripper system for assembly of detachable surface-micromachined MEMS. Sens. Actuators, A **179**, 328–336 (2012)
91. Wang, D., Yang, Q., Dong, H.: A monolithic compliant piezoelectric-driven microgripper: design, modeling, and testing. IEEE/ASME Trans. Mechatron. **18**(1), 138–147 (2011)
92. Wang, F., Liang, C., Tian, Y., Zhao, X., Zhang, D.: Design of a piezoelectric-actuated microgripper with a three-stage flexure-based amplification. IEEE/ASME Trans. Mechatron. **20**(5), 2205–2213 (2014)

93. Wang, F., Liang, C., Tian, Y., Zhao, X., Zhang, D.: Design and control of a compliant microgripper with a large amplification ratio for high-speed micro manipulation. IEEE/ASME Trans. Mechatron. **21**(3), 1262–1271 (2016)
94. Wang, F., Shi, B., Huo, Z., Tian, Y., Zhang, D.: Control and dynamic releasing method of a piezoelectric actuated microgripper. Precis. Eng. **68**, 1–9
95. Wang, F., Shi, B., Tian, Y., Huo, Z., Zhao, X., Zhang, D.: Design of a novel dual-axis micromanipulator with an asymmetric compliant structure. IEEE/ASME Trans. Mechatron. **24**(2), 656–665 (2019)
96. Wang, P., Xu, Q.: Design and modeling of constant-force mechanisms: a survey. Mech. Mach. Theory **119**, 1–21 (2018)
97. Wei, Y., Xu, Q.: A survey of force-assisted robotic cell microinjection technologies. IEEE Trans. Autom. Sci. Eng. **16**(2), 931–945 (2018)
98. Wenjie, C., Wei, L.: Design of a flexure-based gripper used in optical fiber handling. In: IEEE Conference on Robotics, Automation and Mechatronics, 2004., vol. 1, pp. 83–88 (2004)
99. Wu, Z., Li, Y.: Design and control of a novel micro-gripper using adaptive backstepping slide mode control method. Microsyst. Technol. 1–13 (2021)
100. Wu, Z., Xu, Q.: Survey on recent designs of compliant micro-/nano-positioning stages. Actuators **7**(1), 5 (2018)
101. Xiao, S., Li, Y., Zhao, X.: Design and analysis of a novel micro-gripper with completely parallel movement of gripping arms. In: 2011 6th IEEE Conference on Industrial Electronics and Applications, pp. 2127–2132 (2011)
102. Xiao, S., Li, Y., Zhao, X.: Optimal design of a novel micro-gripper with completely parallel movement of gripping arms. In: 2011 IEEE 5th International Conference on Robotics, Automation and Mechatronics (RAM), pp. 35–40 (2011)
103. Xing, Q.: Design of asymmetric flexible micro-gripper mechanism based on flexure hinges. Adv. Mech. Eng. **7**(6), 1687814015590331 (2015)
104. Xing, Q., Ge, Y.: Parametric study of a novel asymmetric micro-gripper mechanism. J. Adv. Mech. Design Syst. Manufact. **9**(5), JAMDSM0075–JAMDSM0075 (2015)
105. Xu, Q.: Mechanism design and analysis of a novel 2-DOF compliant modular microgripper. In: 2012 7th IEEE Conference on Industrial Electronics and Applications (ICIEA), pp. 1966–1971 (2012)
106. Xu, Q.: A new compliant microgripper with integrated position and force sensing. In: 2013 IEEE/ASME International Conference on Advanced Intelligent Mechatronics, pp. 591–596 (2013)
107. Xu, Q.: New robust position and force regulation for a compliant microgripper. In: 2013 IEEE International Conference on Automation Science and Engineering (CASE), pp. 801–806 (2013)
108. Xu, Q.: Precision position/force interaction control of a piezoelectric multimorph microgripper for microassembly. IEEE Trans. Autom. Sci. Eng. **10**(3), 503–514 (2013)
109. Xu, Q.: Robust impedance control of a compliant microgripper for high-speed position/force regulation. IEEE Trans. Industr. Electron. **62**(2), 1201–1209 (2014)
110. Xu, Q.: Design of a constant-force microgripper mechanism for biological micromanipulation. In: 2017 IEEE 12th International Conference on Nano/Micro Engineered and Molecular Systems (NEMS), pp. 418–421 (2017)
111. Xu, Q., Li, Y.: Analytical modeling, optimization and testing of a compound bridge-type compliant displacement amplifier. Mech. Mach. Theory **46**(2), 183–200 (2011)
112. Yamanishi, Y., Sakuma, S., Arai, F.: On-chip cell manipulation by magnetically modified soft microactuators. In: 2008 IEEE International Conference on Robotics and Automation, pp. 899–904 (2008)
113. Yang, S., Xu, Q.: A review on actuation and sensing techniques for MEMS-based microgrippers. J. Micro-Bio Robot. **13**(1), 1–14 (2017)
114. Yang, Y.l., Wei, Y.d., Lou, J.q., Tian, G., Zhao, X.w., Fu, L.: A new piezo-driven microgripper based on the double-rocker mechanism. Smart Mater. Struct. **24**(7), 075031 (2015)

115. Yang, Y.l., Wei, Y.d., Lou, J.q., Xie, F.r., Fu, L.: Development and precision position/force control of a new flexure-based microgripper. J. Micromech. Microeng. **26**(1), 015005 (2015)
116. Yao, S., Li, H., Pang, S., Yu, L., Fatikow, S., Zhang, X.: Motion measurement system of compliant mechanisms using computer micro-vision. Opt. Express **29**(4), 5006–5017 (2021)
117. Ye, G., Li, W., Wang, Y.q., Yang, X.f., Yu, L.: Kinematics analysis of bridge-type micro-displacement mechanism based on flexure hinge. In: The 2010 IEEE International Conference on Information and Automation, pp. 66–70 (2010)
118. Ye, T., Ling, J., Kang, X., Feng, Z., Xiao, X.: A novel two-stage constant force compliant microgripper. J. Mech. Design **143**(5) (2021)
119. Yu, J., Bi, S., Zong, G., Liu, X.J.: On the design of compliant-based micro-motion manipulators with a nanometer range resolution. In: Proceedings 2003 IEEE/ASME International Conference on Advanced Intelligent Mechatronics (AIM 2003), vol. 1, pp. 149–154 (2003)
120. Zhang, D., Zhang, Z., Gao, Q., Xu, D., Liu, S.: Development of a monolithic compliant SPCA-driven micro-gripper. Mechatronics **25**, 37–43 (2015)
121. Zhang, J., Lu, K., Chen, W., Jiang, J., Chen, W.: Monolithically integrated two-axis microgripper for polarization maintaining in optical fiber assembly. Rev. Sci. Instrum. **86**(2), 025105 (2015)
122. Zhang, R., Chu, J., Wang, H., Chen, Z.: A multipurpose electrothermal microgripper for biological micro-manipulation. Microsyst. Technol. **19**(1), 89–97 (2013)
123. Zhang, X., Xu, Q.: Design and analysis of a 2-DOF compliant gripper with constant-force flexure mechanism. J. Micro-Bio Robot. **15**(1), 31–42 (2019)
124. Zhang, Y., Xu, Q.: Adaptive sliding mode control with parameter estimation and kalman filter for precision motion control of a piezo-driven microgripper. IEEE Trans. Control Syst. Technol. **25**(2), 728–735 (2016)
125. Zhao, Y., Huang, X., Liu, Y., Wang, G., Hong, K.: Design and control of a piezoelectric-driven microgripper perceiving displacement and gripping force. Micromachines **11**(2), 121 (2020)
126. Zhu, B., Zhang, X., Zhang, H., Liang, J., Zang, H., Li, H., Wang, R.: Design of compliant mechanisms using continuum topology optimization: a review. Mech. Mach. Theory **143**, 103622 (2020)
127. Zhu, J., Hao, G.: Design and test of a compact compliant gripper using the Scott-Russell mechanism. Arch. Civil Mech. Eng. **20**(3), 1–12 (2020)
128. Zubir, M.N.M., Shirinzadeh, B.: Development of a high precision flexure-based microgripper. Precis. Eng. **33**(4), 362–370 (2009)
129. Zubir, M.N.M., Shirinzadeh, B., Tian, Y.: Development of a novel flexure-based microgripper for high precision micro-object manipulation. Sens. Actuators, A **150**(2), 257–266 (2009)
130. Zubir, M.N.M., Shirinzadeh, B., Tian, Y.: A new design of piezoelectric driven compliant-based microgripper for micromanipulation. Mech. Mach. Theory **44**(12), 2248–2264 (2009)

Chapter 3
Design of a Compliant Gripper Based on the Scott-Russell Mechanism

Abstract This chapter presents the structural design and analysis of a compliant microgripper based on a Scott-Russell (SR) mechanism. The integration of the SR mechanism with a parallelogram guiding structure enables pure translational motion at the gripper tips, a critical feature for precision micromanipulation and microassembly applications. To validate the design, comprehensive kinematic, static, and dynamic models of the gripper mechanism were established and verified through finite element analysis (FEA) simulations. A prototype was subsequently fabricated for experimental testing. The results confirm not only the feasibility of the proposed SR-based gripper design but also demonstrate its promising performance when driven by piezoelectric stack actuators. Furthermore, several structural variations of the gripper are also presented to illustrate its design flexibility.

Keywords Microgripper · Compliant mechanism · Mechanism design · Scott-Russell mechanism · FEA simulation

3.1 Introduction

Robotic micro- and nano-handling systems are essential for enabling automated manipulation and assembly at micro- and nanometer scales [3]. As a crucial component of these systems, microgrippers have attracted significant attention from both academia and industry. Various microgripper designs have been proposed based on different actuation principles, including electrostatic [1], electrothermal [8], and piezoelectric drives [9]. Among these, piezoelectric actuators are particularly advantageous due to their rapid response speed and ultrahigh positioning resolution [15, 19]. Consequently, a piezoelectric stack actuator (PSA) is employed as the driving mechanism in this research.

In terms of structural design, compliant mechanisms have been widely adopted in microgripper development. Unlike traditional bearings, compliant mechanisms generate motion through the elastic deformation of materials, offering notable benefits such as backlash-free operation, friction elimination, and no need for lubrication [6, 10, 13, 17, 21]. However, most existing microgrippers utilize two-arm

Q. Xu, *Piezoelectric-Driven Flexible Robotic Microgrippers*,
https://doi.org/10.1007/978-981-95-4685-5_3

configurations based on rotary motion principles [16]. A significant limitation arises when gripping objects with curved surfaces: the reaction force at the contact point may push the object away from the gripper tips [7]. Therefore, microgrippers capable of pure translational motion at their tips are highly desirable [12, 20]. Although such parallel-motion grippers exist, they often feature complex structures [7, 20], which complicate both analysis and fabrication processes.

The present research aims to develop a new parallel-motion gripper with a simplified architecture. Specifically, a Scott-Russell (SR) mechanism combined with a parallelogram mechanism is adopted for the design of each gripper arm. The SR mechanism has been previously utilized in micropositioning systems; for example, the work [2] employed two SR mechanisms for two-level displacement amplification in a micropositioner, while [4] presented an optimal design for maximizing PSA displacement amplification. More recently, [5] developed an ultraprecision rotary micropositioning stage using an SR mechanism. Despite these advancements, the application of SR mechanisms in gripper design remains largely unexplored.

This chapter introduces a compliant microgripper based on an SR mechanism. The use of the SR mechanism enables a simpler structural design compared to traditional microgrippers. Additionally, the in-plane space is efficiently utilized, resulting in a compact overall dimension. The feasibility and performance of the proposed gripper are validated through finite-element analysis (FEA) simulations and experimental studies.

3.2 Mechanism Design of the Microgripper

Figure 3.1 presents a CAD model of the devised compliant microgripper. The mechanism is constructed using flexure hinges and is actuated by two piezoelectric stack actuators (PSAs). Two preloading screws are incorporated to adjust the preload force applied between each PSA and the gripper structure.

The gripper architecture integrates two Scott-Russell (SR) mechanisms to amplify the input displacement from the PSAs. Parallelogram mechanisms are connected to each SR mechanism to ensure pure translational motion of the gripper tips. A second-stage lever amplifier is inserted between each SR mechanism and its corresponding parallelogram mechanism to provide additional displacement amplification for each gripper arm. Notably, the PSAs are fully embedded within the gripper structure, enabling efficient use of the in-plane space and resulting in a highly compact overall design.

In this design, all flexure hinges employ a right-circular shape. As illustrated in Fig. 3.2, each flexure hinge can be modeled as a revolute joint combined with a torsional spring. Compared to other hinge profiles, the right-circular geometry minimizes the center shift of the instantaneous rotational center. Based on the pseudo-rigid-body model (PRBM) framework, the displacement and stiffness models of the gripper mechanism are established in the following sections.

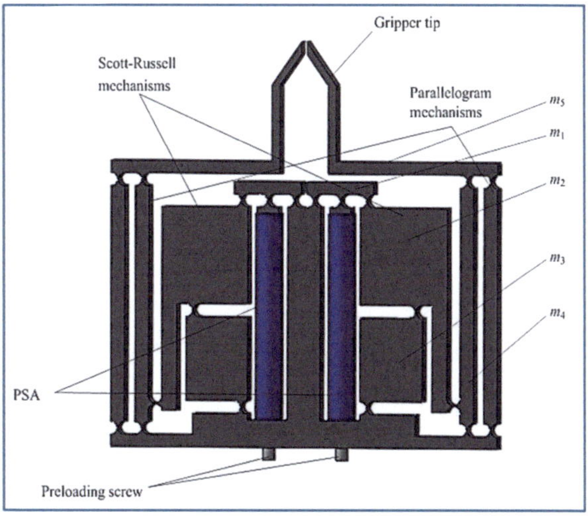

Fig. 3.1 Computer-aided design (CAD) model of the compliant microgripper mechanism, featuring dual piezoelectric stack actuators (PSAs) for actuation and preloading screws for force adjustment

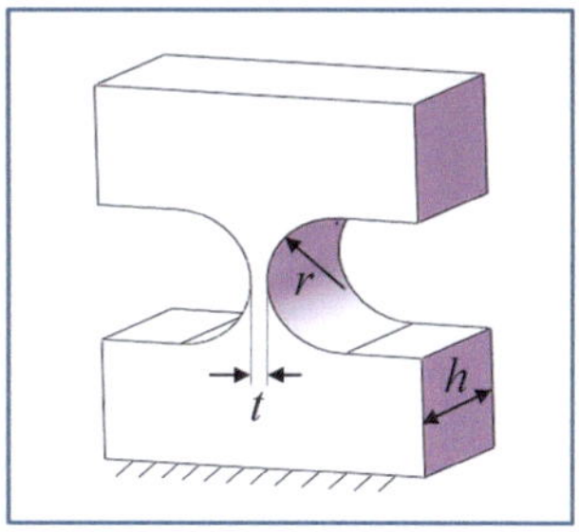

Fig. 3.2 CAD model of a right-circular flexure hinge, illustrating its key geometric parameters

3.2.1 Scott-Russell Mechanism

The schematic of a flexure hinge-based Scott-Russell mechanism is depicted in Fig. 3.3, where point O represents a fixed pin joint, point A is the input point driven by a piezoelectric stack actuator (PSA), and point B is the output point.

Given the geometric constraints $|AC| = |BC| = |OC| = l$ and $\angle OAB = \theta$, the coordinates of output point B are expressed as:

$$x_B = \sqrt{4l^2 - y_A^2} \tag{3.1}$$

$$y_B = 0 \tag{3.2}$$

where y_A denotes the y-coordinate of point A.

For a Scott-Russell mechanism satisfying $|AC| = |BC| = |OC|$, when point A undergoes a displacement Δy_A along the y-axis, point B exhibits a straight-line

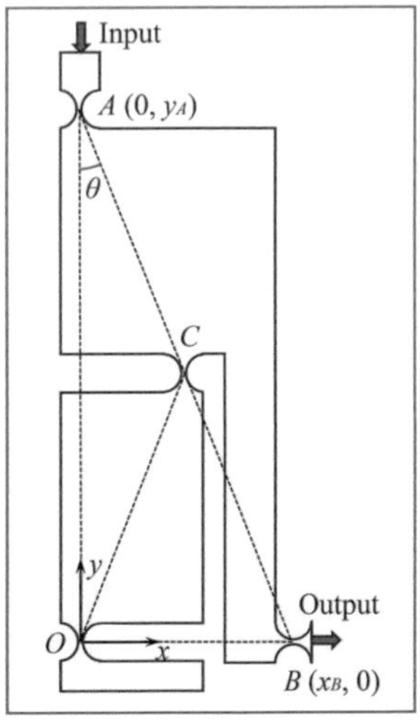

Fig. 3.3 Schematic diagram of a Scott-Russell (SR) compliant mechanism implemented with right-circular flexure hinges, showing the geometric relationships for displacement amplification and orthogonal motion conversion

translational displacement Δx_B along the x-axis. The kinematic relationship is governed by:

$$x_B^2 + y_A^2 = (x_B + \Delta x_B)^2 + (y_A + \Delta y_A)^2. \tag{3.3}$$

Solving this equation yields the nonlinear input-output displacement relation:

$$\Delta x_B = -x_B + \sqrt{x_B^2 - 2y_A \Delta y_A - \Delta y_A^2}. \tag{3.4}$$

For small displacements ($\Delta y_A \to 0$), the relationship simplifies to the linearized form:

$$\eta_1 = \frac{\Delta x_B}{\Delta y_A} \approx \frac{dx_B}{dy_A} = -\frac{y_A}{\sqrt{4l^2 - y_A^2}} = -\frac{y_A}{x_B} = -\cot\theta, \tag{3.5}$$

as derived from Eq. (3.1).

Equation (3.5) reveals that a small input displacement Δy_A along the y-axis produces an amplified output displacement $\Delta x_B = -\Delta y_A \cot\theta$ along the x-axis when $0 < \theta < \pi/4$. The negative sign indicates that a decrease in Δy_A results in an increase in Δx_B. Thus, for small-motion applications, the Scott-Russell mechanism functions as a straight-line motion amplifier with a fixed amplification ratio $\eta_1 = -\cot\theta$.

3.2.2 Lever Mechanism

The second-stage lever mechanism is illustrated in Fig. 3.4. The amplification ratio of this mechanism is given by:

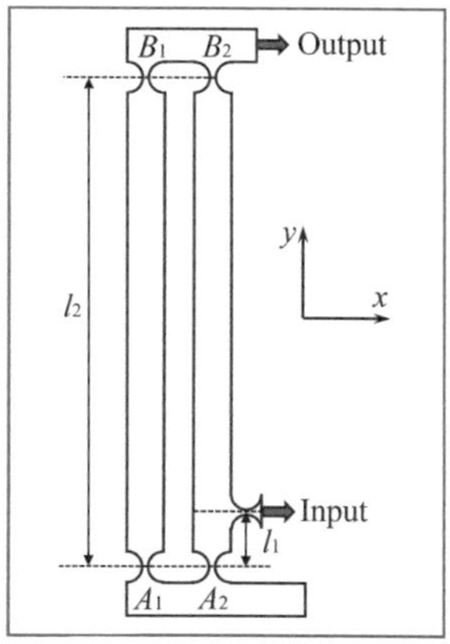

Fig. 3.4 Geometric parameters and kinematic relationships of the second-stage lever amplification mechanism

$$\eta_2 = \frac{l_2}{l_1} \tag{3.6}$$

where l_1 and l_2 represent the input and output lever arms, respectively.

By combining the amplification ratios of both the Scott-Russell mechanism (Eq. (3.5)) and the lever mechanism (Eq. (3.6)), the overall input-output displacement relationship of the gripper is derived as:

$$A_S = \frac{x_{\text{out}}}{y_{\text{in}}} = \frac{l_2 \cot\theta}{l_1} \tag{3.7}$$

This equation defines the total displacement amplification ratio between the gripper's output tip displacement (x_{out}) and the input displacement (y_{in}) from the piezoelectric actuator.

3.3 Kinematics Modeling

Given the input displacements (q_1 and q_2) of the two piezoelectric stack actuator (PSA) actuators, the resulting gripper output motions (d_{x1} and d_{x2}) and the required actuator input forces (F_{in1} and F_{in2}) are governed by the following kinematic and static equations:

$$\begin{bmatrix} d_{x1} \\ d_{x2} \end{bmatrix} = \begin{bmatrix} A_S & 0 \\ 0 & A_S \end{bmatrix} \begin{bmatrix} q_1 \\ q_2 \end{bmatrix} \tag{3.8}$$

$$\begin{bmatrix} F_{\text{in1}} \\ F_{\text{in2}} \end{bmatrix} = \begin{bmatrix} K_{\text{in}} & 0 \\ 0 & K_{\text{in}} \end{bmatrix} \begin{bmatrix} q_1 \\ q_2 \end{bmatrix} \tag{3.9}$$

where A_S is the displacement amplification ratio, and K_{in} is the input (actuation) stiffness of the gripper structure.

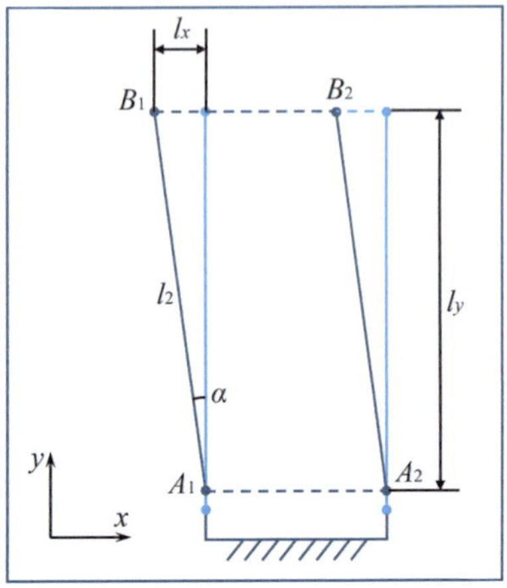

Fig. 3.5 PRBM of the displacement amplifier and its parameters

These relationships allow the kinematics and statics problems to be solved by calculating the structure's amplification ratio and input stiffness, respectively. This chapter establishes a pseudo-rigid-body model (PRBM) for the structure by modeling each flexure hinge as a 1-DOF revolute joint paired with a torsional spring of stiffness K_r. The stiffness equation for K_r with the highest accuracy, as compared in [18], is adopted. Let F_x, Δx, and Δy represent the input force, input displacement, and output displacement, respectively.

In this PRBM, compliance is considered only for the flexure hinges. This means each hinge is assumed to possess 1-DOF rotational compliance due to bending deformation, while all other components are treated as rigid bodies.

The free-body diagram of one amplifier leg is shown in Figs. 3.4 and 3.5. Under equilibrium conditions, the moment equation about point A_1 is derived as:

$$2M_r + \frac{F_x l_y}{2} = \frac{F_y l_x}{2} \tag{3.10}$$

where the moment M_r is given by:

$$M_r = K_r \Delta\alpha \tag{3.11}$$

Here, K_r and $\Delta\alpha$ represent the rotational stiffness and angular deformation of a notch hinge, respectively.

Differentiating the displacement relation $l_x = l_2 \sin\alpha$ (where both l_x and α vary during operation) with respect to time, yields:

$$\Delta y = l_2 \cos\alpha, \quad \Delta\alpha = l_y \Delta\alpha \tag{3.12}$$

The translational stiffness K_1 of the compound parallelogram mechanism originates from the rotational stiffness of its four notch hinges. Based on a potential energy analysis, it is calculated as:

$$K_1 = \frac{K_r}{l_2^2} \tag{3.13}$$

where l_2 is the length of the limb leg, as shown in Fig. 3.4. Consequently, the force F_x can be expressed as:

$$F_x = K_1 \Delta y \tag{3.14}$$

Substituting Eqs. (3.11)–(3.14) into Eq. (3.10) gives the relation between F_x and $\Delta\alpha$:

$$\Delta\alpha = \frac{F_y l_x}{4K_r + \dfrac{K_r l_y^2}{l_2^2}} \tag{3.15}$$

Furthermore, applying the principle of virtual work gives:

$$F_y \Delta x - F_x \Delta y = M_r \Delta\alpha \tag{3.16}$$

This equation states that the work done by external forces equals the work done by internal moments.

Substituting Eqs. (3.11)–(3.15) into Eq. (3.16) yields a relation solely between F_x and Δx. From this, the input displacement is obtained as:

$$\Delta x = \frac{F_y l_x^2}{K_r + \dfrac{K_r l_y^2}{l_2^2}} \tag{3.17}$$

The input stiffness is then derived from Eq. (3.17) as:

$$K_{\text{in}} = \frac{K_r + \dfrac{K_r l_y^2}{l_2^2}}{l_x^2} \tag{3.18}$$

The main parameters of the compliant gripper designed in this work are listed in Table 3.1.

Table 3.1 Key design parameters and their values for the developed compliant micro-gripper mechanism.

Parameter	Value	Unit
θ	24.6	°
l	39.0	mm
l_1	10.0	mm
l_2	190.0	mm
r	2.5	mm
t	0.5	mm
h	10.0	mm
l_y	7.5	mm

3.4 Stiffness, Stress, and Dynamics Analysis

3.4.1 Stiffness and Compliance Analysis

The in-plane stiffness (K_x) of the gripper structure relates the in-plane external forces (F_x) applied to the gripper to the induced deflections (t_x). Conversely, compliance ($C_x = 1/K_x$), being the inverse of stiffness, reflects the linear deflection under an external load as follows:

$$t_x = C_x F_x \tag{3.19}$$

The compliance characterizes the output compliance (or equivalently, the output stiffness) of the structure under an external load.

Owing to the symmetry between the left and right arms of the gripper structure, it can be deduced that $C_{x1} = C_{x2} = C_x$ in theory. In the following derivation, the output compliance C_x of the stage is analyzed by considering an x-axis external force F_x applied to the output platform. The resulting rotation at the hinge can be related to the output displacement by:

$$\theta = \frac{t_x}{2l_2} \tag{3.20}$$

The potential energy of the structure, which arises from the elastic deformation of the material, is stored in the two limbs. This energy can be expressed in two equivalent forms:

$$\frac{1}{2}K_x t_x^2 = 8 \times \frac{1}{2}K_r \theta^2 \tag{3.21}$$

where K_x is the output stiffness of the stage in the x-direction, and K_r represents the rotational stiffness of each flexure hinge around its working axis.

Substituting Eq. (3.20) into Eq. (3.21) yields an expression for the output stiffness:

$$K_x = \frac{2K_r}{l_2^2} \tag{3.22}$$

which implies that the output compliance is:

$$C_x = \frac{l_2^2}{2K_r}. \tag{3.23}$$

3.4.2 Stress and Load Capacity Analysis

The in-plane load capacity, denoted $F_x^{\max}$, is defined as the maximum external force that can be applied to the structure in the x-direction without causing material failure (yielding).

For in-plane operation, the analysis focuses on the bending stress to determine the load limit. This is justified because the axial tensile or compressive stress in the flexure hinge is negligible compared to the maximum bending stress. The failure condition is reached when the maximum stress $\sigma^{\max}$, which occurs at the outermost surface of the hinge's thinnest cross-section, equals the material's yield stress σ_y. At this point, the hinge experiences its maximum allowable angular displacement $\theta^{\max}$.

The relationship between the maximum bending stress and the maximum rotational deformation for a right-circular flexure hinge is given by [11]:

$$\sigma^{\max} = \frac{E(1+\beta)^{9/20}}{\beta^2 f(\beta)}\theta^{\max} \tag{3.24}$$

where E is the Young's modulus of the material, $\beta = t/(2r)$ is a dimensionless geometry factor, and $f(\beta)$ is a dimensionless compliance factor defined as:

$$f(\beta) = \frac{1}{2\beta+\beta^2}\left[\frac{3+4\beta+\beta^2}{(1+\beta)(2\beta+\beta^2)} + \frac{6(1+\beta)}{(2\beta+\beta^2)^{3/2}}\tan^{-1}\left(\sqrt{\frac{2+\beta}{\beta}}\right)\right] \tag{3.25}$$

All flexure hinges in the mechanism are assumed to rotate by the same angle. This maximum rotation is related to the stage's maximum linear deformation $t_x^{\max}$ by the kinematic equation:

$$\theta^{\max} = \frac{t_x^{\max}}{2l_2} \tag{3.26}$$

where l_2 is the length of the output lever arm.

Substituting this kinematic relation into the stress equation (3.24) yields the condition to avoid yielding:

$$\sigma^{\max} = \frac{E(1+\beta)^{9/20}}{2l_2\beta^2 f(\beta)}t_x^{\max} \leq \sigma_y \tag{3.27}$$

Solving this inequality provides the upper limit for the safe linear deformation:

$$t_x^{\max} \leq \frac{2l_2\beta^2 f(\beta)\sigma_y}{E(1+\beta)^{9/20}} \tag{3.28}$$

Finally, the maximum in-plane load capacity is obtained by multiplying this maximum safe deformation by the structure's effective stiffness in the x-direction, K_x:

$$F_x^{\max} = K_x \cdot t_x^{\max}. \tag{3.29}$$

3.4.3 Dynamics and Resonance Frequency Calculation

For the gripper structure, the in-plane motion is described using the generalized coordinate vector $\mathbf{q} = [q_1, q_2]^T$. Considering the kinematic relation in Eq. (3.1), the kinetic energy (T) and potential energy (V) of the stage, stored in each gripper arm, are expressed in terms of these coordinates.

The equations of motion are derived using Lagrange's equation:

$$\frac{d}{dt}\left(\frac{\partial T}{\partial \dot{q}_i}\right) - \frac{\partial T}{\partial q_i} + \frac{\partial V}{\partial q_i} = F_i \tag{3.30}$$

for $i = 1, 2$. This leads to the dynamic equation for the free motion of the stage:

$$\mathbf{M}\ddot{\mathbf{q}} + \mathbf{K}\mathbf{q} = 0 \tag{3.31}$$

Here, $\mathbf{M}$ and $\mathbf{K}$ are the 2×2 equivalent mass and stiffness matrices, respectively. Given the system's symmetry, these matrices are diagonal: $\mathbf{M} = \text{diag}\{M, M\}$ and $\mathbf{K} = \text{diag}\{K, K\}$, where the equivalent mass (M) and stiffness (K) are given by:

$$M = m_1 + \frac{1}{2}m_2 + \frac{1}{25}m_3 + \frac{88}{5}m_4 + 40m_5 \tag{3.32}$$

$$K = K\text{in} \tag{3.33}$$

The masses m_1 to m_5, corresponding to different components of the gripper arm, are illustrated in Fig. 3.1.

Consequently, the in-plane fundamental resonant frequency of the stage is obtained as:

$$f = \frac{1}{2\pi}\sqrt{\frac{K}{M}}. \tag{3.34}$$

3.5 Finite-Element Analysis Simulation Study

A case study is conducted by designing an SR-based compliant gripper with the parameters specified in Table 3.1. The gripper is fabricated from Al-7075 alloy, which has a Young's modulus $E = 7.1 \times 10^{11}$ Pa, Poisson's ratio $\mu = 0.31$, density $\rho = 2810$ kg/m^3, and a yield strength $\sigma_y = 524$ MPa. For actuation, a model P-840.30 piezoelectric stack actuator (PSA) from Physik Instrumente (PI) GmbH & Co. KG is selected. This actuator is 68 mm in length and offers a stroke of 45 μm and a maximum blocking force of 1000 N. Based on the kinematic model (Eq. (3.7)), the displacement amplification ratio is predicted to be $A_S = 13.1$.

3.5.1 Static FEA Simulation Results

A static structural FEA was performed by applying an input force of 1 N to the gripper's input end. The resulting deformed shape is shown in Fig. 3.6a. The displacement amplification ratio, calculated from the extracted displacement values, is $A_S = 12.6$.

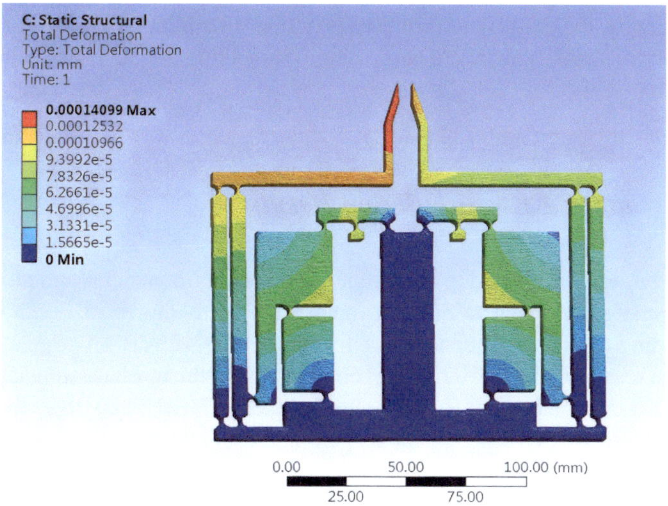

(a)

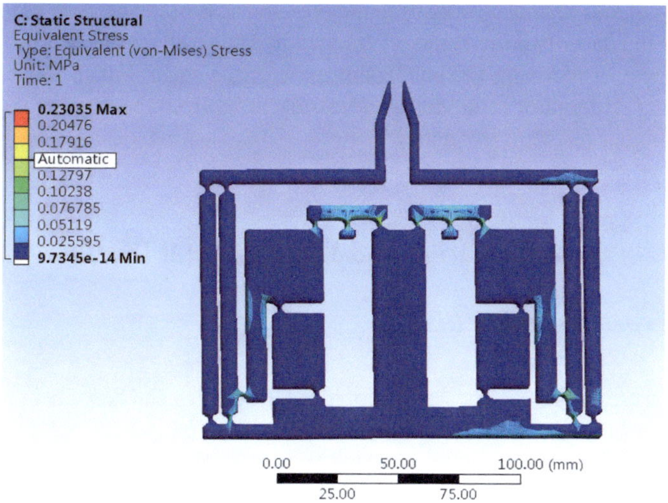

(b)

Fig. 3.6 Finite element analysis (FEA) results for the compliant gripper under a static load. **a** Total deformation shape; **b** von Mises stress distribution

Compared to the FEA result, the kinematic model overestimates the amplification ratio by 3.98%. This discrepancy occurs because the kinematic model, based on the PRBM approach, only considers the bending deformations of the flexure hinges. In contrast, the FEA reveals that other deformations (e.g., in the links between hinges) also occur, which reduces the overall amplification. Consequently, a more comprehensive nonlinear model that accounts for all deformations in the gripper mechanism would improve the accuracy of the kinematic predictions.

Additionally, the equivalent (von Mises) stress distribution is shown in Fig. 3.6b. The maximum stress is observed at the hinge connecting the Scott-Russell and lever mechanisms.

3.5.2 Dynamic FEA Simulation Results

A modal FEA was conducted to determine the resonant frequencies and mode shapes of the gripper structure. The first six mode shapes are shown in Fig. 3.7, and the corresponding resonant frequencies are listed in Table 3.2.

The first resonant mode (102.74 Hz) corresponds to the in-phase translation of the two parallelogram mechanisms. The second mode (107.12 Hz) involves their out-of-phase translation, which represents the opening/closing working mode of the gripper tips. Modes three to five are characterized by out-of-plane deformations, while the sixth mode is caused by an in-plane bending deformation of a single gripper arm.

The resonant frequency calculated by the dynamic model (Eq. (3.34)) is 108.83 Hz. Compared to the FEA result, the dynamic model overestimates the frequency with a deviation of less than 6%, validating the effectiveness of the derived model. Equation (3.34) indicates that the resonant frequency can be increased by enhancing the input stiffness or reducing the equivalent mass of the stage. For instance, employing a thinner material or removing unnecessary mass from the moving parts would achieve a higher resonant frequency.

3.6 Prototype Fabrication and Experimental Results

3.6.1 Experimental Setup

This section presents the fabricated prototype of the gripper and describes preliminary open-loop tests conducted to evaluate its performance.

The microgripper prototype, fabricated from Al-7075 alloy using a wire-cutting process, is shown in Fig. 3.8. For actuation, two PSAs (model P-840.30, Physik Instrumente GmbH & Co. KG) with a stroke of 45 μm are used to drive the mechanism; their specifications are provided in Table 3.3.

A PCI-based D/A board (National Instruments NI-9215 module) generates analog control signals. These signals are amplified by a three-axis voltage amplifier (PI

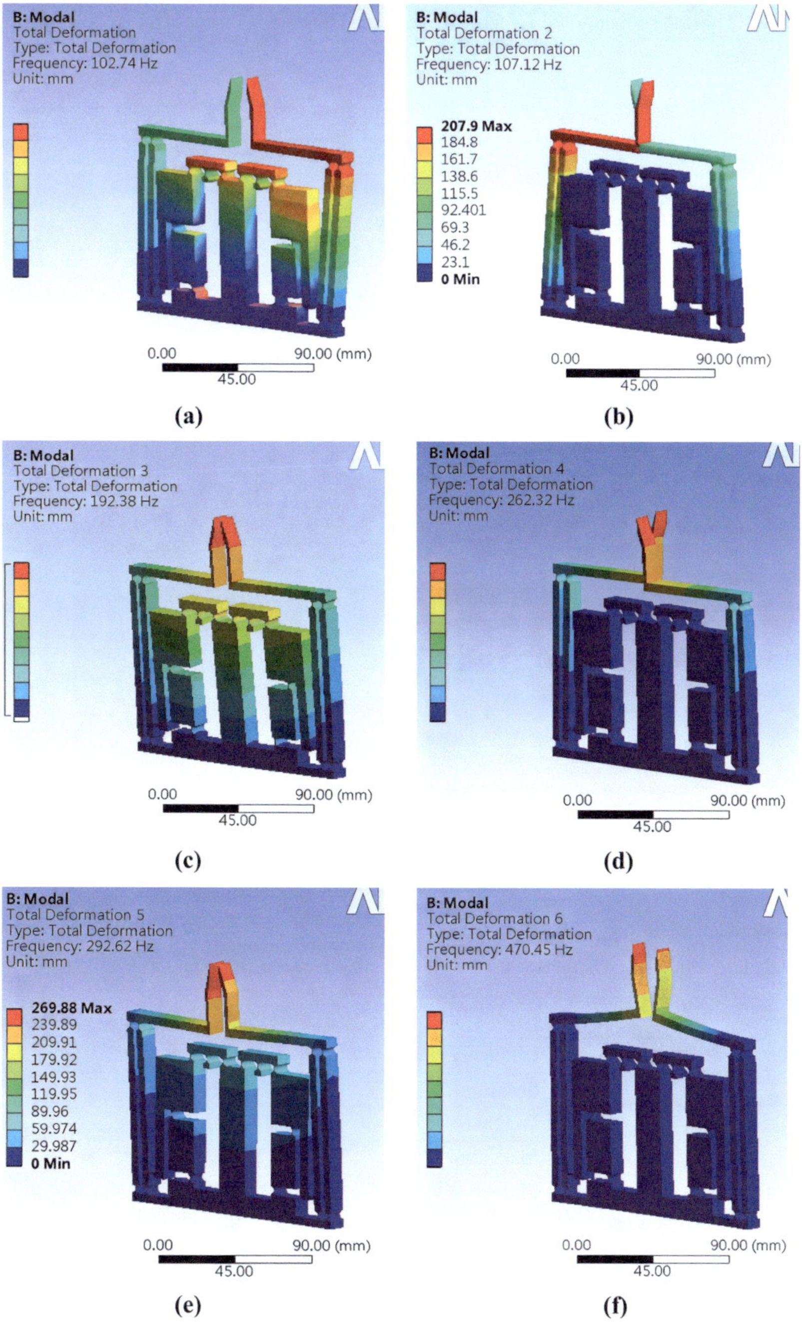

Fig. 3.7 Results of the modal finite element analysis (FEA) showing the first six eigenmodes (**a**–**f**) of the compliant gripper structure

Table 3.2 The first six natural frequencies and corresponding mode shapes of the compliant gripper.

Mode sequence	Frequency (Hz)
1	102.74
2	107.12
3	192.38
4	262.32
5	292.62
6	470.45

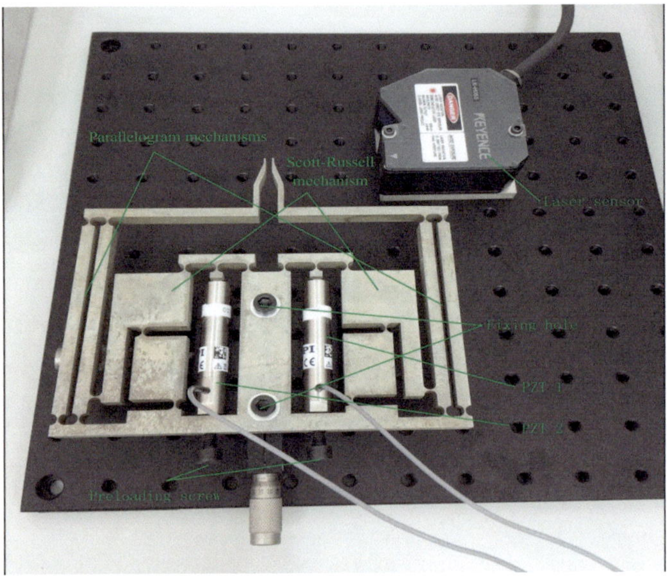

Fig. 3.8 Experimental prototype of the developed compliant microgripper

Table 3.3 Specifications of the employed piezoelectric stack actuator (PSA)

Parameter (Unit)	Value
Size (mm)	$\Phi 12 \times 68$
Weight (g)	46
Max. stroke (μm)	45
Max. trust force (N)	1000
Output stiffness (N/μm)	19

model E-503.00) to provide the 0–100 V drive voltage required by the PSAs. The output displacement of the moving platform is measured using a laser triangulation displacement sensor (Keyence Corp., model LK-H055). The sensor's analog output is acquired by a data acquisition (DAQ) board (NI-9263 ADC module) and read

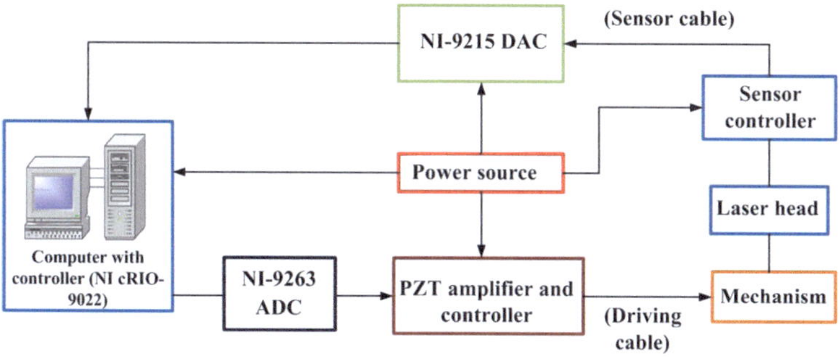

Fig. 3.9 Schematic of the experimental setup hardware connections

into a personal computer. A real-time controller (NI cRIO-9022) programmed with LabVIEW software is used to implement the control system, as illustrated in Fig. 3.9.

Furthermore, each PSA is inserted into the mechanical amplifier and preloaded using a screw mechanism. This creates an interference fit between the actuator and the amplifier, eliminating operational clearances through the elastic deformation of the material.

3.6.2 Open-Loop Experimental Test

The open-loop static characteristics of the gripper were first experimentally evaluated. A low-frequency (0.1 Hz) sinusoidal voltage signal ranging from 0 to 100 V was applied to drive PSA 1. Under this open-loop voltage-driven strategy, the PSA exhibits significant nonlinearity, primarily due to hysteresis effects.

This hysteresis negatively impacts the motion accuracy of the mechanism. Figure 3.10 shows the dominant hysteresis loops for the x-axis motion under different input voltage rates applied to PSA 1, indicating a gripping range of approximately 1000 μm for each tip. Similar results were observed when driving PSA 2.

Given the PSA's nominal stroke of 45 μm, the experimentally determined displacement amplification ratio is 22.2. This value is larger than the result obtained from FEA. The discrepancy is primarily attributed to the preloading effect from the PSA mounting. Each PSA is inserted into its mechanical amplifier and preloaded using an adjustment screw. This preloading alters the initial architectural parameters of the flexure mechanism and the effective input displacement of the PSA by creating an interference fit between the actuator and the amplifier. Consequently, the measured amplification ratio exceeds the nominal design value.

The frequency response of the gripper was experimentally identified via a swept-sine excitation. The result, presented in Fig. 3.11, shows a fundamental natural frequency of approximately 70 Hz. The discrepancy between this measured value and the higher FEA prediction is attributed to the unmodeled mass of the two mounted PSAs in the simulation.

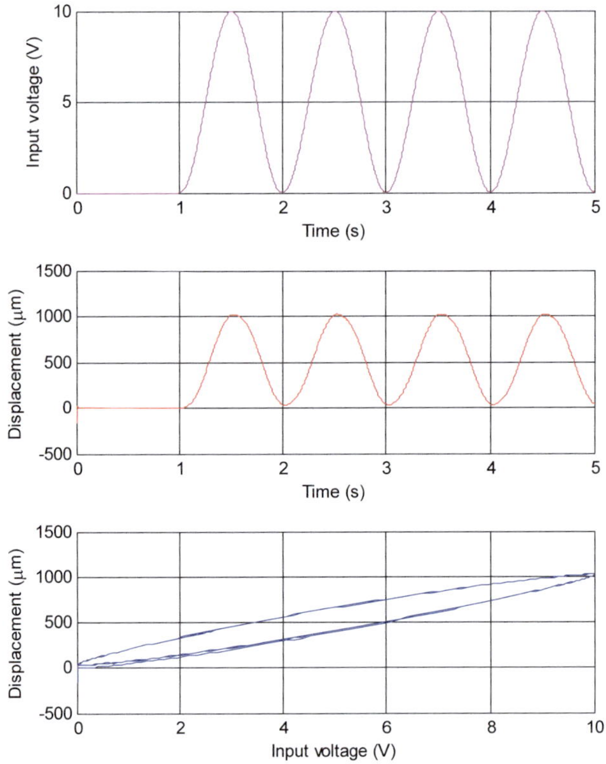

Fig. 3.10 Output displacement and hysteresis characteristics of the gripper tip in response to a 1-Hz actuation voltage applied to PSA 1

Fig. 3.11 Bode diagrams of the system model

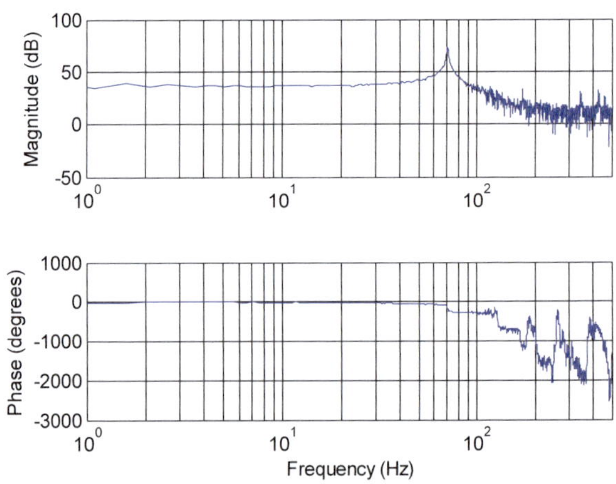

3.7 Discussion and Future Work

Experimental studies have verified the feasibility of the proposed gripper design based on the Scott-Russell (SR) mechanism. The accuracy of the kinematic model could be enhanced by accounting for all structural deformations, not just those at the flexure hinges. Furthermore, the resonant frequency of 70 Hz could be increased through the optimal design of stage parameters to reduce mass and increase stiffness.

The gripper design can also be extended to multiple tips, as illustrated in Fig. 3.12. This design allows the three tips to be actuated independently, enabling the grasping of complex micro-objects and facilitating more dexterous and reliable micromanipulation. Using a single gripper arm as a basic module, a modular multi-tip gripper can be readily configured.

To achieve precise microscale positioning, the detrimental effects of piezoelectric hysteresis must be compensated through closed-loop control. Both hysteresis and other error sources currently limit the open-loop positioning accuracy of the stage. Future work will, therefore, focus on the design and implementation of an advanced controller to mitigate these effects and enable precise manipulation.

Furthermore, for target applications in micromanipulation and microassembly, force control at the gripper tips is crucial for the safe handling of fragile objects [13, 14]. Future developments will integrate force sensors to measure gripping forces directly, enabling hybrid position/force control strategies. The practical feasibility of the gripper will be demonstrated by manipulating various objects with controlled, application-specific grasp forces.

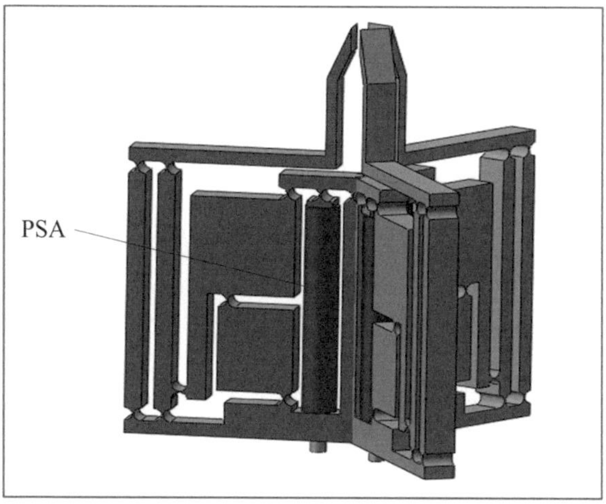

Fig. 3.12 CAD model of a modular compliant gripper featuring three independently actuated tips

3.8 Conclusion

This chapter presented a compliant microgripper based on the Scott-Russell mechanism to amplify the input displacement from piezoelectric stack actuators (PSAs) and achieve parallel motion at the gripper tips. An analytical model was established to facilitate the design of the gripping range and was validated through finite element analysis. Simulation results confirmed that the gripper arms undergo a pure translational motion. Furthermore, the gripper structure exhibits a high resonant frequency, enabling rapid dynamic response. A prototype was fabricated, and experimental tests were conducted. Compared to traditional microgrippers, the proposed Scott-Russell-based design offers a simpler structure and a more compact form factor by efficiently utilizing the in-plane area. Experimental results demonstrate that the developed gripper achieves a resonant frequency of 70 Hz and a large displacement amplification ratio of 22.2. Future work will focus on implementing force sensing and control to ensure the safe and reliable microhandling of fragile objects.

References

1. Beyeler, F., Neild, A., Oberti, S., Bell, D., Sun, Y., Dual, J., Nelson, B.: Monolithically fabricated microgripper with integrated force sensor for manipulating microobjects and biological cells aligned in an ultrasonic field. J. Microelectromech. Syst. **16**, 7–15 (2007)
2. Chang, S., Chang, B.: A precision piezodriven micropositioner mechanism with large travel range. Rev. Sci. Instrum. **69**, 1785–1791 (1998)
3. Fatikow, S., Eichorn, V.: Nanohandling automation: Trends and current developments. Proc. Inst. Mech. Eng. Part C-J. Mech. Eng. Sci. **222**, 1353–1369 (2008)
4. Ha, J., Kuang, Y., Hu, S., Fung, R.: Optimal design of a micro-positioning scott-russell mechanism by taguchi method. Sens. Actuator A-Phys. **125**, 565–572 (2006)
5. Hwang, D., Byun, J., Jeong, J., Lee, M.: Robust design and performance verification of an in-plane xy micropositioning stage. IEEE Trans. Nanotechnol. **10**, 1412–1423 (2011)
6. Kemper, M.: Development of a tactile low-cost microgripper with integrated force sensor. IEEE Trans. Robot. 1461–1466 (2004)
7. Keoschkerjan, R., Wurmus, H.: A novel microgripper with parallel movement of gripping arms. In: Proceedings of 8th International Conference on New Actuators, pp. 321–324 (2002)
8. Kim, K., Liu, X., Zhang, Y., Sun, Y.: Nanonewton force-controlled manipulation of biological cells using a monolithic mems microgripper with two-axis force feedback. J. Micromech. Microeng. **18**, 055013 (2008)
9. Menciassi, A., Eisinberg, A., Carrozza, M., Dario, P.: Force sensing microinstrument for measuring tissue properties and pulse in microsurgery. IEEE/ASME Trans. Mechatron. **8**, 10–17 (2003)
10. Nah, S., Zhong, Z.: A microgripper using piezoelectric actuation for micro-object manipulation. Sens. Actuator A-Phys. **133**, 218–224 (2007)
11. Smith, S.: Flexures: Elements of Elastic Mechanisms. Gordon and Breach, New York (2000)
12. Xu, Q.: Mechanism design and analysis of a novel 2-dof compliant modular microgripper. In: Proceedings of the 7th IEEE Conference on Industrial Electronics and Applications, pp. 1966–1971 (2012)
13. Xu, Q.: New flexure parallel-kinematic micropositioning system with large workspace. IEEE Trans. Robot. **28**, 478–491 (2012)

14. Xu, Q.: Precision position/force interaction control of a piezoelectric multimorph microgripper for microassembly. IEEE Trans. Autom. Sci. Eng. **10**, 503–514 (2012)
15. Xu, Q.: Identification and compensation of piezoelectric hysteresis without modeling hysteresis inverse. IEEE Trans. Ind. Electron. **60**, 3927–3937 (2013)
16. Xu, Q.: A new compliant microgripper with integrated position and force sensing. In: Proceedings IEEE/ASME International Conference on Advanced Intelligent Mechatronics, pp. 591–596 (2013)
17. Xu, Q.: Design and development of a compact flexure-based xy precision positioning system with centimeter range. IEEE Trans. Ind. Electron. **61**, 893–903 (2014)
18. Yong, Y., Lu, T., Handley, D.: Review of circular flexure hinge design equations and derivation of empirical formulations. Precis. Eng. **32**, 63–70 (2008)
19. Zhang, G., Zhang, C., Gu, J.: A memory-based hysteresis model in piezoelectric actuators. J. Control Sci. Eng. **2012**, 7 (2012)
20. Zubir, M., Shirinzadeh, B., Tian, Y.: Development of a novel flexure-based microgripper. Precis. Eng. **33**, 362–370 (2009)
21. Zubir, M., Shirinzadeh, B., Tian, Y.: Development of a novel flexure-based microgripper for high precision micro-object manipulation. Sens. Actuator A-Phys. **150**, 257–266 (2009)

Chapter 4
Design and Testing of a Large-Range Compliant Microgripper

Abstract This chapter introduces a new microgripper designed with a flexible hinge structure to achieve a large gripping range and nearly straight-line motion. It uses a two-stage amplification system–a modified Scott-Russell mechanism and a parallel mechanism–to significantly increase the motion range of the piezoelectric actuator. The modified Scott-Russell design alone provides 3.56 times more amplification than conventional versions. The gripper offers a gripping range of over 720 μm, a resonant frequency of 70.7 Hz, and minimal sideways motion. With an overall amplification ratio of 16.13, the design was validated through simulations and physical tests using an aluminum prototype. This microgripper is suitable for precision tasks such as handling optical fibers in microassembly applications.

Keywords Microgripper · Compliant mechanism · Scott-russell mechanism · FEA simulation · Piezoelectric stack actuator

4.1 Introduction

With the rapid development of micro-electromechanical systems (MEMS) and precision engineering, significant research efforts have been devoted to advancing automatic micromanipulation and microassembly technologies. As fundamental end-effectors in these domains, microgrippers serve as essential devices for micro- and nanomanipulation tasks [9, 12]. A typical microgripper consists of an actuator and a guiding mechanism. Various types of microgrippers have been developed, including adhesion-based and mechanical grippers [2, 16]. However, many applications are limited by the restricted gripping range of these devices. To address this limitation, displacement amplification mechanisms–such as bridge-type amplifiers–have been employed to enhance output displacement [1]. These advancements are of considerable practical importance for enabling reliable microassembly technologies.

Flexure-based mechanisms have been widely adopted to achieve ultra-high precision displacement [6, 15, 17]. Additionally, many micro- and nanogrippers utilize smart actuators, such as piezoelectric stack actuators, which have been successfully applied in biological cell manipulation and MEMS [3, 7]. Piezoelectric (PZT) actu-

Q. Xu, *Piezoelectric-Driven Flexible Robotic Microgrippers*,
https://doi.org/10.1007/978-981-95-4685-5_4

ators offer several advantages, including high resolution, fast response, large force output, and low power consumption [11, 18, 20]. Nevertheless, their limited output displacement often proves insufficient for many grasping applications. This shortcoming can be mitigated through the use of displacement amplification mechanisms.

Compliant mechanisms, which generate motion via elastic deformation, have been increasingly employed in gripper design. These mechanisms offer benefits such as frictionless operation and elimination of lubrication requirements. However, most existing microgrippers produce rotary motion at the tips. When such grippers contact curved objects, reaction forces may cause the object to be pushed away. The Scott-Russell (SR) mechanism has previously been used in micromanipulation systems due to its simple structure and ability to provide rectilinear output motion [5, 8, 10]. A notable drawback of conventional SR mechanisms, however, is their limited displacement amplification ratio, which can further reduce the already small output displacement of PZT actuators. Thus, a key motivation of this work is to overcome this limitation through an improved SR mechanism design.

This chapter presents the mechanical design, fabrication, and testing of a novel large-range compliant microgripper based on a flexure-hinge structure, as illustrated in Fig. 4.1. The gripper's main advantage lies in its ability to provide both large-range output displacement and nearly rectilinear motion, owing to minimal displacement in non-working directions. The device is actuated by a single PZT actuator mounted at its center, offering a resolution of 0.45 nm (open-loop) and a stroke of 45 μm. A dual-stage flexure amplifier is designed to magnify the actuator's stroke. The first stage consists of a modified Scott-Russell mechanism positioned along the hypotenuse, using circular notch hinges [14] to enable both motion amplification and rectilinear output. This modified design achieves an amplification ratio 3.56 times greater than existing SR mechanisms. The second stage incorporates a parallelogram mechanism that suppresses non-working directional deformations, a concept previously applied in metamorphic grippers for microassembly [4].

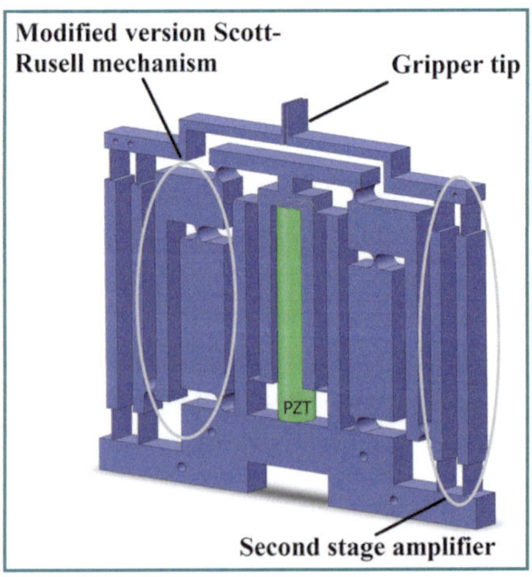

Fig. 4.1 Overall mechanical architecture of the designed compliant microgripper, illustrating the integration of Scott-Russell mechanisms, parallelogram guides, and lever amplifiers

The proposed gripper achieves a total displacement amplification ratio of 16 and a gripping range of 720 μm. For experimental validation, a mesoscale prototype was developed using a PZT actuator with a length of 68 mm. The gripper's size can be further reduced by adopting alternative actuation methods. Detailed design procedures and experimental results are presented in the subsequent sections of this chapter.

4.2 Mechanism Design and Modeling

This section details the mechanism design of the microgripper, encompassing the guiding mechanism, the first-stage and second-stage displacement amplifiers, and the dynamics modeling of the overall system.

4.2.1 Guiding Mechanism Design

The actuation displacement is transmitted to the gripper through a guiding mechanism comprising four leaf flexures. The computer-aided design (CAD) model of this mechanism is depicted in Fig. 4.2. When the output force from the piezoelectric (PZT) actuator is applied, these guiding flexures constrain the motion to a pure translation.

Each leaf flexure is modeled as a fixed-guided beam. Considering the boundary conditions–where the rotation angles at both ends are zero and the guided end undergoes a deformation δ_x–the following equations govern its behavior:

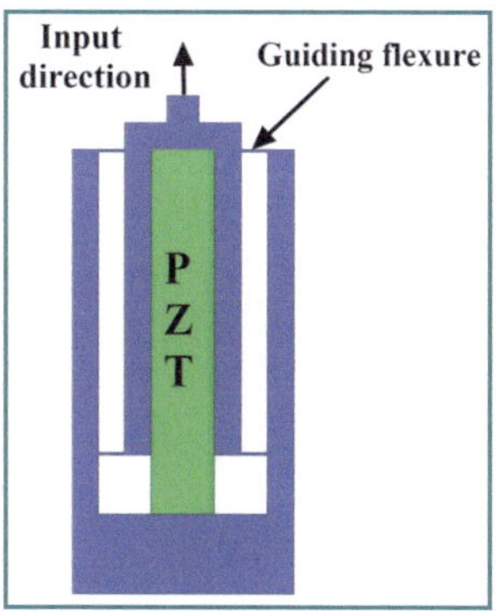

Fig. 4.2 CAD model of the guiding mechanism

$$0 = \frac{Fl^2}{2EI} - \frac{Ml}{EI} \tag{4.1}$$

$$\delta_x = \frac{Fl^3}{2EI} - \frac{Ml^2}{EI} \tag{4.2}$$

where l is the length of the leaf flexure, F is the applied force, M is the moment, and E is the elastic modulus of the material.

The moment of inertia for the beam is given by:

$$I_b = \frac{bh^3}{12} \tag{4.3}$$

where b is the width and h is the thickness of the leaf flexure.

Solving Eqs. (4.1) and (4.2), yields

$$M = \frac{Fl}{2} \tag{4.4}$$

$$\delta_x = \frac{Fl^3}{12EI} \tag{4.5}$$

From the kinetic energy and kinematic relationships of the mechanism, the equivalent mass M_1 is derived as:

$$M_1 = m_a \left(\frac{1}{4}\right)^2 + 4m_b \left(\frac{1}{8}\right)^2 + 4I_b \frac{1}{(8l)^2} \tag{4.6}$$

where m_a is the mass of the moving platform (without the leaf flexures), and m_b is the mass of a single leaf flexure, as illustrated in Fig. 4.6 later.

The potential energy stored in the four guiding flexures is:

$$V = 4\left(\frac{1}{2}K_s\theta_z^2\right) = \frac{2K_s q^2}{(8l)^2} \tag{4.7}$$

where K_s is the rotational stiffness of a flexure, θ_z is its angular deflection, and q is the linear deformation at the guided end. For further details, refer to [18].

The rotational stiffness K_s for the presented flexure is given by [18]:

$$K_s = 2\gamma K_\theta \frac{EI}{l} \tag{4.8}$$

where $\gamma = 0.85$ and $K_\theta = 2.6686$.

Consequently, the natural frequency of the guiding mechanism is:

$$f = \frac{1}{2\pi}\sqrt{\frac{K}{M_1}} \tag{4.9}$$

where the equivalent stiffness K is:

$$K = \frac{K_s}{(4l)^2} \tag{4.10}$$

4.2.2 First-Stage Amplifier Design

The first-stage amplifier is implemented as a modified Scott-Russell (SR) mechanism. First, the fundamental model of the standard SR mechanism is introduced. As established in the literature, when the lengths satisfy AB = BC = BD = L, the structure constitutes a classical SR mechanism, as illustrated in Fig. 4.3.

To derive the relationship between the input displacement at point A and the output displacement at point C, we assume the horizontal displacement component Δy_A at point A is negligible post-input. Based on the geometric constraints within the right triangle formed by the mechanism, the fundamental kinematic relationship is given by Eq. (4.11). Solving this trigonometric relationship yields the output displacement and the amplification ratio, as presented in Eqs. (4.12) and (4.13), respectively.

$$(\Delta y_A + l_{AD})^2 + (\Delta x_c + l_{CD})^2 \approx (l_{AD}^2 + l_{CD}^2) \tag{4.11}$$

$$\Delta x_C = (l_{DC}^2 - 2L_{AD}\Delta y_A - \Delta y_A^2)^{\frac{1}{2}} - l_{DC} \tag{4.12}$$

$$R_1 \approx \frac{-[(2L)^2 - l_{AD}^2]^{\frac{1}{2}}}{l_{AD}} = -\tan\alpha \tag{4.13}$$

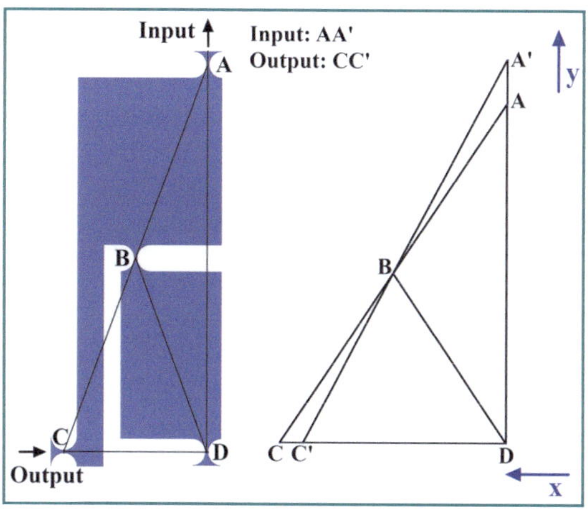

Fig. 4.3 Schematic diagram of the Scott-Russell (SR) compliant mechanism

where $L = 38.84$ mm and $l_{AD} = 72.75$ mm in this work.

From the expression for the amplification ratio in Eq. (4.13), it is observed that the output displacement of the Scott-Russell (SR) mechanism decreases when the length of segment CD is less than that of AD. This implies that $\tan\alpha$ is always less than unity. Consequently, although the conventional SR mechanism provides a rectilinear output, its amplification ratio is inherently less than one, representing a significant limitation.

In this design, the angle α is set to 20.47°. With this configuration, the amplification ratio of the conventional SR mechanism is calculated as:

$$R_1 \approx \frac{-[(2L)^2 - l_{AD}^2]^{\frac{1}{2}}}{l_{AD}} = -\tan(20.47^\circ) = -0.37 \tag{4.14}$$

where the negative sign indicates that the output displacement occurs in the direction opposite to the input.

The modified SR mechanism addresses this limitation by employing unequal lengths for segments AB and BC. This design modification not only enhances the amplification ratio but also maintains an approximately rectilinear output. The improved amplification ratio is derived as:

$$R_2 \approx \frac{(l_{AB} + l_{BC})\tan\alpha}{l_{AB}} \tag{4.15}$$

A CAD model of this modified amplifier and its key geometric parameters are presented in Fig. 4.4. The input from the piezoelectric (PZT) actuator is applied at point A along the y-axis. Substituting the design values $l_{AB} = 22.01$ mm and $l_{BC} = 55.89$ mm yields:

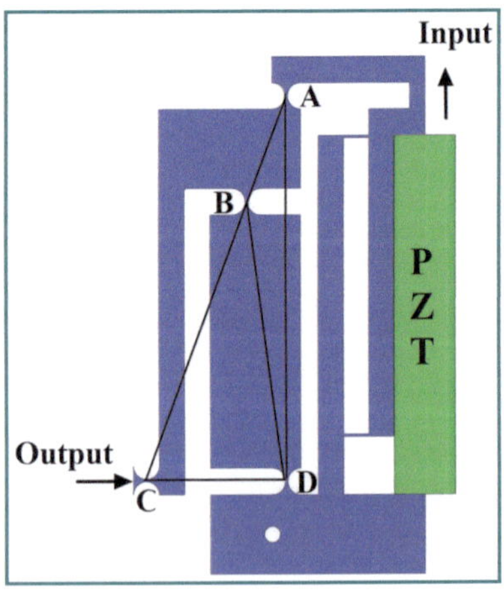

Fig. 4.4 Modified Scott-Russell (SR) compliant mechanism with integrated right-circular flexure hinges, illustrating the enhanced kinematic design for improved motion guidance and reduced parasitic rotation

$$R_2 \approx \frac{(22.01 + 55.89)\tan(20.4°)}{22.01} = 1.32 \tag{4.16}$$

The improvement factor η, representing the ratio of the new amplification ratio to that of the conventional SR mechanism, is obtained as:

$$\eta = \frac{R_2}{R_1} \approx \frac{(l_{AB} + l_{BC})}{l_{AB}} \tag{4.17}$$

Substituting the numerical values:

$$\eta = \frac{1.32}{0.37} \approx \frac{(22.01 + 55.89)}{22.01} = 3.56 \tag{4.18}$$

This result demonstrates that the modified SR mechanism achieves a 3.56-fold increase in amplification ratio compared to the conventional design.

4.2.3 Second-Stage Amplifier Design

The second-stage amplifier is implemented using a parallelogram mechanism constructed with leaf flexures. Although the first amplification stage provides initial displacement gain, its ratio remains insufficient for many applications. The primary objectives of this second module are to: (i) provide additional displacement amplification, and (ii) constrain motion to suppress parasitic displacements in non-working directions. The structure and key parameters of this mechanism are illustrated in Fig. 4.5.

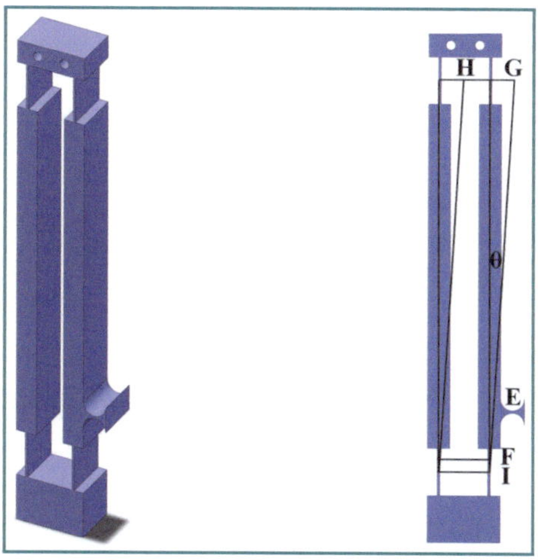

Fig. 4.5 Model of the second-stage amplifier, featuring a parallelogram mechanism designed to suppress non-working directional deformations and enhance motion stability

By cascading the two amplification stages, the overall performance of the microgripper is significantly enhanced. The total amplification ratio for a single gripper arm is calculated as:

$$R_3 \approx \frac{(l_{AB} + l_{BC})l_{IG} \tan \alpha}{l_{AB}l_{IE}} \tag{4.19}$$

Substituting the design parameters yields:

$$R_3 \approx \frac{(22.01 + 55.89) \times 83 \times \tan(20.4^\circ)}{22.01 \times 12.75} = 8.59 \tag{4.20}$$

This result indicates an amplification ratio of 8.59 for one gripper arm. During operation, both left and right arms move symmetrically. Since a single actuator drives both arms through the mechanism, the total system amplification ratio reaches 17.18.

The rotational stiffness at point I is determined by [1]:

$$K_r = \left(\frac{F_y l_{HG}}{4 + \frac{l_{FG}^2}{l_{IG}^2}} \right)^{\frac{1}{2}} \tag{4.21}$$

The input stiffness of the mechanism is subsequently derived as:

$$K = \frac{K_r + \frac{K_r}{l_{IG}^2}}{l_{HG}^2} \tag{4.22}$$

The equivalent mass of the gripper mechanism is given by:

$$M = M_1 + \frac{9}{5}m_2 + \frac{18}{125}m_3 + \frac{1287}{25}m_4 + \frac{4797}{40}m_5 \tag{4.23}$$

where M_1 is obtained from Eq. (4.6), and the mass distribution components (m_2–m_5) are identified in Fig. 4.6.

Finally, the natural frequency (in Hz) is calculated as:

$$f = \frac{1}{2\pi}\sqrt{\frac{K}{M}} \tag{4.24}$$

4.3 Finite Element Analysis Simulation

Finite element analysis (FEA) was performed utilizing the ANSYS software suite. The primary objectives of this simulation were to validate the analytical models developed in the preceding subsection and to conduct a comprehensive evaluation

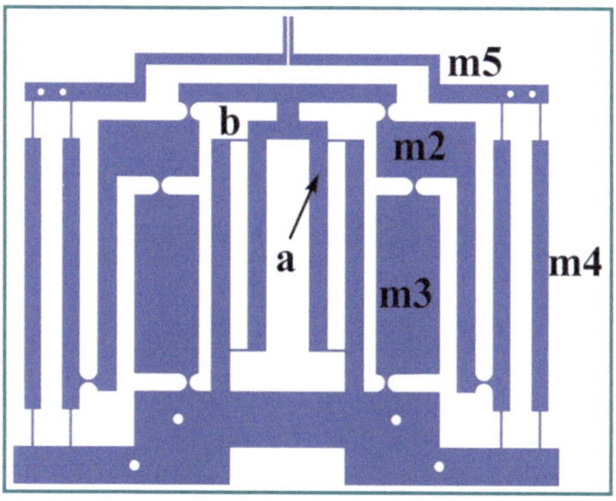

Fig. 4.6 Mass distribution analysis of the microgripper, showing the inertial properties and center of mass location for dynamic performance evaluation

Table 4.1 Material properties of aluminum 7075-T6 alloy used in the microgripper fabrication

Parameter	Value	Unit
Elongation	11	%
Density	2810	kg/m^3
Modulus of elasticity	72	GPa
Yield strength	455	MPa
Tensile strength	524	MPa
Poisson's ratio	0.33	–

of the microgripper's performance. The assessment focused on three key metrics: the output displacement at the gripper jaws, the parasitic displacement in the non-working direction, and the modal characteristics of the structure. Given its critical role in predicting performance prior to costly prototype fabrication, FEA serves as an indispensable step in the design verification process.

The gripper structure was modeled using the material properties of Aluminum 7075-T6 alloy. This material was selected for its superior mechanical properties, which include a high strength-to-weight ratio, excellent performance at low temperatures, and notable resistance to stress corrosion cracking. The complete set of material properties employed in the simulation is summarized in Table 4.1.

4.3.1 Static Structural Analysis

The static structural performance of the gripper was analyzed under an applied input displacement. As illustrated in Fig. 4.7a, the gripper achieves a maximum output displacement of 364.4 μm at its jaws under a full input displacement of 45 μm.

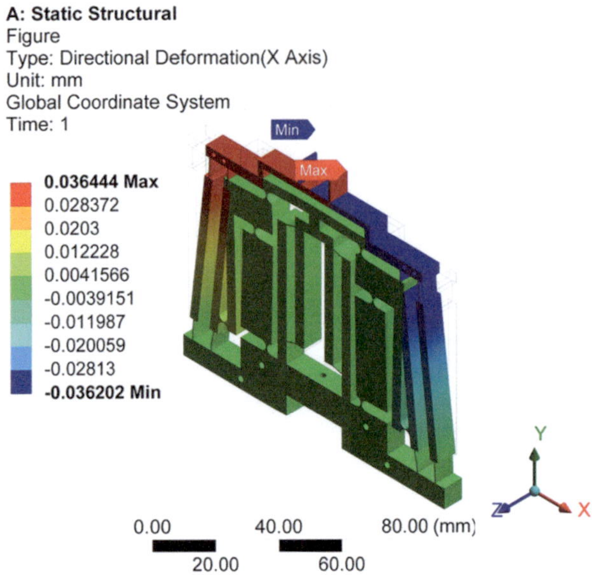

(a) Working direction displacement

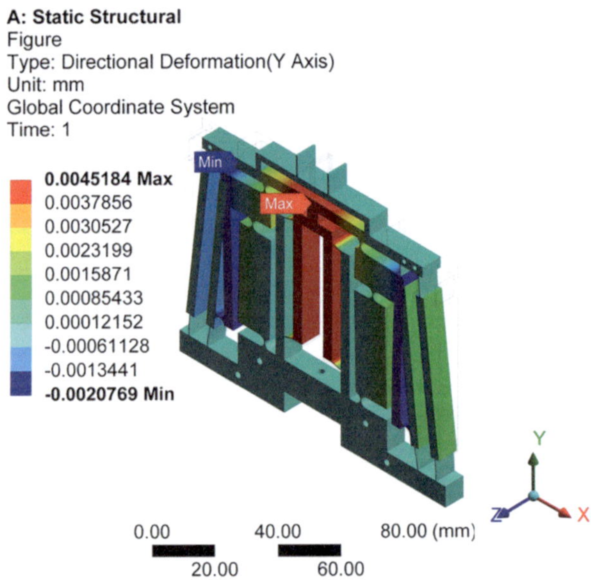

(b) Non-working direction displacement

Fig. 4.7 Finite element analysis (FEA) simulation results: **a** total displacement output at the gripper jaws, and **b** parasitic displacement in the non-working direction

This input value corresponds to the maximum stroke of the piezoelectric actuator designated for use in the physical prototype.

Furthermore, the displacement in the non-working (parasitic) direction was scrutinized to evaluate the quality of the motion. Figure 4.7b demonstrates that the magnitude of this parasitic displacement is merely 61.2 nm. This value is negligible when compared to the primary output displacement, confirming the gripper's ability to produce a nearly rectilinear motion.

4.3.2 *Modal Analysis*

A modal analysis was conducted to evaluate the dynamic behavior of the gripper structure. This analysis identified the resonant mode shapes and their corresponding natural frequencies, which are critical for understanding the system's response to dynamic inputs and for avoiding operational resonance.

The first six resonant mode shapes are presented in Fig. 4.8. Their corresponding natural frequencies are listed in Table 4.2. The first natural frequency of the gripper is found to be 70.7 Hz, indicating a high fundamental resonance and, consequently, a potential for high response speed. The analysis of the mode shapes reveals that Modes 3 and 4 correspond to vertical translations along the y-axis, while Modes 5 and 6 are a result of bending of the gripper arm about the z-axis.

4.3.3 *Summary of Simulation Results*

The FEA simulation results conclusively validate that the design objectives have been met. The compliant microgripper provides a significant output displacement range while maintaining minimal parasitic motion. Coupled with a high first natural frequency, these characteristics render the design highly suitable for the precise manipulation of various microparticles in practical microassembly applications.

4.4 Experimental Results and Validation

Following the successful validation through FEA simulations, a physical prototype of the microgripper was fabricated. A comprehensive experimental evaluation was subsequently conducted to verify its practical performance and to correlate the results with the analytical and simulation predictions. The primary objective of this testing was to accurately measure the output displacement at the gripper tips under actuation.

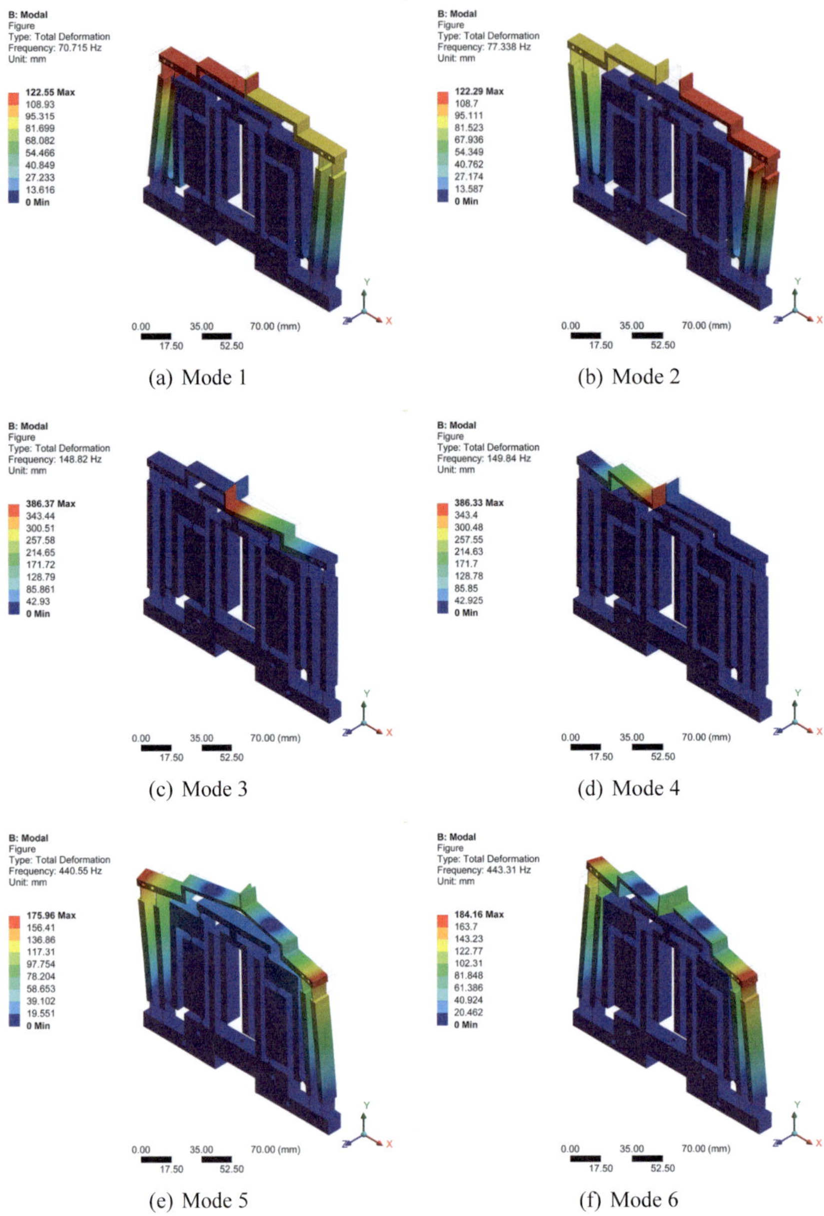

Fig. 4.8 Finite element analysis results showing the first six resonant mode shapes **a–f** of the compliant microgripper structure

Table 4.2 First six natural frequencies and their corresponding mode shapes for the compliant microgripper structure

Mode	Frequency (Hz)
1	70.715
2	77.338
3	148.82
4	149.84
5	440.55
6	443.31

4.4.1 Experimental Setup

The experimental test setup, depicted in Fig. 4.9, was configured for precise performance characterization. The core of the displacement measurement system was a high-resolution laser displacement sensor (Keyence LK-H008), which offers a resolution of 25 nm. The microgripper was driven by a piezoelectric stack actuator (Physik Instrumente P-840.30), chosen for its high resolution and force output. The technical specifications of the actuator are provided in Table 4.3. The actuator was operated within a voltage range of 0–100 V, corresponding to its maximum nominal displacement of 45 μm.

4.4.2 Displacement Performance

The output displacement of the gripper tips was measured at the maximum driving voltage of 100 V. The results from four consecutive tests, as recorded by the laser displacement sensor, are summarized in Table 4.4. The measured displacements show a high degree of consistency, with an average output displacement of approximately

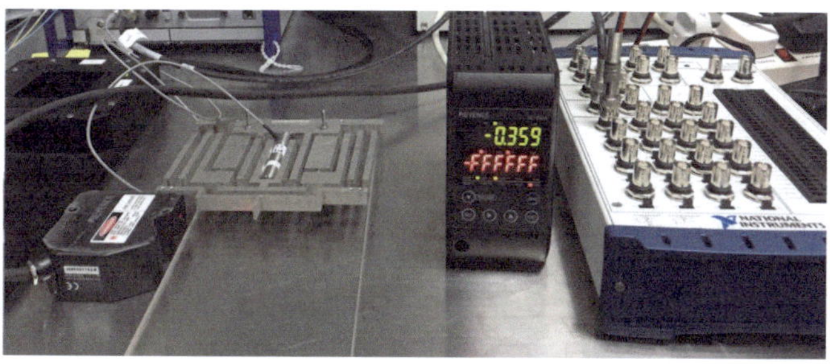

Fig. 4.9 Experimental test setup for microgripper performance characterization, showing the integrated measurement systems and positioning stages

Table 4.3 Technical specifications of the piezoelectric stack actuator (Physik Instrumente P-840.30)

Parameter	Value	Unit
Model	P-840.30	–
Open-loop travel @100 V	±45	μm
Blocking force	1000	N
Tension force	50	N
Outer diameter	12	mm
Length	68	mm
Resolution (approx.)	<1.0	nm
Stiffness	25	N/μm

Table 4.4 Experimental results of output displacement per gripper arm at the maximum driving voltage of 100 V

Test #	1	2	3	4
Output displacement (μm)	362	363	360	363
Average	362 μm			

Fig. 4.10 Gripper tip positions: **a** in the neutral state and **b** during active actuation at 100 V, demonstrating the displacement range

362 μm per arm. Figure 4.10 provides a visual comparison of the gripper tip positions in the neutral state and during full actuation, clearly demonstrating the significant displacement range achieved.

Table 4.5 Comparison of output displacement results from analytical, FEA, and experimental methods

Method	Output displacement (μm)	Error versus expt. avg.	Notes
Analytical model	388.0	7.2%	
FEA simulation	364.4	0.7%	
Experimental (average)	362.0	–	Average of tests 1–4

4.4.3 Model Validation and Discussion

A direct comparison between the analytical, FEA simulation, and experimental results is presented in Table 4.5. The close agreement among these values validates the accuracy of the developed kinematic model and the FEA simulations. The minor discrepancies observed can be attributed to factors such as manufacturing tolerances, slight variations in material properties, and the inherent assumptions within the analytical model.

The experimental results conclusively confirm the effectiveness of the proposed compliant microgripper design. The implemented modified Scott-Russell (SR) mechanism successfully amplifies the input displacement while the integrated parallelogram module effectively minimizes parasitic motion. The FEA simulations predicted a non-working directional displacement of only 61.2 nm, a value that is negligible for most practical applications and could potentially be reduced further through future structural optimization.

The dual-stage amplification mechanism achieves a calculated amplification ratio of 8.066 per arm. Consequently, the total gripping range, resulting from the symmetric displacement of both arms, is 16.13 times the input displacement of the piezoelectric actuator. This performance surpasses that of conventional single-arm actuated grippers, highlighting a significant advantage of the proposed two-arm, dual-stage kinematic design.

4.4.4 Discussion and Future Work

For practical micromanipulation applications, precise grasp force regulation is critical to safely handle delicate objects such as biological cells or micro-optical components without causing damage [13, 19]. While this study successfully validated the kinematic and mechanical performance of the gripper, future work will focus on the integration of force sensing capabilities. The subsequent implementation of a closed-loop force control system will be essential to enable accurate and reliable force output during manipulation tasks. Finally, application-specific validation in real-world micromanipulation scenarios will be conducted to demonstrate the gripper's full operational potential.

4.5 Conclusion

This chapter has presented the design, analysis, fabrication, and experimental validation of a novel large-displacement compliant microgripper. Each module of the gripper was systematically designed and optimized. Following the development of an analytical model, finite-element analysis (FEA) simulations were conducted to verify the performance of the proposed design. A prototype was subsequently manufactured and tested experimentally. The results demonstrate that the gripper achieves a gripping range of 0–720 μm with a displacement amplification ratio of 16.13, providing sufficient motion for manipulating a wide variety of microparticles. The modified Scott-Russell (SR) mechanism achieved a 3.56-fold improvement in amplification ratio compared to conventional designs.

Future work will focus on implementing precision motion and force control modules to enable accurate and stable manipulation, further enhancing the gripper's applicability in micromanipulation and microassembly tasks.

References

1. Ai, W., Xu, Q.: New structural design of a compliant gripper based on the Scott-Russell mechanism. Int. J. Adv. Robot. Syst. **11** (2014). https://doi.org/10.5772/59655
2. Arai, F., Fukuda, T.: Adhesion-type micro end effector for micromanipulation. In: Proceedings of 1997 IEEE International Conference on Robotics and Automation, vol. 2, pp. 1472–1477. IEEE (1997)
3. Bergander, A., Driesen, W., Varidel, T., Breguet, J.M.: Development of miniature manipulators for applications in biology and nanotechnologies. In: Proceedings of IEEE/RSJ International Conference on Intelligent Robots and Systems (IROS 2003), pp. 11–35. Las Vegas, NV, USA (2003)
4. Bruzzone, L., Bozzini, G.: A flexible joints microassembly robot with metamorphic gripper. Assem. Autom. **30**(3), 240–247 (2010)
5. Chang, S., Du, B.: A precision piezodriven micropositioner mechanism with large travel range. Rev. Sci. Instrum. **69**(4), 1785–1791 (1998)
6. Dunning, A.G., Tolou, N., Herder, J.: Inventory of platforms towards the design of a statically balanced six degrees of freedom compliant precision stage. Mech. Sci. **2**, 157–168 (2011)
7. Götze, H., Pagel, L.: Development of a micro-manipulator based on piezoelectric-technology. Microelectron. Eng. **84**(5), 1333–1336 (2007)
8. Ha, J.L., Kung, Y.S., Hu, S.C., Fung, R.F.: Optimal design of a micro-positioning Scott-Russell mechanism by Taguchi method. Sens. Actuators, A **125**(2), 565–572 (2006)
9. Han, K., Lee, S.H., Moon, W., Park, J.S.: Fabrication of the micro-gripper with a force sensor for manipulating a cell. In: Proceedings of SICE-ICASE International Joint Conference, pp. 5833–5836. IEEE, Bexco, Busan, Korea (2006)

10. Hwang, D., Byun, J., Jeong, J., Lee, M.G.: Robust design and performance verification of an in-plane xyθ micropositioning stage. IEEE Trans. Nanotechnol. **10**(6), 1412–1423 (2011)
11. Khadraoui, S., Rakotondrabe, M., Lutz, P.: Optimal design of piezoelectric cantilevered actuators with guaranteed performances by using interval techniques. IEEE/ASME Trans. Mechatron. **19**(5), 1660–1668 (2014)
12. Lamers, A., Sánchez, J.A.G., Herder, J.L.: Design of a statically balanced fully compliant grasper. Mech. Mach. Theory **92**, 230–239 (2015)
13. Liu, X., Kim, K., Zhang, Y., Sun, Y.: Nanonewton force sensing and control in microrobotic cell manipulation. Int. J. Robot. Res. **28**(8), 1065–1076 (2009)
14. Merken, P., Smal, O., Debongnie, J.F., Raucent, B.: Design and test of a circular notch hinge. In: Proceedings of the International Precision Assembly Seminar (IPAS 2004), pp. 51–56. Bad Hofgastein, Austria (2004)
15. Merriam, E.G., Jones, J.E., Magleby, S.P., Howell, L.L.: Monolithic 2 DOF fully compliant space pointing mechanism. Mech. Sci. **4**, 381–390 (2013)
16. Tanikawa, T., Arai, T., Masuda, T.: Development of micro manipulation system with two-finger micro hand. In: Proceedings of IEEE/RSJ International Conference on Intelligent Robots and Systems (IROS 96), vol. 2, pp. 850–855. IEEE, Osaka, Japan (1996)
17. Tian, Y., Liu, C., Liu, X., Wang, F., Li, X., Qin, Y., Zhang, D., Shirinzadeh, B.: Design, modelling and characterization of a 2-DOF precision positioning platform. Trans. Inst. Meas. Control. **37**(3), 396–405 (2015)
18. Xu, Q.: Design, testing and precision control of a novel long-stroke flexure micropositioning system. Mech. Mach. Theory **70**, 209–224 (2013)
19. Xu, Q.: Design and development of a novel compliant gripper with integrated position and grasping/interaction force sensing. IEEE Trans. Autom. Sci. Eng. **14**(3), 1415–1428 (2017)
20. Zhang, Y., Lu, T.F., Al-Sarawi, S.: Formulation of a simple distributed-parameter model of multilayer piezoelectric actuators. J. Intell. Mater. Syst. Struct. **27**(11), 1485–1491 (2015)

Chapter 5
Design of a Piezoelectric-Actuated Symmetric Compliant Microgripper

Abstract This chapter presents a piezoelectrically actuated compliant microgripper designed to enhance both static and dynamic performance. The design combines a bridge mechanism, an L-shaped mechanism, and a levered parallelogram to achieve large gripping motion. Structural parameters were optimized using response surface methodology. Finite element analysis and experimental testing demonstrated a gripping stroke of 312.8 μm, a natural frequency of 786 Hz, and resolutions of ±0.6 μm (motion) and ±1.69 mN (force). The microgripper successfully manipulated a 200 μm optical fiber and 100 μm metal sheet using position and force control, confirming its practical capability for micromanipulation tasks.

Keywords Mircogripper · Piezoelectric actuator · Compliant mechanism · Displacement amplifier · Micromanipulation

5.1 Introduction

Micro/nano-manipulation robots represent a critical advancement in robotic automation technology, enabling precise operations at micro/nano-meter scales. These systems are extensively employed across biomedical engineering, precision manufacturing, and advanced optics for applications requiring micromanipulation and microassembly capabilities [28]. As a fundamental end-effector in these domains, microgrippers play a pivotal role in the grasping and release of micro-objects [18]. Various actuation principles–including electrostatic, electrothermal, electromagnetic, piezoelectric, and shape memory alloy (SMA) technologies [4, 19]–have been explored for microgripper designs. Among these, piezoelectric-actuated microgrippers have garnered significant attention due to their compact form factor, high displacement resolution, rapid response, and substantial force output. Compliant mechanisms are typically integrated as displacement amplification and transmission systems between the piezoelectric actuator (PEA) and the microgripper jaw, offering distinct advantages such as frictionless operation, elimination of lubrication requirements, vacuum compatibility, and ultra-precise motion resolution [14, 17].

Q. Xu, *Piezoelectric-Driven Flexible Robotic Microgrippers*,
https://doi.org/10.1007/978-981-95-4685-5_5

Given that the output displacement of a PEA is inherently constrained (typically $\approx$0.1% of its length), a primary design challenge for piezoelectric-actuated microgrippers lies in achieving a large gripping range. A substantial gripping stroke enhances the gripper's adaptability to handle objects of diverse sizes when the jaw gap is predetermined. Researchers have consequently employed various compliant amplification mechanisms–including lever and bridge-type architectures–in the development of micromanipulators [6, 9, 23]. Concurrently, high natural frequency represents another critical performance metric, as it extends the control bandwidth, thereby improving the responsiveness of the control system [26]. Furthermore, high-frequency vibration capabilities can mitigate adhesion forces between biological specimens and gripper jaws [3]. Thus, an ideal microgripper must concurrently exhibit both a large gripping stroke and a high natural frequency.

Previous research has yielded numerous piezoelectric-actuated compliant microgripper designs [1, 5, 7, 8, 10–13, 15, 16, 21, 22, 24, 25, 27, 29, 30]. For instance, the integration of pseudo-rigid-body model (PRBM) and finite element analysis (FEA) techniques enabled the development of microgrippers with gripping ranges of approximately 100 μm [30]. Subsequent designs incorporating lever, Scott–Russell (SR), and parallelogram mechanisms achieved strokes up to 1000 μm [1], while optimized SR mechanisms yielded displacements exceeding 720 μm [16]. However, these designs often suffered from limited natural frequencies (e.g., 70–70.7 Hz) and bulky structures. Recent advances include a symmetrical microgripper combining lever, bridge, and parallelogram mechanisms to attain a natural frequency of 1044 Hz [11], albeit with a restricted gripping displacement of 93.52 μm. Another design achieved a resonant frequency of 854 Hz with a 184.04 μm stroke [8]. Despite these advances, few designs have successfully reconciled large gripping ranges with high natural frequencies, highlighting a persistent gap in the field.

This chapter contributes a piezoelectric-actuated microgripper design that simultaneously addresses the dual objectives of large gripping stroke and high natural frequency. The proposed architecture comprises a piezoelectric actuator, a three-stage compliant amplification/guiding mechanism, and dual jaws. Computational simulations–incorporating response surface analysis and FEA–were employed to optimize critical structural parameters and validate performance. Experimental characterization through open-loop and closed-loop testing quantified key metrics including gripping stroke, force output, natural frequency, tracking precision, and motion resolution. Practical functionality was demonstrated via gripping experiments on metal micro-wires and micro-shims, confirming the design's viability for real-world micromanipulation applications.

5.2 Mechanism Design and Modeling

This section details the compliant mechanism design and kinematic modeling of the proposed piezoelectric-actuated microgripper.

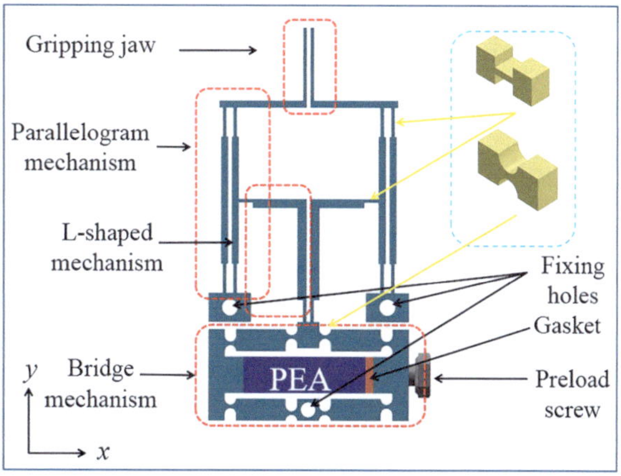

Fig. 5.1 Computer-aided design (CAD) model of the proposed piezoelectric-actuated microgripper, showing the three-stage compliant amplification mechanism (bridge, L-shaped, and levered parallelogram components) and integrated piezoelectric actuator

5.2.1 *Mechanism Design*

Figure 5.1 presents a computer-aided design (CAD) model of the microgripper, which utilizes a symmetrical compliant mechanism to balance internal stresses and enhance displacement amplification. A piezoelectric actuator (PEA) is embedded within a bridge-type mechanism that provides bidirectional actuation. The overall mechanism integrates three key components: (i) a bridge mechanism for initial displacement amplification and force transmission, (ii) an L-shaped mechanism for motion redirection, and (iii) a levered parallelogram mechanism for parallel jaw motion.

The bridge mechanism, positioned at the gripper's base, employs double-notch right circular flexure hinges to achieve high precision and compactness. It converts the PEA's horizontal displacement into vertical motion. The L-shaped mechanism transmits this motion to the parallelogram mechanism, which ensures parallel jaw movement through rectangular flexure hinges. These hinges were selected for their simplicity and suitability for unidirectional bending. Three mounting holes enable integration with external positioning systems, while pre-tightening screws facilitate PEA installation and initial jaw gap adjustment.

5.2.2 *Kinematic Modeling*

The displacement amplification performance was analyzed using a pseudo-rigid-body model (PRBM) approach. Figure 5.2 shows the half-model representation, where flexure hinges are modeled as revolute joints with torsional springs. Based on the velocity instantaneous center method, the kinematic relationships are derived as follows:

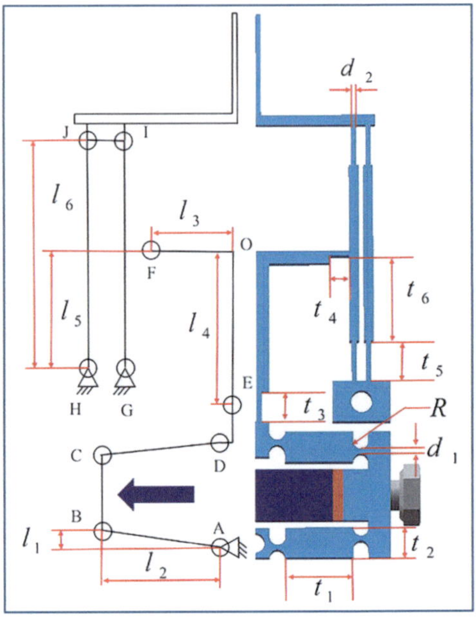

Fig. 5.2 Pseudo-rigid-body model (PRBM) of the microgripper showing critical geometric parameters and kinematic relationships

$$v_C = \omega_1 \times l_1 \tag{5.1}$$

$$v_D = \omega_1 \times l_2 \tag{5.2}$$

$$v_E = v_D = \omega_2 \times l_3 \tag{5.3}$$

$$v_F = \omega_2 \times l_4 = \omega_3 \times l_5 \tag{5.4}$$

$$v_I = \omega_3 \times l_6 \tag{5.5}$$

where ω_1, ω_2, and ω_3 denote the angular velocities of linkages CD, O_2E, and GI, respectively (see Fig. 5.3). The displacement amplification ratio R_m is then obtained as:

$$R_m = \frac{v_I}{v_C} = \frac{l_2}{l_1} \cdot \frac{l_4}{l_3} \cdot \frac{l_6}{l_5}. \tag{5.6}$$

This model provides a theoretical foundation for optimizing the geometric parameters l_1 to l_6 to achieve the desired amplification performance.

The analysis reveals that the displacement amplification of the microgripper exhibits a linear relationship with a specific subset of structural parameters, highlighting the mechanism's deterministic behavior under actuation. When the bridge mechanism is displaced, the entire structure undergoes elastic deformation, with rotation angles at flexure hinges derived as:

$$\varphi_A = \varphi_B = \varphi_C = \varphi_D = \theta_1 = \frac{l_2 x_{\text{in}}}{2l_1\sqrt{l_1^2 + l_2^2}} \tag{5.7}$$

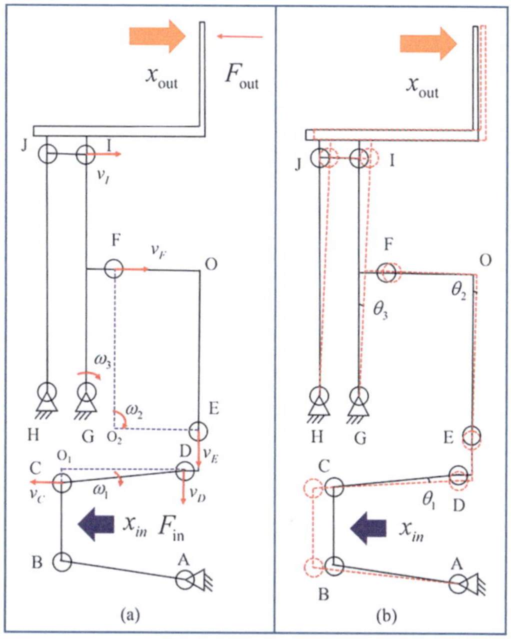

Fig. 5.3 Kinematic analysis of the microgripper mechanism: **a** velocity vector diagram illustrating instantaneous centers and transmission paths; **b** displacement relationship diagram showing geometric constraints and amplification principles

$$\varphi_E = \varphi_F = \theta_2 = \frac{l_2 x_{\text{in}}}{l_1 l_3} \tag{5.8}$$

$$\varphi_G = \varphi_H = \varphi_I = \varphi_J = \theta_3 = \frac{l_2 l_4 x_{\text{in}}}{l_1 l_3 l_5} \tag{5.9}$$

where θ_1, θ_2, and θ_3 denote the angular deflections of linkages CD, EO_1, and GI, respectively.

The work done by external forces is given by:

$$W = \frac{1}{2} F_{\text{in}} x_{\text{in}} \tag{5.10}$$

where F_{in} and x_{in} are the input force and displacement.

Ignoring gravity, the potential energy stored in the flexure hinges is:

$$E_p = \frac{1}{2} \sum_{i=A}^{J} K_{ri} \varphi_i^2. \tag{5.11}$$

Expanding this expression yields:

$$E_p = \frac{K_{rA} l_2^2 x_{\text{in}}^2}{2(l_1^2 + l_2^2) l_1^2} + \frac{K_{rE} l_2^2 x_{\text{in}}^2 (1 + \eta F)}{2l1^2 l_3^2} + 2K_{rE} \eta G x_{\text{in}}^2 \left(\frac{l_2 l_4}{l_1 l_3 l_5} \right)^2 \tag{5.12}$$

where K_{rA} and K_{rE} are the rotational stiffnesses of hinges A and E, and η_F, η_G are stiffness ratios.

The rotational stiffness for right circular and rectangular hinges is analytically estimated as:

$$K_{ri} = \frac{2Eb_t t_i^{5/2}}{9\pi r_i^{1/2}}, \qquad (i = A, B, C, D) \tag{5.13}$$

$$K_{ri} = \frac{Eb_t t_i^3}{12 l_i}, \qquad (i = E, F, \ldots, J) \tag{5.14}$$

where E is the elastic modulus, and t_i, b_t, l_i, r_i are hinge dimensions.

The input stiffness K_{in} is derived as:

$$K_{\text{in}} = \frac{K_{rA} l_2^2}{(l_1^2 + l_2^2) l_1^2} + \frac{K_{rE} l_2^2 (1 + \eta F)}{l 1^2 l_3^2} + 4 K_{rE} \eta G \left(\frac{R_m}{l_6}\right)^2. \tag{5.15}$$

The total kinetic energy, including rotational and translational components, is:

$$E_K = I_{AB}\omega_1^2 + \frac{1}{2}(IEO + I_{FO})\omega_2^2 + IGI\omega_3^2 + \frac{1}{2}(m_{\text{in}} + R_m^2)\dot{x}_{\text{in}}^2, \tag{5.16}$$

where m_{in} is the input mass, and I_{AB}, I_{EO}, I_{FO}, I_{GI} are rotational inertias.

Applying Lagrange's equation yields the equation of motion:

$$M\ddot{x}_{\text{in}} + K\dot{x}_{\text{in}} = F_{\text{in}} - R_m F_{\text{out}}, \tag{5.17}$$

where M and K are the equivalent mass and stiffness:

$$M = \frac{I_{AB} l_2^2}{2 l_1^2 (l_1^2 + l_2^2)} + \frac{(I_{EO} + I_{FO}) l_2^2}{l_1^2 l_3^2} + 2 I_{GI} \left(\frac{R_m}{l_6}\right)^2 + m_{\text{in}} + R_m^2 m_{\text{out}}, \tag{5.18}$$

$$K = K_{\text{in}}. \tag{5.19}$$

The natural frequency is then:

$$f = \frac{1}{2\pi}\sqrt{\frac{K}{M}}. \tag{5.20}$$

This modeling framework provides a comprehensive basis for optimizing the microgripper's performance, balancing displacement amplification, structural stiffness, and dynamic response.

5.3 Computational Simulation Study

While the kinematic and dynamic models provide fundamental insights, a comprehensive performance evaluation requires advanced computational techniques. This section details response surface analysis (RSA) and finite element analysis (FEA) conducted using ANSYS Workbench to optimize and validate the microgripper's design.

5.3.1 *Response Surface Analysis*

Response surface methodology (RSM) was employed to model the relationship between design parameters and performance objectives through a series of deterministic simulations. This statistical approach efficiently captures the effects of multiple variables on system behavior, making it ideal for optimizing complex compliant mechanisms. For foundational RSM principles, refer to [2, 20].

The geometric parameters $[l_1, l_2, l_3, l_4, l_5, l_6]$ and $[t_1, t_2, t_3, t_4, t_5, t_6]$ defined in Fig. 5.2 were analyzed. Given that the l_i parameters are interdependent with the t_i array, optimization focused on the thickness parameters $[t_1, t_2, t_3, t_4, t_5, t_6]$. The overall gripper dimensions were constrained to 54 mm $\times$ 28 mm $\times$ 5 mm to ensure compactness and integrability. The bridge mechanism's double-notch hinges had a radius $R = 0.8$ mm to maximize compliance, while minimum thicknesses $d_1 = d_2 = 0.5$ mm respected manufacturing constraints.

The optimization objectives were to maximize the output displacement S and natural frequency F, while maintaining a safety factor $Q > 1$ to prevent material yielding. Sensitivity analysis of the thickness parameters on S yielded values $[-0.40, 0.02, 0.05, 0.003, 0.40, 0.21]$, identifying t_1 (negative correlation), t_5, and t_6 (positive correlations) as dominant factors, as depicted in Fig. 5.4. Response surface plots (see Fig. 5.5) confirmed that S increases with t_3 and t_5 but decreases with t_1. The optimized parameters, finalized in Table 5.1, balance performance and structural integrity.

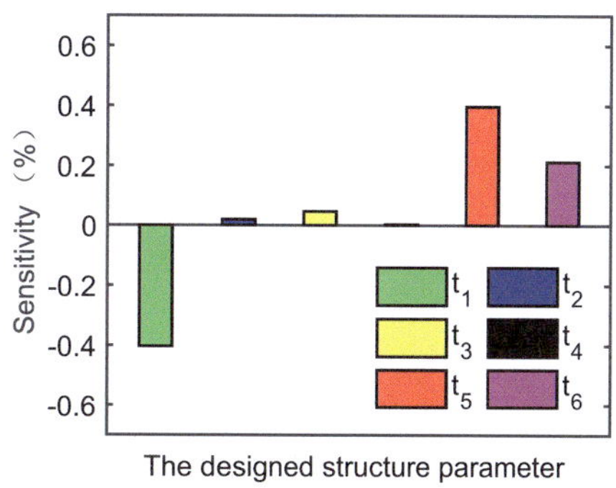

Fig. 5.4 Sensitivity analysis of design variables (t_1 to t_6) on maximum output displacement, showing dominant influence of parameters t_1, t_5, and t_6 with sensitivity values of -0.40, 0.40, and 0.21, respectively

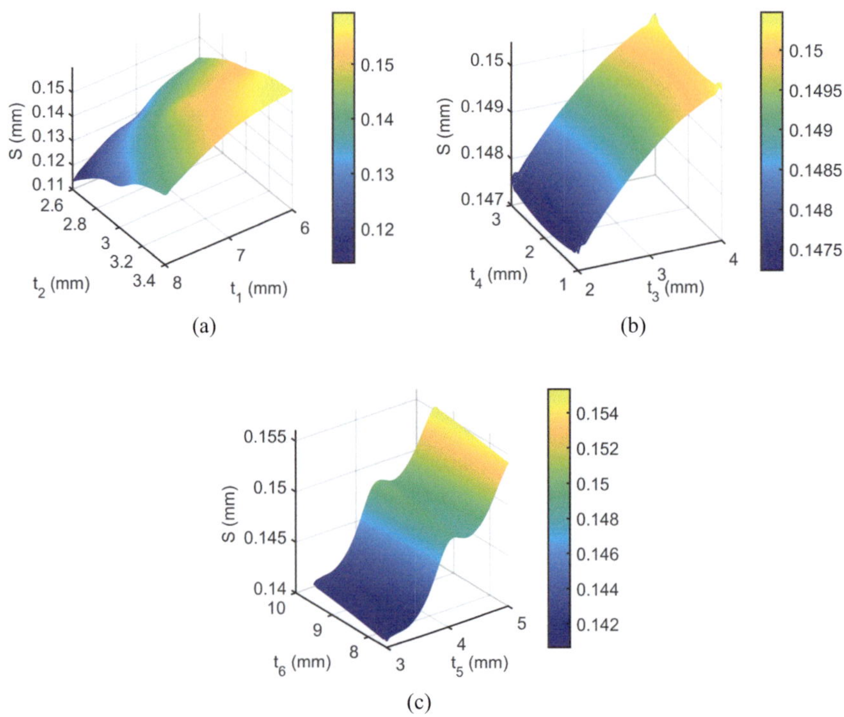

Fig. 5.5 Response surfaces illustrating the relationship between maximum output displacement and structural parameters: **a** t_1–t_2 interaction showing dominant negative influence of t_1; **b** t_3–t_4 interaction demonstrating moderate positive effect of t_3; **c** t_5–t_6 interaction revealing strong positive correlation with output displacement

Table 5.1 Optimized structural parameters of the micro-gripper obtained through response surface methodology and sensitivity analysis

Parameter	Value (mm)
t_1	7.0
t_2	3.0
t_3	3.0
t_4	2.0
t_5	4.0
t_6	8.5

5.3.2 FEA Simulation Result

To validate the performance of the optimized microgripper design, a finite element analysis (FEA) was conducted using ANSYS Workbench. The gripper was modeled with Al-7075 alloy, selected for its favorable combination of low density

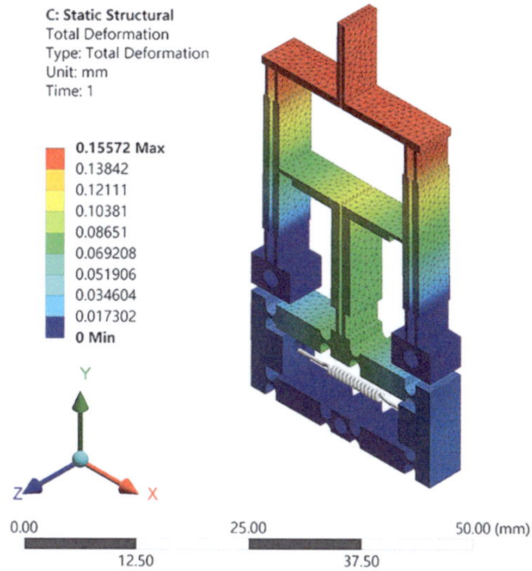

Fig. 5.6 Finite element analysis (FEA) simulation results showing total deformation distribution under maximum actuation force (900 N) with 155.72 μm jaw displacement

(2810 kg/m^3), high Young's modulus (71 GPa), and yield strength (524 MPa). The piezoelectric actuator (PEA model RP150/55/20) was simulated as a linear spring with equivalent stiffness, representing its nominal displacement of 20 μm and maximum blocking force of 900 N.

Under an applied force of 900 N at both ends of the bridge mechanism, the simulation revealed a maximum unilateral input displacement of 8.6 μm. The resulting deformation analysis, shown in Fig. 5.6, demonstrated a maximum jaw displacement of 155.72 μm in the horizontal direction. Due to the symmetrical architecture, the total gripping range reached 311.44 μm, confirming the design's capability for handling micro-objects across a substantial range. The displacement amplification ratio was calculated as 18.1, consistent with theoretical predictions. Stress analysis confirmed that von Mises stresses remained below the material yield limit, with a safety factor exceeding 2.5 throughout the structure.

Figure 5.7 demonstrates the linear relationship between input and output displacement, confirming an amplification ratio of 16.72 under simulated loading conditions. The equivalent (von Mises) stress distribution reveals stress concentration at the compliant hinges of the bridge mechanism's upper section, with a peak value of 162.11 MPa. This remains well below the yield strength of Al-7075 alloy (524 MPa), resulting in a minimum safety factor of 3.53 at maximum actuation force (900 N). These results verify the structural integrity of the design during full-range operation, ensuring no material yielding occurs at the maximum gripping displacement.

Modal analysis results for the first three resonant modes are presented in Fig. 5.8. The fundamental mode (843 Hz) exhibits coordinated rotational motion of the parallelogram, L-shaped, and bridge mechanisms about the fixed mounting points. The second mode (933 Hz) represents the operational working mode, characterized by symmetrical horizontal motion of both gripping jaws–confirming the mechanism's

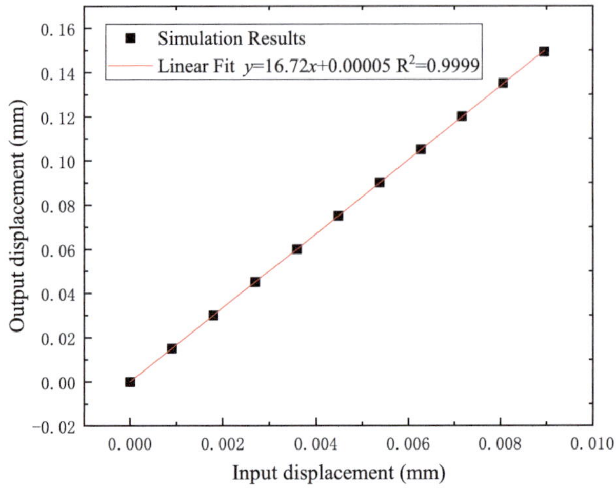

Fig. 5.7 Simulation results demonstrating the linear relationship between output and input displacements: input-output displacement correlation with amplification ratio of 16.72

ability to produce pure translational motion for gripping operations. The third mode (1665 Hz) involves out-of-plane bending of the jaw structures, representing a higher-order deformation pattern outside the primary working plane.

These results validate the microgripper's dynamic performance, particularly demonstrating that the fundamental operational mode (933 Hz) achieves the desired translational motion while maintaining a high natural frequency suitable for high-speed applications. The clear separation between operational modes (in-plane) and parasitic modes (out-of-plane) confirms effective structural design for precision manipulation tasks.

5.4 Prototype Fabrication and Experimental Tests

This section details the fabrication process and experimental validation of the micro-gripper prototype. The performance characteristics–including output displacement, natural frequency, gripping force, and motion resolution–were systematically evaluated through a series of experimental tests. The experimental setup and measurement methodologies are described, followed by presentation and discussion of the results.

5.4.1 Prototype Fabrication and Experimental Setup

The microgripper prototype was fabricated from Al-7075 alloy using wire electrical discharge machining (WEDM), achieving precise dimensional accuracy and surface quality. The material properties include: Young's modulus = 71 GPa, Poisson's ratio

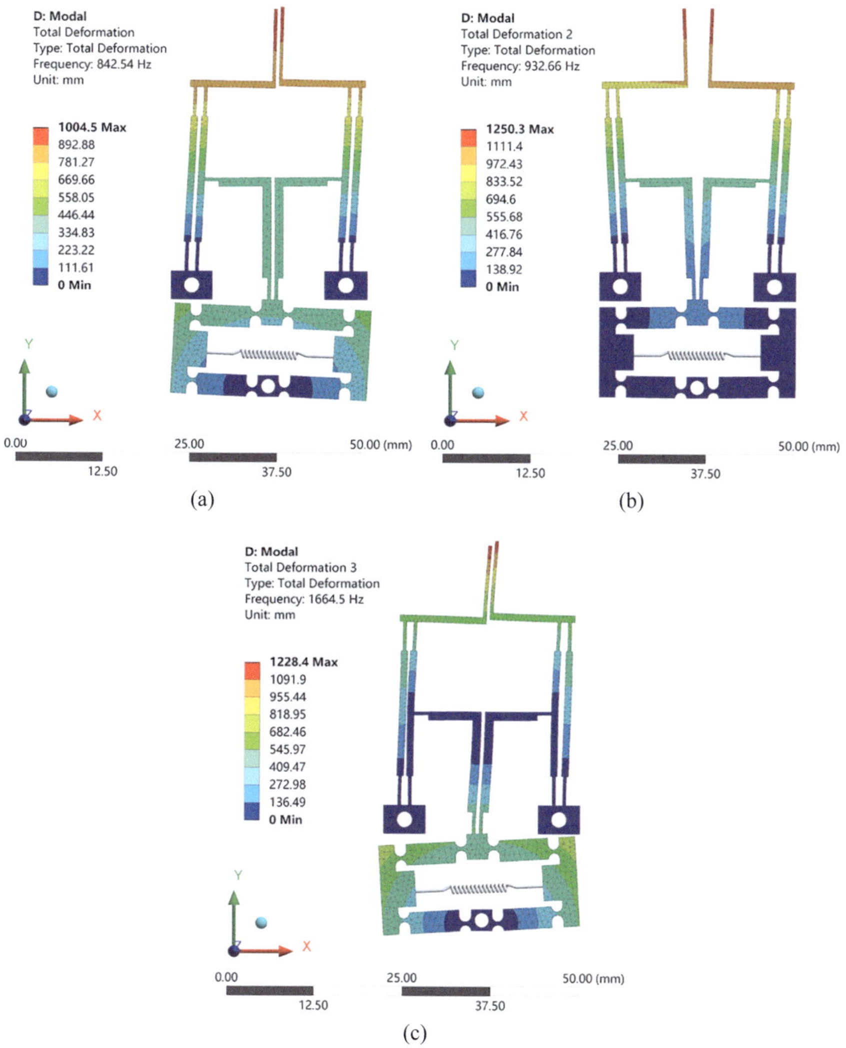

Fig. 5.8 Finite element simulation results of the first three mode shapes of the microgripper: **a** first mode (843 Hz) showing coordinated rotational motion of amplification mechanisms; **b** second mode (933 Hz) demonstrating in-plane operational gripping motion; **c** third mode (1665 Hz) exhibiting out-of-plane bending displacement

= 0.33, density = 2810 kg/m^3, and yield strength = 524 MPa. The fabricated prototype (Fig. 5.9a) maintains the compact design (54 mm × 28 mm × 5 mm) optimized through computational studies.

An integrated experimental system was developed to characterize the microgripper's performance (Fig. 5.9b, c). The prototype was mounted on a high-precision XYZ translational stage (model: HTML25-NS, from Hangzhou SPL Photonics Co.,

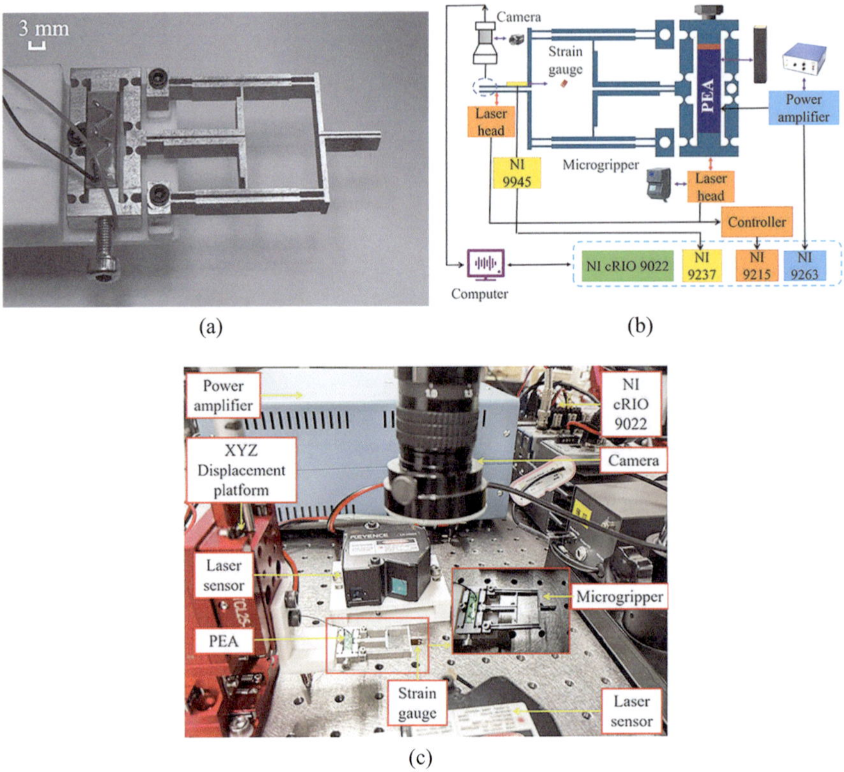

Fig. 5.9 Experimental setup for microgripper performance evaluation: **a** fabricated Al-7075 prototype; **b** schematic diagram showing laser sensors, force measurement, and control systems; **c** photograph of complete apparatus with vibration isolation

Ltd., China) using a custom fixture. Displacement measurements were obtained using two laser sensors (model: LK-H055, from Keyence Corp., Japan; resolution: 25 nm, range: ± 10 mm) positioned to simultaneously capture input and output displacements. Force sensing was implemented through a strain gauge (model: BFH 350-3AA, from Chengtec Electronics Ltd., China; sensitivity factor: 2.0) bonded to the gripper jaw and connected to a quarter Wheatstone bridge circuit with signal conditioning.

Actuation was provided by a voltage amplifier (model: EPA-104, from Piezo Systems, Inc., USA) delivering 0–150 V to the piezoelectric actuator. Visual monitoring was accomplished with an industrial camera (model: TD-UV20S, from Sanqtid Co., China) mounted above the workspace. A real-time controller (model: NI cRIO 9022, from National Instruments Corp., USA) with analog I/O modules managed all signal generation, data acquisition, and control algorithms at 1 kHz sampling rate. The entire system was isolated from environmental vibrations using an optical table (model: RS2000, from Newport Corp.) to minimize external disturbances.

This comprehensive setup enabled simultaneous measurement of kinematic, dynamic, and force characteristics while ensuring experimental repeatability under controlled conditions.

5.4.2 Open-Loop Test of the Microgripper Displacement

Open-loop displacement tests were conducted by applying a linearly varying voltage (0–150 V) to the piezoelectric actuator. The experiments were repeated three times to evaluate repeatability and characterize hysteresis effects.

Figure 5.10 presents the experimental results, showing the input voltage, input displacement, and output displacement. At the maximum driving voltage of 150 V, the unilateral jaw displacement reached 156.4 μm. The input displacement signal (Fig. 5.10b) exhibited noticeable noise due to the small displacement-to-voltage ratio inherent to piezoelectric actuation. The total output displacement for both jaws was measured at 312.8 μm, yielding an experimental amplification ratio of 15.64.

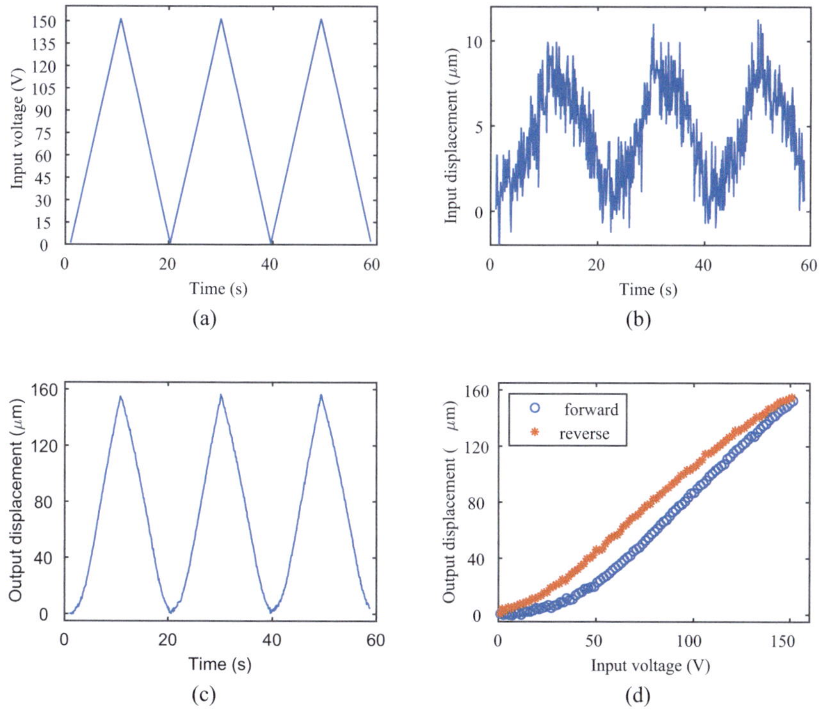

Fig. 5.10 Experimental results of gripping displacement characterization: **a** input voltage signal (0–150 V); **b** measured input displacement; **c** measured output displacement; **d** output displacement versus input voltage relationship showing 16.72 amplification ratio

This result shows good agreement with the FEA simulation result (16.72), with the minor discrepancy (6.9%) attributable to manufacturing tolerances and assembly variations. The theoretical model prediction (21.02) exceeded both experimental and simulation values due to two primary factors: (i) the theoretical model assumed ideal hinge deformation without accounting for energy dissipation in rigid segments, and (ii) it neglected the thickness effects of structural components. The hysteresis characteristic observed in Fig. 5.10d confirms the nonlinear piezoelectric behavior, highlighting the need for closed-loop control strategies to achieve precise positioning.

5.4.3 Open-Loop Test of the Natural Frequency

A dynamic characterization experiment was performed to identify the natural frequencies and corresponding mode shapes of the microgripper. The system was excited using a sinusoidal sweep signal with the following parameters: frequency range = 1–1000 Hz, amplitude = 1 V, sweep rate = 1 Hz/s, and frequency increment = 0.1 Hz. The excitation was applied to the embedded piezoelectric actuator (PEA) to induce structural vibrations. Data acquisition was performed using an NI cRIO 9022 controller operating in FPGA mode with a sampling frequency of 3000 Hz. A laser displacement sensor (Keyence LK-H055) captured the displacement response at the jaw tip with nanometer-level resolution.

The frequency response function (FRF) obtained from the test is shown in Fig. 5.11. The fundamental natural frequency was identified at 786 Hz, corresponding to the peak amplitude in the FRF. This experimental value is 6.8% lower than the FEA-predicted value (843 Hz) and 14.8% lower than the theoretical model prediction (922 Hz). The discrepancies are primarily attributed to unmodeled mass contributions from the PEA (20 g), mounting bolts, and gaskets in the physical system. According to Eq. (5.20), the natural frequency $f = \frac{1}{2\pi}\sqrt{K/M}$ is inversely proportional to the square root of the equivalent mass M. The additional mass components not accounted for in the models thus reduce the experimental natural frequency. Other

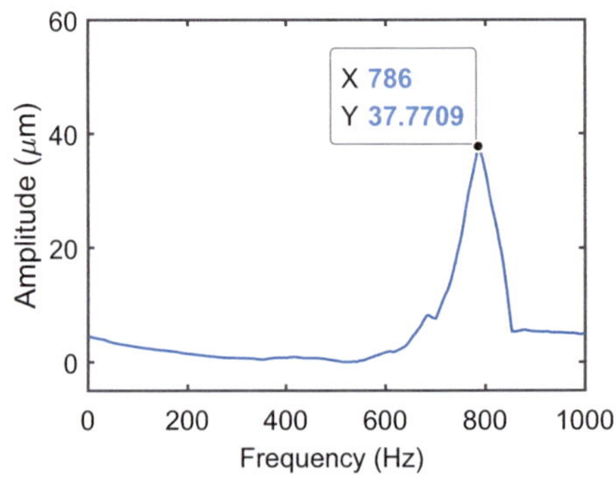

Fig. 5.11 Frequency response test result of the gripping displacement showing magnitude response with first natural frequency at 786 Hz

contributing factors include slight variations in material properties and boundary conditions between the ideal models and physical prototype.

5.4.4 Open-Loop Test of the Microgripper Force

The strain-gauge force sensor was calibrated using precision weights (1 and 2 g) through sequential loading and unloading cycles, as illustrated in Fig. 5.12a. The initial sensor output voltage under no-load conditions was measured at 0.585 mV. Figure 5.12b shows the calibration curve, where applied weights caused a linear decrease in output voltage during loading, with symmetric recovery during unloading. The calibration established a force-to-voltage sensitivity of −1185.2 mN/mV.

A gripping experiment was performed on a 350-μm diameter metal micro-wire, with results shown in Fig. 5.13. The process was segmented into five distinct phases:

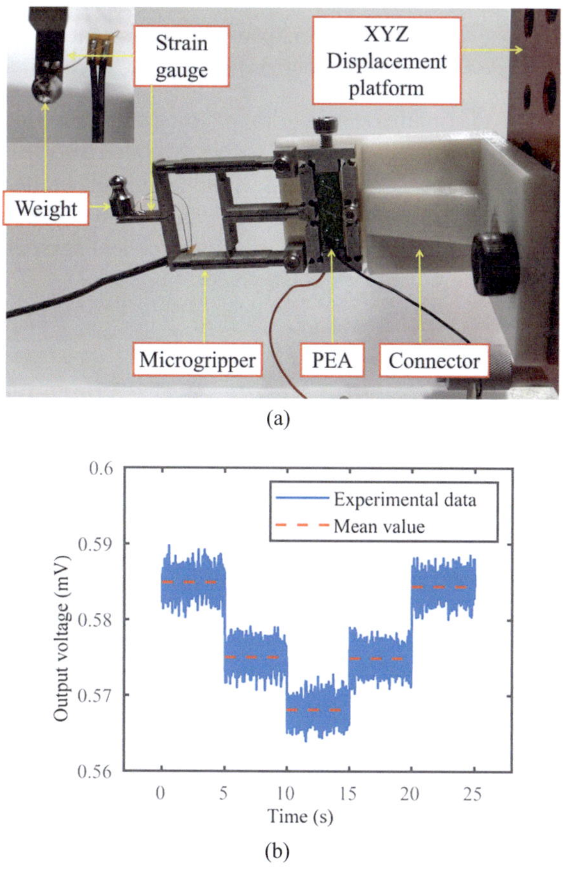

Fig. 5.12 Experimental results of strain-gauge force sensor characterization: **a** calibration setup using standard weights (1 and 2 g); **b** calibration curve showing linear force-voltage relationship with conversion coefficient of −1185.2 mN/mV

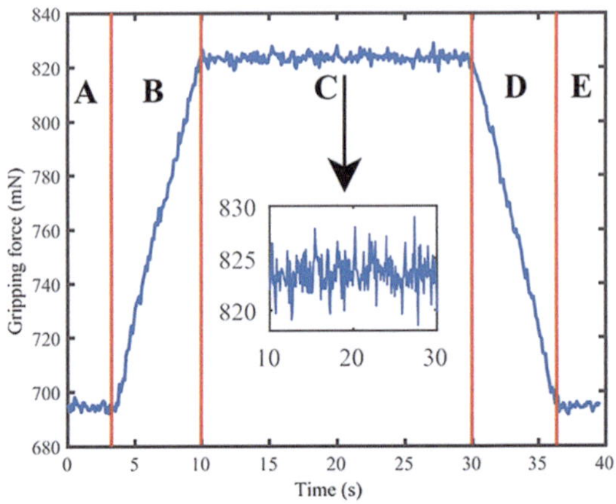

Fig. 5.13 Experimental results of gripping force measurement when grasping a 350-μm diameter metal micro-wire: **a** five-phase gripping process (A–E) showing contact transition and stable gripping; **b** force sensor noise distribution with ±1.69 mN resolution (2σ)

- **Phases A and E**: Non-contact states showing baseline force offset of 692.91 mN
- **Phase B**: Contact and force increase phase during jaw closure
- **Phase C**: Stable gripping phase with constant 823.71 mN force
- **Phase D**: Force release phase during jaw opening.

The maximum relative gripping force was measured at 130.80 mN during the stable gripping phase. The force sensor exhibited a noise level of 1.69 mN (2σ), indicating 95% of measurements fell within ±1.69 mN of the mean value. This corresponds to a force resolution of ±1.69 mN, which is sufficient for delicate micromanipulation tasks. The experimental results show excellent agreement with FEA predictions and follow theoretically expected physical behavior.

5.5 Experimental Results of the Microgripper with Closed-Loop Control

The inherent nonlinearities in piezoelectric actuation, particularly hysteresis and creep, necessitate closed-loop control to achieve precise position tracking of the microgripper jaws. This work implements a Proportional-Integral (PI) control strategy to compensate for these effects and demonstrate accurate motion control.

A pure proportional controller results in steady-state error due to the system's nonlinear characteristics. The incorporation of an integral term eliminates this error by continuously accumulating and compensating for past deviations. The PI control law is formulated as:

$$u(t) = K_P \left[e(t) + \frac{1}{T_I} \int_0^t e(\tau) d\tau \right] \tag{5.21}$$

where K_P represents the proportional gain, T_I the integral time constant, $e(t) = r(t) - y(t)$ the tracking error between reference $r(t)$ and output $y(t)$, and $u(t)$ the control signal.

Through systematic experimental tuning, the optimal controller parameters were determined to be $K_P = 35$ and $K_I = 0.00005$ (where $K_I = K_P/T_I$), providing a balance between response speed and stability. The closed-loop system operated at a sampling frequency of 1 kHz, ensuring adequate bandwidth for precise trajectory tracking.

5.5.1 Experimental Results of Motion Tracking

The microgripper's tracking performance was systematically evaluated through three experimental campaigns: amplitude variation tests, frequency response tests, and step resolution tests.

5.5.1.1 Amplitude Variation Tests

Sinusoidal trajectories with amplitudes of 20, 40, 60, and 80 μm at 0.25 Hz were applied. The maximum tracking errors were measured as 1.38, 2.16, 2.88, and 3.81 μm, respectively, corresponding to 6.9%, 5.4%, 4.6%, and 4.7% of the respective output displacements. The errors exhibited a normal distribution with a 2σ value of 2.54 μm, indicating 95% of errors fell within ±2.54 μm (Fig. 5.14).

5.5.1.2 Frequency Response Tests

Using a fixed amplitude of 60 μm, frequencies of 0.25, 0.5, 0.75, and 1 Hz were tested. Maximum tracking errors increased to 2.81, 5.61, 8.07, and 11.06 μm (4.6–19.3% of the reference amplitude). The error distribution showed a 95% confidence interval of ±9.88 μm (Fig. 5.15), reflecting the system's bandwidth limitations.

5.5.1.3 Step Resolution Tests

A staircase signal with 1 μm steps and 2 s duration was applied. The laser sensor noise ($2\sigma = 0.5978$ μm) determined the closed-loop resolution, with 95% of readings within ±0.6 μm (Fig. 5.16). This confirms the microgripper's capability for sub-micrometer precision motion.

These results validate the PI controller's effectiveness in achieving precise motion control across varying operational conditions, with the microgripper demonstrating robust performance suitable for high-precision micromanipulation tasks.

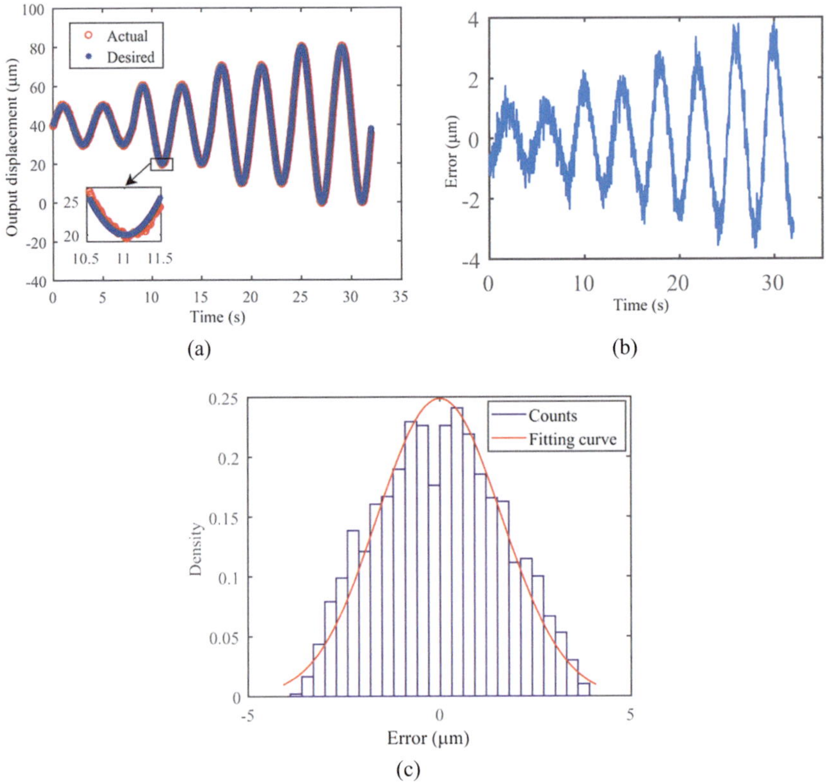

Fig. 5.14 Experimental results of position tracking with PI controller for sine wave inputs with varying displacement amplitudes (20–80 μm): **a** reference and actual displacement trajectories; **b** position tracking error; **c** histogram of tracking error distribution showing ±2.54 μm (2σ) precision

5.5.2 Experimental Results of Gripping Test

The gripping capability of the designed microgripper was experimentally validated through a series of tests involving different micro-objects and operational scenarios.

5.5.2.1 Material Handling and Docking Demonstration

Initial experiments demonstrated successful manipulation of an optical fiber with a diameter of 200 μm and a metal sheet with a thickness of 100 μm. Snapshots capturing the clamping states during these operations are presented in Fig. 5.17.

The gripping performance was further evaluated through a docking experiment utilizing a square shim with a thickness of 200 μm. The docking procedure, illus-

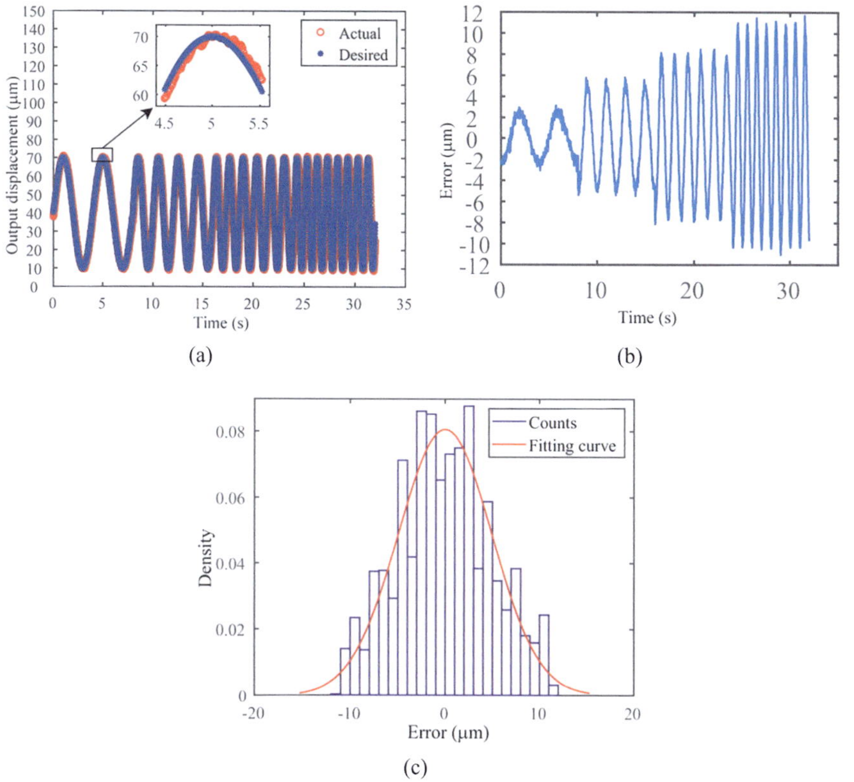

Fig. 5.15 Experimental results of position tracking with PI controller for sine wave inputs with varying frequencies (0.25–1 Hz): **a** reference and actual displacement trajectories; **b** position tracking error; **c** histogram of tracking error distribution showing ±9.88 μm (2σ) precision at 1 Hz

Fig. 5.16 Experimental results of motion resolution test for the microgripper jaw, showing step response with 1-μm increments and measured resolution of ±0.6 μm (2σ)

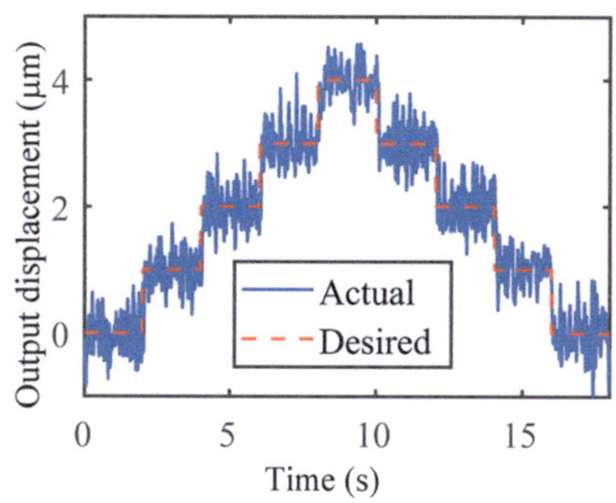

trated in Fig. 5.18, involved coordinated operation between two micrograppers. The left gripper initially clamped the shim and advanced toward the right gripper. Precise alignment between the two grippers was achieved through manual adjustment of their

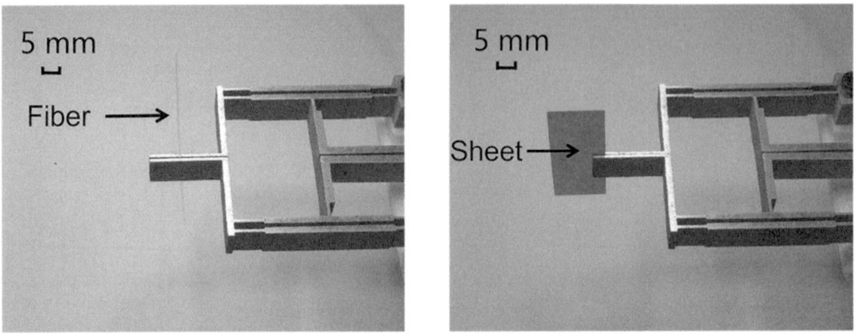

(a) (b)

Fig. 5.17 Snapshots of gripping experiments: **a** optical fiber with 200-μm diameter; **b** metal sheet with 100-μm thickness

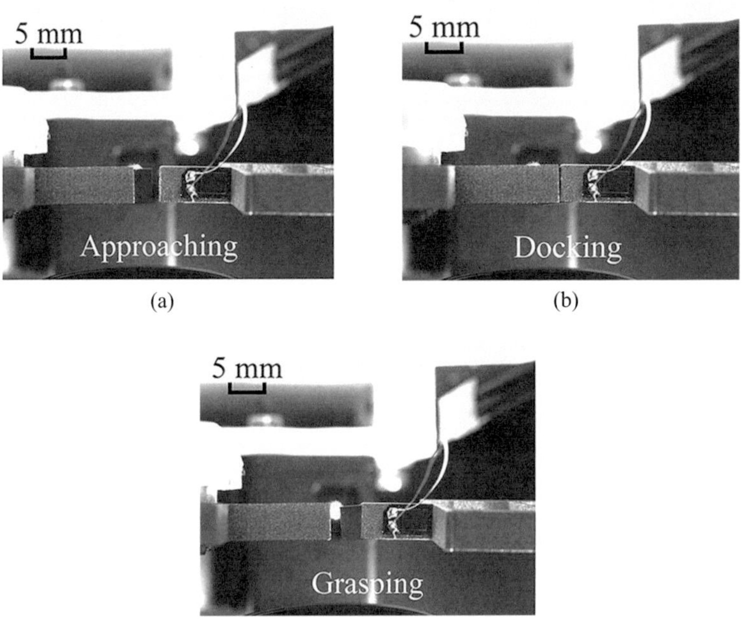

(c)

Fig. 5.18 Snapshots of the docking experiment using two microgrippers: **a** approaching phase; **b** docking alignment; **c** successful grasping by the designed microgripper

relative position. Subsequently, the right gripper was activated to engage the shim while the left gripper simultaneously released its grip and retracted. The successful transfer resulted in the shim being securely suspended by the right gripper–the microgripper developed in this study. This experiment confirms the reliable performance and operational adaptability of the proposed microgripper for practical micromanipulation applications.

5.5.2.2 Force-Controlled Manipulation

To ensure damage-free manipulation of fragile micro-objects, precise control of gripping force is essential during handling operations. This work implements a position/force switch control strategy to achieve non-destructive precision gripping of micro-wires. The approach begins with position control during the jaw approach phase, followed by automatic switching to force control upon contact detection while maintaining position control in a deactivated state.

Prior to implementing the hybrid control strategy, an open-loop characterization experiment was conducted using full-scale driving voltage to establish the relationship between gripping displacement and force. Figure 5.19 displays the displacement and force signals recorded during open-loop gripping of a 350-μm diameter micro-wire. The gripping force is represented by the dotted curve, while the jaw displacement is denoted by the dashed curve. The process is divided into three distinct segments based on the gripping dynamics: Segments 1 and 3 exhibit rapid displacement changes while maintaining nearly constant gripping force, whereas Segment 2 demonstrates the opposite behavior with significant force variations occurring while displacement remains stabilized. The transition points between these segments provide critical reference information for implementing the switch control strategy.

The results of the closed-loop gripping experiment utilizing position/force switching control are presented in Fig. 5.20. The process is similarly divided into three segments corresponding to different control variables. Segments 1 and 3 represent position control phases where step position signals facilitate the grasp and release operations of the metal wire. The experimental results confirm that both gripping

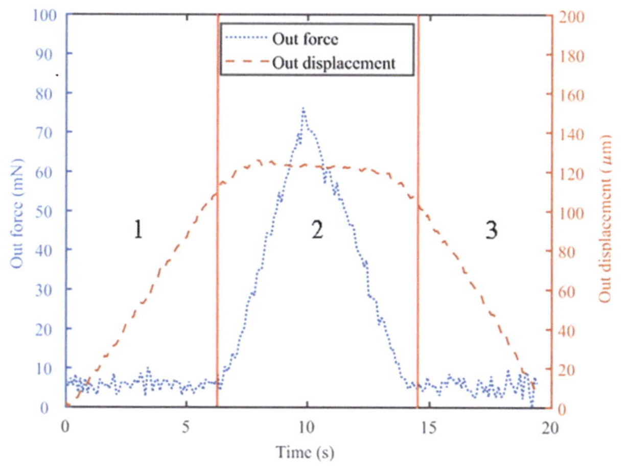

Fig. 5.19 Experimental results of open-loop gripping process test showing displacement (dashed line) and force (dotted line) signals during manipulation of a 350-μm diameter micro-wire, with three distinct operational segments identified

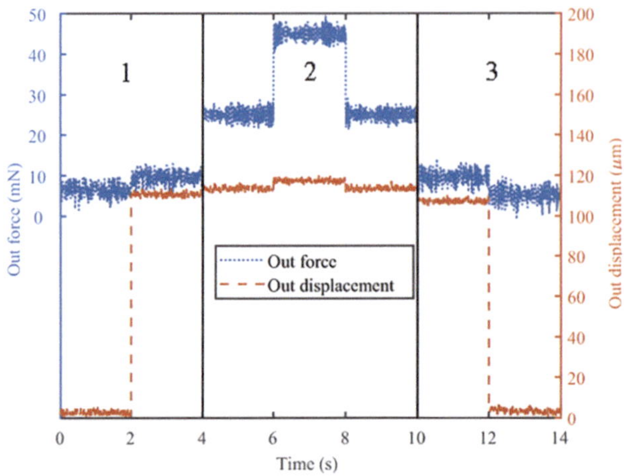

Fig. 5.20 Experimental results of closed-loop gripping process test with position/force switch control, demonstrating precise transition between position control (Segments 1 and 3) and force control (Segment 2) for non-destructive manipulation

position and force are effectively regulated using the PI control strategy, enabling precise and non-destructive manipulation of delicate micro-objects.

5.5.3 Performance Comparison

A comprehensive performance comparison was conducted between the fabricated microgripper and existing piezoelectric-actuated microgrippers reported in the literature. Key performance metrics, including displacement amplification ratio, gripping stroke, and natural frequency, are systematically summarized in Table 5.2. The

Table 5.2 Performance comparison of typical piezoelectric-actuated microgrippers, highlighting the proposed design's superior comprehensive performance metric (stroke × frequency)

Reference	Amplification ratio	Stroke (μm)	Frequency (Hz)	Stroke × Frequency (μm × Hz)
[29]	2.85	100	–	–
[13]	20.2	101	130	13,130
[1]	22.2	2000	70	140,000
[16]	16.13	720	70.7	50,904
[11]	12.76	93.52	1044	97,634.88
[8]	12.05	184.04	854	157,170.16
This work	15.64	312.8	786	245,860.8

comparative analysis reveals that the proposed microgripper achieves an exceptional balance between large gripping stroke and high natural frequency relative to previous designs.

To quantitatively assess the overall performance, a composite metric–defined as the product of working stroke and resonant frequency–is introduced. This parameter effectively captures the critical trade-off between motion range and dynamic response characteristics, with higher values indicating superior comprehensive capability for high-speed micromanipulation applications. As demonstrated in the final column of Table 5.2, the proposed gripper achieves the highest value of this performance metric among all compared devices, confirming its advanced capabilities for applications requiring both substantial motion range and rapid dynamic response.

The performance superiority is particularly evident when comparing with previous designs that typically excel in either stroke capacity (e.g., 1000 μm @70 Hz) or dynamic response (e.g., 93.52 μm @1044 Hz), but rarely achieve both simultaneously. The proposed design's ability to maintain a high natural frequency (786 Hz) while achieving a substantial gripping stroke (312.8 μm) represents a significant advancement in microgripper design methodology. This performance enhancement is attributed to the optimized three-stage amplification mechanism and carefully balanced structural dynamics achieved through the response surface methodology-based optimization process.

By examining the relationship between amplification ratio and gripping stroke across various microgripper designs, it is evident that the amplification ratio does not exhibit direct proportionality to the achievable gripping stroke. This discrepancy arises due to differences in input displacement capabilities of piezoelectric actuators, structural constraints, and energy efficiency across designs. Consequently, the gripping stroke–rather than the amplification ratio alone–is adopted as the primary design objective for evaluating and optimizing microgripper performance, as it more directly reflects the functional capability in practical applications.

5.6 Conclusion

This chapter has presented the development and validation of a novel compliant microgripper driven by a piezoelectric actuator, specifically designed for precision micromanipulation and microassembly applications. The microgripper incorporates a three-stage flexible amplification mechanism that enables large-stroke parallel motion of both jaws while maintaining compact dimensions. A comprehensive mathematical model was established to characterize the displacement amplification behavior, and response surface methodology was employed to optimize critical structural parameters for enhanced performance. Finite element analysis simulations verified the design's capabilities, confirming both substantial gripping range and high natural frequency. Experimental characterization of the fabricated prototype demonstrated a gripping stroke of 311.44 μm and a first natural frequency of 786 Hz. Closed-loop control experiments utilizing a proportional-integral (PI) control strategy achieved precise tracking of sinusoidal reference trajectories and demonstrated a motion resolution of $\pm 0.6\ \mu m$ through step response tests. The practical utility of the design was further validated through successful manipulation of delicate components, including a 200-μm diameter optical fiber and a 100-μm thick metal sheet.

Future research directions will focus on the implementation of advanced hybrid position/force control strategies to enable delicate manipulation of an expanded range of micro-objects, including metal wires and optical fibers. Additional work will explore integration with automated microassembly systems and the development of vision-based control algorithms to enhance the gripper's applicability in complex micromanipulation scenarios.

References

1. Ai, W., Xu, Q.: New structural design of a compliant gripper based on the Scott-Russell mechanism. Int. J. Adv. Rob. Syst. **11**(12), 192 (2014)
2. Allaix, D.L., Carbone, V.I.: An improvement of the response surface method. Struct. Saf. **33**(2), 165–172 (2011)
3. Avci, E., Hattori, T., Kamiyama, K., Kojima, M., Horade, M., Mae, Y., Arai, T.: Piezo-actuated parallel mechanism for biological cell release at high speed. Biomed. Microdevice **17**(5), 1–10 (2015)
4. Cauchi, M., Grech, I., Mallia, B., Mollicone, P., Sammut, N.: Analytical, numerical and experimental study of a horizontal electrothermal MEMS microgripper for the deformability characterisation of human red blood cells. Micromachines **9**(3), 108 (2018)
5. Chen, X., Deng, Z., Hu, S., Gao, J., Gao, X.: Design of a flexible piezoelectric microgripper based on combined amplification principles. Nanotechnol. Precis. Eng. **2**(3), 138–143 (2019)
6. Chen, Z., Li, Z., Jiang, X., Zhang, X.: Strain-based multimode integrating sensing for a bridge-type compliant amplifier. Meas. Sci. Technol. **30**(10), 105106 (2019)
7. Das, T.K., Shirinzadeh, B., Al-Jodah, A., Ghafarian, M., Pinskier, J.: Computational parametric analysis and experimental investigations of a compact flexure-based microgripper. Precis. Eng. **66**, 363–373 (2020)
8. Das, T.K., Shirinzadeh, B., Al-Jodah, A., Ghafarian, M., Pinskier, J.: A novel compliant piezoelectric actuated symmetric microgripper for the parasitic motion compensation. Mech. Mach. Theory **155**, 104069 (2021)
9. Das, T.K., Shirinzadeh, B., Ghafarian, M., Al-Jodah, A.: Design, analysis, and experimental investigation of a single-stage and low parasitic motion piezoelectric actuated microgripper. Smart Mater. Struct. **29**(4), 045028 (2020)
10. Das, T.K., Shirinzadeh, B., Ghafarian, M., Al-Jodah, A., Pinskier, J.: Characterization of a compact piezoelectric actuated microgripper based on double-stair bridge-type mechanism. J. Micro-Bio Robot. **16**(1), 79–92 (2020)
11. Das, T.K., Shirinzadeh, B., Ghafarian, M., Al-Jodah, A., Zhong, Y., Smith, J.: Design, analysis and experimental investigations of a high precision flexure-based microgripper for micro/nano manipulation. Mechatronics **69**, 102396 (2020)
12. Das, T.K., Shirinzadeh, B., Ghafarian, M., Pinskier, J.: A flexure-based 2-DOF microgripper for handling micro-objects. In: Proceedings of 2018 International Conference on Manipulation, Automation and Robotics at Small Scales (MARSS), pp. 1–6 (2018)
13. Feng, F.Y., Cui, Y.G., Xue, F., Wu, L.E.: Design of a new piezo-electric micro-gripper based on flexible magnifying mechanism. Appl. Mech. Mater. **201**, 907–911 (2012)
14. Howell, L.L., Magleby, S.P., Olsen, B.M., Wiley, J.: Handbook of Compliant Mechanisms. Wiley Online Library (2013)

15. Liang, C., Wang, F., Shi, B., Huo, Z., Zhou, K., Tian, Y., Zhang, D.: Design and control of a novel asymmetrical piezoelectric actuated microgripper for micromanipulation. Sens. Actuators, A **269**, 227–237 (2018)
16. Liu, Y., Xu, Q.: Mechanical design, analysis and testing of a large-range compliant microgripper. Mech. Sci. **7**(1), 119–126 (2016)
17. Lyu, Z., Wu, Z., Xu, Q.: Design and development of a novel piezoelectrically actuated asymmetrical flexible microgripper. Mech. Mach. Theory **171**, 104736 (2022)
18. Lyu, Z., Xu, Q.: Recent design and development of piezoelectric-actuated compliant microgrippers: a review. Sens. Actuators, A **331**, 113002 (2021)
19. Ruiz, D., Díaz-Molina, A., Sigmund, O., Donoso, A., Bellido, J.C., Sánchez-Rojas, J.L.: Optimal design of robust piezoelectric unimorph microgrippers. Appl. Math. Model. **55**, 1–12 (2018)
20. Shi, C., Dong, X., Yang, Z.: A microgripper with a large magnification ratio and high structural stiffness based on a flexure-enabled mechanism. IEEE/ASME Trans. Mechatron. **26**(6), 3076–3086 (2021)
21. Sun, X., Chen, W., Fatikow, S., Tian, Y., Zhou, R., Zhang, J., Mikczinski, M.: A novel piezo-driven microgripper with a large jaw displacement. Microsyst. Technol. **21**(4), 931–942 (2015)
22. Wang, D., Yang, Q., Dong, H.: A monolithic compliant piezoelectric-driven microgripper: design, modeling, and testing. IEEE/ASME Trans. Mechatron. **18**(1), 138–147 (2011)
23. Wang, F., Liang, C., Tian, Y., Zhao, X., Zhang, D.: Design of a piezoelectric-actuated microgripper with a three-stage flexure-based amplification. IEEE/ASME Trans. Mechatron. **20**(5), 2205–2213 (2014)
24. Wang, F., Liang, C., Tian, Y., Zhao, X., Zhang, D.: Design and control of a compliant microgripper with a large amplification ratio for high-speed micro manipulation. IEEE/ASME Trans. Mechatron. **21**(3), 1262–1271 (2016)
25. Xu, Q.: A new compliant microgripper with integrated position and force sensing. In: Proceedings of 2013 IEEE/ASME International Conference on Advanced Intelligent Mechatronics, pp. 591–596 (2013)
26. Yang, S., Xu, Q.: Design of a microelectromechanical systems microgripper with integrated electrothermal actuator and force sensor. Int. J. Adv. Rob. Syst. **13**(5), 1729881416663375 (2016)
27. Yang, Y.L., Wei, Y.D., Lou, J.Q., Xie, F.R., Fu, L.: Development and precision position/force control of a new flexure-based microgripper. J. Micromech. Microeng. **26**(1), 015005 (2015)
28. Zimmermann, S., Tiemerding, T., Fatikow, S.: Automated robotic manipulation of individual colloidal particles using vision-based control. IEEE/ASME Trans. Mechatron. **20**(5), 2031–2038 (2014)
29. Zubir, M.N.M., Shirinzadeh, B.: Development of a high precision flexure-based microgripper. Precis. Eng. **33**(4), 362–370 (2009)
30. Zubir, M.N.M., Shirinzadeh, B., Tian, Y.: A new design of piezoelectric driven compliant-based microgripper for micromanipulation. Mech. Mach. Theory **44**(12), 2248–2264 (2009)

Chapter 6
Design of a Piezoelectrically Actuated Asymmetrical Flexible Microgripper

Abstract This chapter presents a novel asymmetric piezoelectric microgripper designed to overcome limitations of symmetric configurations. It features a three-stage flexure mechanism that enables precise, friction-free motion and separates driving and sensing arms to prevent interference. Through modeling and optimization, key parameters were refined to enhance performance. Simulations and experiments validated its working stroke, clamping force, and resonant frequency. The gripper successfully manipulated micro-objects of varying sizes and shapes, demonstrating its suitability for practical micromanipulation and microassembly tasks. The design ensures reliable operation and eliminates resonant mode issues common in symmetric designs.

Keywords Microgripper · Flexure mechanism · Compliant mechanism · Asymmetric gripper · Symmetric gripper · Micromanipulation

6.1 Introduction

Microgrippers represent sophisticated tools essential for micromanipulation and microassembly tasks, with widespread applications in semiconductor technology, biomedical engineering, and optical engineering [5, 30, 32]. As critical devices bridging macro- and micro-scale systems, microgrippers play a vital role in micro/nano-scale operations [22, 24]. These devices typically feature a front end that interacts directly with micro-components for grasping, transporting, assembling, and releasing operations, while the back end connects to macro-mechanisms to enable multi-degree-of-freedom movement within large workspaces. The diversity of object materials, shapes, and dimensions necessitates various microgripper designs to accommodate different operational requirements [31].

Microgrippers can be categorized by actuation method into vacuum adsorption, electrostatic, electromagnetic, shape memory alloy, and piezoelectric types [10, 12, 13, 19, 27]. Among these, stack piezoelectric ceramic actuation microgrippers have gained prominence in micromanipulation robotics due to their high precision, rapid response, and ease of control. These devices effectively combine piezo-

Q. Xu, *Piezoelectric-Driven Flexible Robotic Microgrippers*,
https://doi.org/10.1007/978-981-95-4685-5_6

Table 6.1 Performance comparison of typical symmetric flexure microgrippers

Reference	1st (Hz)	2nd (Hz)	3rd (Hz)	Working mode
[3]	1062	1344	1707	1st
[8]	815	1092	3025	1st
[11]	1578.9	1637.5	1672.4	1st
[34]	1446.1	1690.5	–	1st
[29]	920.9	1213.5	2047.9	2nd
[6]	185.4	202.7	466.7	2nd
[28]	576.86	622.66	1756.4	2nd
[21]	70.715	77.338	148.82	2nd
[18]	672	920	1676	2nd
[17]	102.74	107.12	192.38	2nd

electric actuators with multi-stage compliant amplification mechanisms to achieve large-stroke clamping performance.

Flexure-based compliant microgrippers utilizing stack piezoelectric ceramic actuator (SPCA) are generally classified into two geometric categories: symmetrical and asymmetrical configurations [20]. Symmetrical grippers offer advantages such as large clamping strokes, balanced internal stresses, and aesthetic appeal. However, they suffer from limitations including sensitivity to manufacturing tolerances and assembly errors, absence of fixed positioning benchmarks, and dynamic performance issues. As documented in Table 6.1, most symmetrical piezoelectric grippers exhibit mismatches between their first resonant modes and working modes [6, 17, 18, 21, 28, 29]. This frequency discrepancy can induce passive resonance during high-frequency operation. Additionally, symmetrical designs often demonstrate dense mode phenomena where closely spaced resonant frequencies complicate control system design–a characteristic inherent to symmetric structures [25]. The integration of microforce sensors (e.g., strain gauges) on compliant hinges further compromises symmetry and affects overall performance.

Asymmetrical grippers present a viable alternative, though they typically provide half the clamping stroke since only one jaw moves while the other remains stationary. Previous developments include Zhang et al.'s lever-parallelogram design achieving 6.0× magnification for 150 μm glass tubes [39], Xing et al.'s optimized asymmetric gripper with 4.16× magnification and 7.5 μm resolution [36], and Liang et al.'s three-stage amplifier offering 13.94× magnification at 531 Hz with 149.4 μm displacement [16]. More complex designs like Chen et al.'s four-stage amplifier achieved 31.6× magnification but required dual actuators, increasing cost and energy consumption [7]. While Ho et al.'s four-bar mechanism produced large displacements, its substantial footprint (75 mm × 113 mm × 5 mm) limited applicability in confined spaces [14].

Recent advances include piezoelectric-actuated asymmetric 2-DOF flexible grippers combining lateral clamping and longitudinal rubbing functions [4, 5, 33]. These designs typically feature working modes aligned with the first two resonant modes, though arm coupling effects during operation remain problematic. Some researchers have addressed this through functional separation between arms, such as Felix et al.'s electrostatic gripper with comb-drive actuation and capacitive sensing [2], Kim et al.'s electrothermal actuator with dual-axis capacitive force sensing [15], and Simon et al.'s monolithic dual-axis tensile tester [23]. Rotary mechanisms have also been incorporated, as demonstrated by Busara et al.'s comb-drive gripper with 90 μm stroke [26] and Xu's electromagnetic design with integrated position/force sensors [37].

This work presents an asymmetrical gripper design based on flexure mechanisms and piezoelectric actuation. We analyze why existing symmetrical grippers exhibit misaligned working and resonant modes, establish new design guidelines, and systematically compare asymmetric and symmetric configurations. Through pseudo-rigid body modeling, we develop comprehensive displacement, static, dynamic, and force measurement models. Finite element analysis and experimental studies demonstrate that while symmetrical grippers provide greater displacement magnification, they suffer from inferior dynamic performance, mode misalignment, and mode density issues. The proposed asymmetrical design overcome these limitations while maintaining suitable working stroke, high resonant frequency, and sufficient clamping force–representing a significant advancement in microgripper technology for precision manipulation applications.

6.2 Mechanism Design and Analytical Modeling

6.2.1 Mechanism Design

Before proceeding with the structural design of the gripper, it is necessary to examine the first non-working mode of the symmetrical microgripper. The first resonant modes of some typical symmetrical grippers are shown in Fig. 6.1. One mode (see Fig. 6.1a) is attributed to the swing of the clamping jaws in the direction perpendicular to the gripper plane. Another mode (see Fig. 6.1b, c) is caused by the two clamping jaws rotating within the gripper plane.

As shown in Fig. 6.1a, the gripping arm of the gripper is regarded as an L-shaped beam to simplify the analysis. The appearance of the former case is due to the length of the BC section (L_B) being close to that of the AB section (L_A). Yoon et al. found that the aspect ratio (L_A/L_B) of the L-shaped cantilever beam has a significant effect on the beam's modal shape [38]. When L_A is much larger than L_B, its first resonant mode will be generated at the vertical beam AB. Hence, the occurrence of the mode shape as shown in Fig. 6.1a is attributed to the instability of the structure, which is caused by the small aspect ratio of the gripper arm.

Another unfavorable mode is caused by the insufficient constraints of the microgripper, which makes the gripper arm equivalent to an overhanging beam or a can-

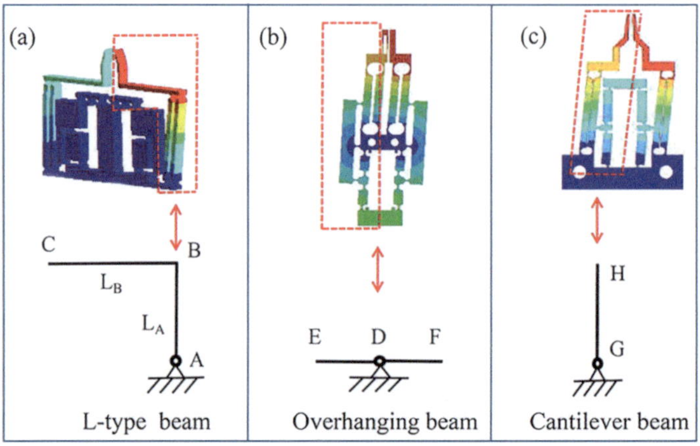

Fig. 6.1 First resonant modes (non-working modes) of representative symmetrical microgrippers reported in literature [1, 11, 17]

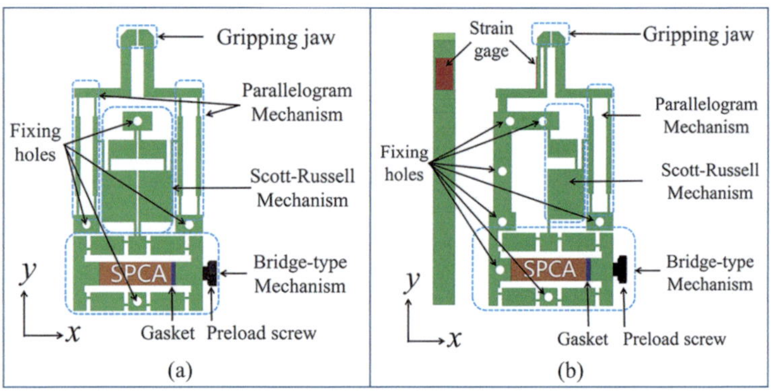

Fig. 6.2 CAD model of the designed flexure microgrippers. **a** Symmetrical gripper; **b** asymmetrical gripper

tilever beam. Obviously, these two beam structures are prone to deflect within the gripper plane. Therefore, to overcome the above shortcomings, it is necessary to consider two issues in the design of the symmetrical microgripper. One is to increase the aspect ratio of the equivalent gripper structure length. The other is to appropriately increase the restraints of the gripper. These two suggestions are also applicable to the architectural design of asymmetric microgrippers.

Figure 6.2 shows the computer-aided design (CAD) model of the designed new microgrippers. While Fig. 6.2a illustrates a symmetrical microgripper, the corresponding asymmetrical design is depicted in Fig. 6.2b. The overall size of the asymmetrical microgripper is 60 mm × 28 mm × 5 mm. The microgripper is composed of a stack piezoelectric ceramic actuator (SPCA), a gasket, a pre-tightening bolt, a three-stage compliant magnification mechanism, and a pair of clamping jaws. The clamping jaws of the microgripper include a stationary jaw and a movable one. The three-stage compliant amplification mechanisms consist of the bridge-type mechanism, Scott-Russell (SR) mechanism, and leveraged parallelogram mechanism.

A SPCA is placed inside the bridge-type magnification mechanism to generate horizontal rightward driving force and displacement. The actuated displacement is transmitted and amplified by a series of compliant rectangular hinges due to their extensive rotational range and high processing accuracy. The gasket is inserted between the SPCA and the pre-tightening bolt to prevent the concentrated load acting on the end face from damaging the SPCA. Simultaneously, a pre-tightening bolt is used to provide pre-tightening force and to adjust the initial gap between the two gripper jaws.

The modified SR mechanism realizes the orthogonal movement directional conversion from the bridge-type mechanism to the parallelogram mechanism. The vertical coupling displacement of the parallelogram mechanism is very small and can be ignored. Hence, it can produce approximately parallel clamping movement, which is beneficial to stable clamping of irregular micro-targets. For practical use, it is necessary to detect and control the clamping force to reduce the damage of tiny targets. The strain gauges attached on a single side will affect the clamping movement's symmetry for a symmetrical microgripper. By contrast, the strain gauge stuck on the asymmetrical microgripper's stationary jaw will eliminate such issue. It realizes the measurement of the clamping force without affecting the elastic deformation of the compliant displacement amplification mechanism.

6.2.2 Kinematic Modeling

Figure 6.3 illustrates the pseudo-rigid-body model (PRBM) and dynamic model of the movable side of the gripper mechanism, where i $(i = A, B, \ldots, N)$ denotes the rotation center of each compliant hinge. In the PRBM approach, each compliant hinge is treated as an ideal single-axis revolute joint equipped with a torsional spring. For kinematic analysis, assuming no drift in the rotation centers and neglecting the stiffness of the compliant hinges, the mechanism interconnected by these hinges is considered as a rigid-body system. The input and output displacements of the microgripper are denoted by x_{in} and x_{out}, respectively. The ideal displacement amplification ratio of the gripper is defined as:

$$R_{\text{amp}} = \frac{x_{\text{out}}}{x_{\text{in}}}. \tag{6.1}$$

The motion vector diagram of the microgripper is shown in Fig. 6.4a. The red straight arrows indicate the working direction of the flexure hinge rotation centers, while the red curved arrows show the rotation direction of the linkages. To simplify the calculation, it is assumed that v_A is directed horizontally to the right and v_C is directed downward. The instantaneous velocities of points A and L are used to

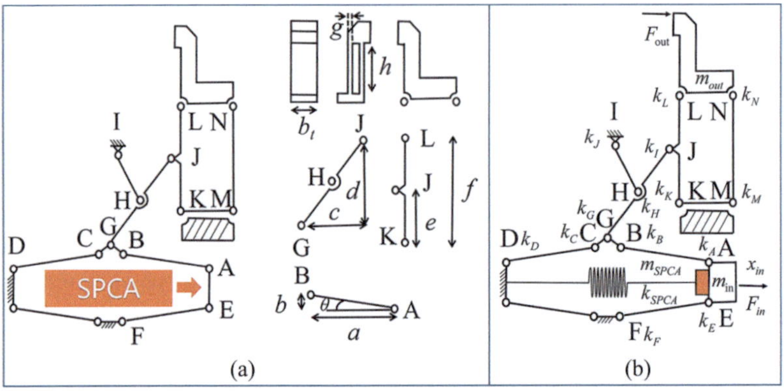

Fig. 6.3 The asymmetrical microgripper. **a** Pseudo-rigid-body model (PRBM) with preliminary geometry parameters; **b** dynamic model scheme showing mass-spring components and their interconnections

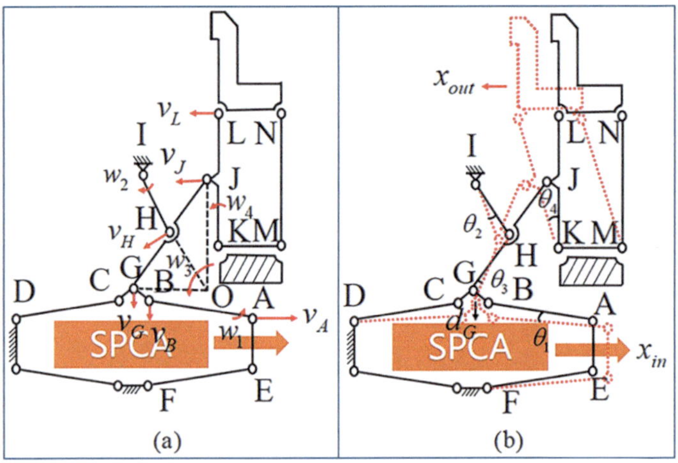

Fig. 6.4 Kinematic model scheme of the designed microgripper. **a** Motion vector diagram showing velocity directions and rotation centers; **b** displacement diagram illustrating input-output motion relationships

determine the amplification ratio. Based on the geometric conditions and kinematic relations shown in Fig. 6.3, the following equations are derived:

$$v_A \cos\theta = v_B \sin\theta \tag{6.2}$$

$$v_B = v_G = \omega_3 \cdot OG \tag{6.3}$$

$$v_J = \omega_3 \cdot OJ = \omega_4 \cdot KJ \tag{6.4}$$

$$v_L = \omega_4 \cdot KL \tag{6.5}$$

where ω_i ($i = 1, 2, 3$, and 4) represent the instantaneous angular velocities of linkages AB, HI, OG, and KL, respectively. Additionally, the quarter-bridge mechanism is fixed, resulting in a halving of its displacement magnification. According to the geometric relationships of the gripper, the displacement amplification ratio can be simplified as:

$$R_{\text{amp}} = \frac{x_{\text{out}}}{x_{\text{in}}} \approx \frac{1}{2}\frac{v_L}{v_A} = \frac{1}{2}\frac{adf}{bce}. \tag{6.6}$$

From the linear displacement relationship in Eq. (6.6), it is clear that the magnification ratio of the microgripper depends solely on the geometric parameters a, b, c, d, e, and f. The geometric meanings of these symbols are illustrated in Fig. 6.3a. In this work, the displacement amplification ratio of the designed gripper is calculated as 9.62. Although these parameters are conceptual within the PRBM framework, they provide essential guidance for selecting design variables in the subsequent optimization process.

6.2.3 Static Modeling

Assume that the input terminal is driven by the SPCA with a small displacement x_{in}. The angle increments θ_1, θ_2, θ_3, and θ_4 of the pivotal moving linkages AB, HI, GH, and JK can be obtained as shown in Fig. 6.3. To simplify the calculation, the linkage BCG is assumed to have a downward displacement d_c. Moreover, the rotational angles ψ_A, ψ_B, ψ_G, ψ_H, ψ_I, ψ_J, and ψ_K of the flexure hinges A, B, G, H, I, J, and K are derived as follows.

$$\psi_A = -\psi_B = \theta_1 = \frac{d_c}{2\sqrt{a^2 + b^2}} \tag{6.7}$$

$$\psi_I = -\psi_G = \theta_2 = -\frac{d_c}{c} \tag{6.8}$$

$$\psi_H = \psi_I - \psi_G = -\frac{2d_c}{c} \tag{6.9}$$

$$\psi_J = \psi_G + \psi_K = \frac{d_c}{c} + \frac{dd_c}{ce} \tag{6.10}$$

$$\psi_K = \theta_4 = \frac{dd_c}{ce} \tag{6.11}$$

where the negative value of the angle indicates clockwise rotation of the flexure hinge.

In the pseudo-rigid-body model (PRBM), each flexure hinge is equivalent to a rigid revolute joint with an associated torsional spring. The stiffness of this torsional spring corresponds directly to the torsional stiffness of the flexure hinge it represents. For a rectangular flexure hinge, the torsional stiffness K_r is calculated as follows:

$$K_i = \frac{E b_t t^3}{12 l_i} = \eta_i K_A, \qquad i = A, B, \ldots, N \tag{6.12}$$

where E represents the elastic modulus of the material. The parameters l, t, and b_t denote the length, in-plane width (minimum dimension), and out-of-plane thickness of the rectangular hinge, respectively. The term η_i represents the stiffness ratio of other hinges relative to hinge A, providing a normalized measure of comparative stiffness across the mechanism's flexure elements.

The torque at the center of rotation of each hinge is expressed as:

$$M_i = -K_i \psi_i \tag{6.13}$$

where $i = A, B, \ldots, N$, and the minus sign indicates that the torque is in the opposite direction of hinge rotation.

The input stiffness of the gripper can be obtained by using Castigliano's first theorem, which is expressed as follows.

$$F_{\text{in}} = \frac{\partial W}{\partial x_{\text{in}}} \tag{6.14}$$

where F_{in} is the input force, x_{in} is the input displacement, and W is the deformation energy. Furthermore, W can be expressed as

$$W = \frac{1}{2} \sum_{i=A}^{N} K_i \psi_i^2. \tag{6.15}$$

Substituting Eqs. (6.7)–(6.11) into (6.15), we can obtain the following equation:

$$W = \frac{K_A a^2 x_{\text{in}}^2}{8b^2} \left[\frac{3}{2(a^2+b^2)} + \frac{\eta_G + \eta_I + 4\eta_H}{c^2} + \left(\frac{2d}{ce}\right)^2 \eta_K + \eta_J \left(\frac{d+e}{ce}\right)^2 \right]. \tag{6.16}$$

The input force can be derived as

$$F_{\text{in}} = \frac{\partial W}{\partial x_{\text{in}}} = \frac{K_A a^2 x_{\text{in}}}{4b^2} \left[\frac{3}{2(a^2+b^2)} + \frac{\eta_G + \eta_I + 4\eta_H}{c^2} + \left(\frac{2d}{ce}\right)^2 \eta_K + \eta_J \left(\frac{d+e}{ce}\right)^2 \right]. \tag{6.17}$$

The input stiffness of the gripper is expressed by

$$K_{\text{in}} = \frac{F_{\text{in}}}{x_{\text{in}}} = \frac{K_A a^2}{4b^2}\left[\frac{3}{2(a^2+b^2)} + \frac{\eta_G + \eta_I + 4\eta_H}{c^2} + \left(\frac{2d}{ce}\right)^2 \eta_K + \eta_J \left(\frac{d+e}{ce}\right)^2\right]. \tag{6.18}$$

Considering that the gripper will be subjected to an external force in practice, the following equation can be obtained:

$$F_{\text{in}} = K_{\text{in}} x_{\text{in}} + R_{\text{amp}} F_{\text{out}}. \tag{6.19}$$

It is observed that the input stiffness of the microgripper is determined by its geometric parameters. The displacement amplification ratio of the gripper exhibits an inverse relationship with the output force at the tip: larger amplification results in smaller output force. When the output force $F_{\text{out}} = 0$ (i.e., no external load), the output displacement of the gripper increases proportionally with the input force. However, once the jaw contacts the manipulated target, the output displacement becomes nearly constant due to the constraint imposed by the object, while the gripping force continues to increase with further input force application.

6.2.4 Dynamic Modeling

The dynamic model of the gripper is established using Lagrange's equation, which is expressed as follows:

$$\frac{d}{dt}\left(\frac{\partial T}{\partial \dot{q}_i}\right) - \frac{\partial T}{\partial q_i} + \frac{\partial V}{\partial q_i} = F_{\text{in}}, \quad i = 1, 2, \ldots, n \tag{6.20}$$

where T represents the total kinetic energy of the system, V denotes the total potential energy, q_i represents the generalized coordinates, $\dot{q}_i$ is the generalized velocity, and n specifies the number of generalized coordinates. Additionally, F_{in} represents the generalized force corresponding to the coordinate q_i.

The kinetic energy of the gripper mechanism can be derived as:

$$T = \frac{1}{2}\left[(m_{\text{in}} + m_{\text{SPCA}})\dot{x}_{\text{in}}^2 + 3I_{AB}\omega_1^2 + I_{HI}\omega_2^2 + I_{GJ}\omega_3^2 + 2I_{KL}\omega_4^2 + m_{\text{out}}\dot{x}_{\text{out}}^2\right] \tag{6.21}$$

where m_{SPCA} denotes the mass of the stack piezoelectric ceramic actuator (SPCA). The terms m_{in} and m_{out} represent the effective masses at the gripper's input and output ends, respectively. The moments of inertia I_{AB}, I_{HI}, I_{GJ}, and I_{KL} correspond to the rotational inertia of linkages AB, HI, GJ, and KL about their respective instantaneous centers of rotation.

The potential energy of the gripper mechanism is calculated by:

$$V = \frac{1}{2}\sum_{i=A}^{N} K_i \psi_i^2 + \frac{1}{2}k_{\text{SPCA}} x_{\text{in}}^2 \tag{6.22}$$

where k_{SPCA} is the stiffness of the SPCA.

The dynamic model of the gripper is given as

$$M\ddot{x}_{\text{in}} + Kx_{\text{in}} = F \tag{6.23}$$

where

$$M = m_{\text{in}} + m_{\text{SPCA}} + \frac{3a^2 I_{AB}}{16b^2(a^2+b^2)} + \frac{a^2(I_{HI}+4I_{GJ})}{4b^2c^2} + \frac{a^2d^2(2I_{KL}+f^2 m_{\text{out}})}{4b^2c^2e^2}, \tag{6.24}$$

$$K = \frac{F_{\text{in}}}{x_{\text{in}}} = k_{\text{SPCA}} + K_{\text{in}}, \tag{6.25}$$

$$F = F_{\text{in}} - R_{\text{amp}} F_{\text{out}}. \tag{6.26}$$

Then, the natural frequency of the gripper system can be calculated by

$$f_g = \frac{1}{2\pi}\sqrt{\frac{K}{M}}. \tag{6.27}$$

It is worth noting that to enhance the dynamic performance of the compliant microgripper, the equivalent mass M should be minimized, while the stiffness K should be maximized. This combination yields a higher natural frequency, as indicated by the relationship $f = \frac{1}{2\pi}\sqrt{\frac{K}{M}}$. Additionally, the stiffness of the SPCA significantly influences the overall system dynamics, as it contributes to the total stiffness K and thus directly affects the gripper's resonant frequency and bandwidth. Optimizing these parameters ensures improved response speed, reduced settling time, and enhanced stability during high-speed micromanipulation tasks.

6.2.5 Clamping Force Measurement

In this study, the clamping force is measured by attaching a strain gauge to the bottom surface of the double flexure beam on the stationary jaw side. The theoretical model for the relationship between strain and clamping force in the flexure hinges is derived as follows.

When the flexure hinge undergoes bending, the strain in the equivalent beam model is given by:

$$\varepsilon = \frac{(\rho + y)d\theta - \rho d\theta}{\rho d\theta} = \frac{y}{\rho} \tag{6.28}$$

where ε denotes the strain of the compliant hinge, ρ is the radius of curvature, y represents the distance from the neutral layer, and $d\theta$ is the angle between two adjacent cross-sections.

When $y = g$, the maximum strain value is given by:

$$\varepsilon_{\max} = \frac{g}{\rho} \tag{6.29}$$

From theoretical mechanics, the relationship between curvature and bending moment is:

$$\frac{1}{\rho} = \frac{M}{EI} \tag{6.30}$$

where $I = \frac{b_t(2g)^3}{12}$ and $M = F_g h$. In addition, g stands for the in-plane width of the flexure hinge, l is the out-of-plane thickness of the hinge, h denotes the length of the flexure beam, and F_g is considered as the reaction to the gripping force.

According to the above equations, the maximum strain at the root of the flexure beam is obtained as follows when the clamping jaw is stressed:

$$F_g = \frac{2Eb_t g^2}{3h} \varepsilon_{\max}. \tag{6.31}$$

It can be observed from Eq. (6.31) that the clamping force exhibits a linear relationship with the strain measured at the bottom of the flexible beam. Therefore, the use of a strain gauge for detecting the clamping force is both theoretically justified and experimentally reliable.

Currently, the models established for the microgrippers remain relatively simplified. A more refined model, accounting for nonlinear deformations and complex interactions within the flexible mechanisms, will be developed in future work to enhance accuracy and predictive capability.

6.3 Design Optimization and FEA Simulation Study

6.3.1 Design Optimization

In this work, the optimization module within ANSYS Workbench software is employed to enhance the gripper's performance. The flowchart of the microgripper optimization process, including the iterative procedure, is displayed in Fig. 6.5. Initially, a parametric model of the microgripper is created in SolidWorks software to define the geometric variables. This CAD model is then imported into the static structural analysis module of ANSYS software to evaluate its total output deformation. The geometric parameters and performance metrics are subsequently imported into the optimization module as design variables and objective functions, respectively. Finally, the optimization model is configured by setting constraints and selecting appropriate optimization algorithms to achieve optimal deformation performance. If the results do not meet the desired objectives, the optimization model is adjusted and the process repeated. A detailed description of the optimization procedure is provided below.

The mathematical formulation of the optimization design includes design variables, constraint functions, and objective functions. The primary goal of the structural optimization is to maximize the gripping stroke of the microgripper. The natural frequency, being an inherent property of the structure, is largely determined once the basic architectural conditions are set. The selected parameters during optimization may have minimal impact on the natural frequency; hence, only the maximum clamping stroke S is chosen as the objective function to achieve large deformations. Significant architectural changes would be required to alter the natural frequency substantially. The design variables are chosen as critical dimensions of the com-

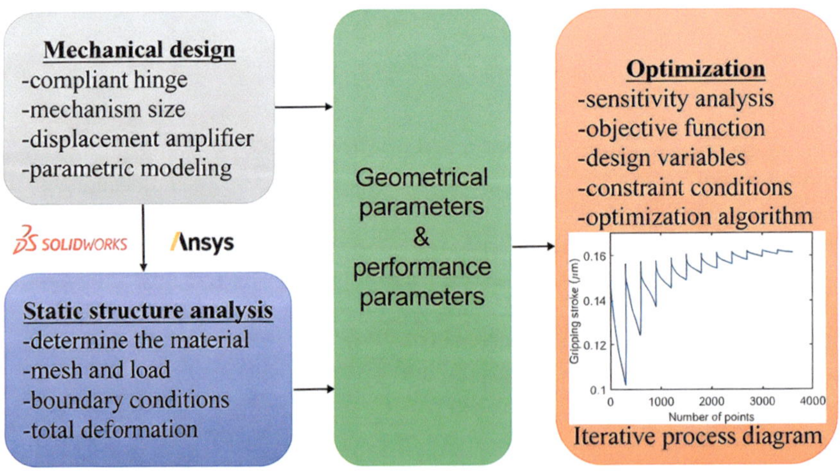

Fig. 6.5 Flowchart of the microgripper optimization process with iterative refinement

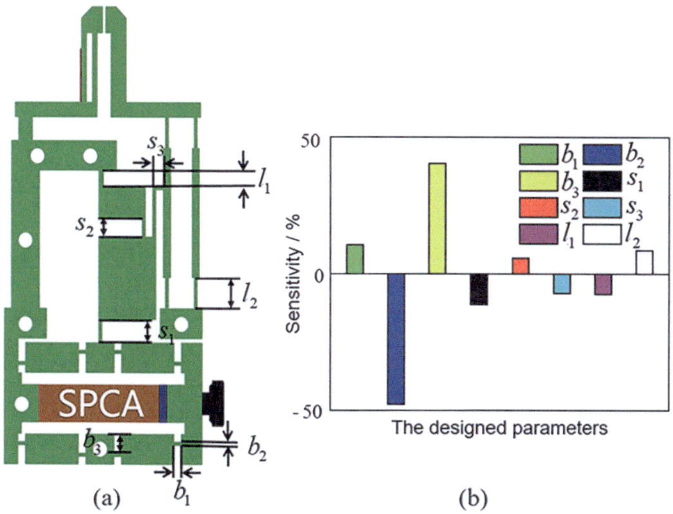

Fig. 6.6 Illustration of **a** geometric parameters of the mechanism to be optimized, and **b** their sensitivities to the clamping stroke

pliant hinges that influence displacement amplification. The selected set of design variables is $[b_1, b_2, b_3, s_1, s_2, s_3, l_1, l_2]$, with their physical meanings illustrated in Fig. 6.6a. The constraints include the spatial limits of the design variables and the requirement that the safety factor Q of the microgripper must be greater than one.

The optimization problem is mathematically described as follows:

$$\text{Maximize:} \quad S = f_1(b_1, b_2, b_3, s_1, s_2, s_3, l_1, l_2) \tag{6.32}$$

$$\text{Subject to} \begin{cases} 0.6\,\text{mm} \le b_1 \le 1.2\,\text{mm} \\ 0.5\,\text{mm} \le b_2 \le 0.9\,\text{mm} \\ 2.0\,\text{mm} \le b_3 \le 2.8\,\text{mm} \\ 3.0\,\text{mm} \le s_1 \le 4.0\,\text{mm} \\ 1.0\,\text{mm} \le s_2 \le 3.0\,\text{mm} \\ 1.5\,\text{mm} \le s_3 \le 2.5\,\text{mm} \\ 2.0\,\text{mm} \le l_1 \le 4.0\,\text{mm} \\ 2.0\,\text{mm} \le l_2 \le 4.0\,\text{mm} \\ Q \ge 1 \end{cases} \tag{6.33}$$

Before performing the optimization, the sensitivities of the design variables to the objective function were obtained through response surface analysis. Sensitivity values reflect the degree of influence each design variable has on the objective function. The sensitivities of the design variables to the clamping stroke are derived as [10.64, −47.753, 40.403, −11.223, 5.8728, −7.1448, −7.4014, 8.6657], as shown in Fig. 6.6b. Notably, parameters b_2 and b_3 demonstrate the strongest influence on the

Table 6.2 Optimization results of key geometric and performance parameters of the microgripper

Variable	b_1	b_2	b_3	s_1	s_2	s_3	l_1	l_2	S
Initial design (mm)	0.9	0.7	2.4	3.5	2.0	2.0	3.0	3.0	0.118
Optimal result (mm)	1.0	0.5	2.5	3	2.5	1.5	2.0	4.0	0.162

gripping stroke, with b_2 exhibiting a large negative sensitivity and b_3 a large positive sensitivity.

In this work, a genetic algorithm (GA) was employed to solve the optimization problem. The GA mimics the natural selection and genetic mechanisms of biological evolution, searching for optimal solutions through simulated evolutionary processes. The optimization model initially generated 400 samples and proceeded through a maximum of 20 iterations, with 300 samples per iteration. As illustrated in Fig. 6.5, the optimization process converged after 3626 generations. The initial and optimized values of the key parameters are summarized in Table 6.2.

6.3.2 *FEA Simulation Results*

Finite element analysis provides an efficient method to validate the performance of the optimized microgripper. The microgripper was assigned aluminum 7075 material properties due to its high elasticity, yield strength, and low density. Fixing constraints were applied to the gripper's mounting holes, and the SPCA was modeled as a linear spring with equivalent stiffness. A driving force of 900 N was applied to the spring connection terminal, with simulations performed using ANSYS software.

As shown in Fig. 6.7a, the maximum deformation at the movable jaw tip reached 162.33 μm. Compared to the initial design's deformation of 118.68 μm (Table 6.2), the optimization yielded a significant 36.8% improvement. The input-output displacement relationship remained linear with a magnification ratio of 8.12, slightly lower than the theoretical value. This discrepancy arises from several factors: (i) idealizing assumptions in the pseudo-rigid-body model (PRBM) that neglect slight deformations in supposedly rigid linkages; (ii) parasitic motion components in velocities v_A and v_C that were omitted in the kinematic model; and (iii) unaccounted effects of flexure hinge thickness on deformation behavior.

Stress analysis revealed maximum concentrations at compliant hinges, with a minimum safety factor of 5.14 under worst-case loading conditions. This substantially exceeds the required threshold of 1.0, ensuring reliable operation without material yielding. The linear force-displacement relationship and high safety margin confirm the optimized design's robustness for practical micromanipulation applications.

The principle of clamping force measurement was further validated through FEA. When a force of 400 mN was applied to the stationary jaw, the maximum strain occurred at the bottom of the flexible beam, as demonstrated in Fig. 6.8a. Both theoret-

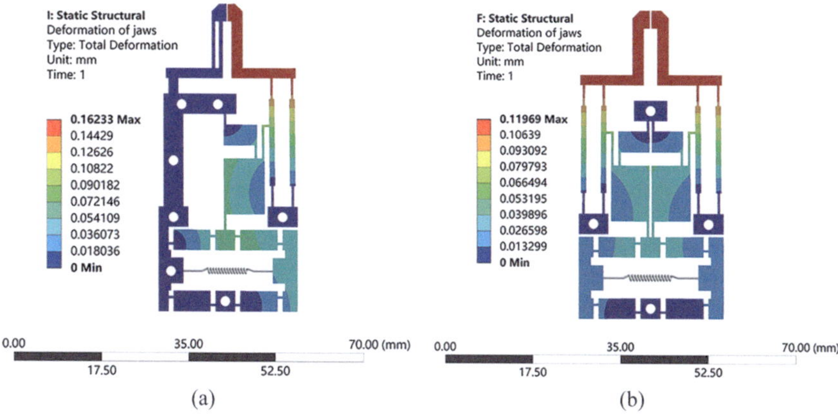

Fig. 6.7 Simulation results of total deformation: **a** asymmetrical gripper; **b** symmetrical gripper

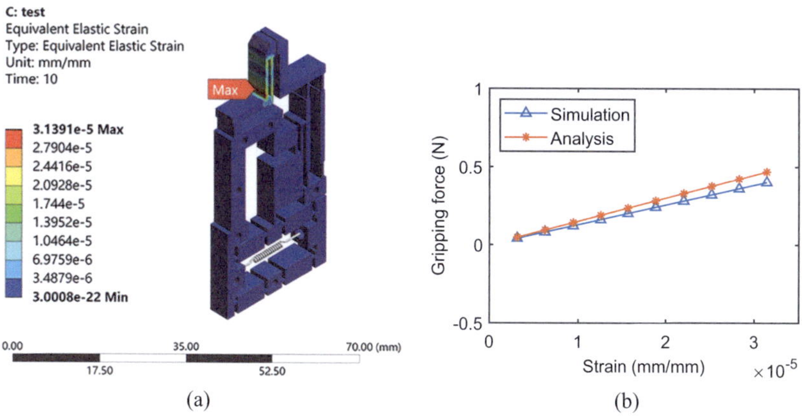

Fig. 6.8 Simulation results of clamping force for the asymmetrical gripper. **a** Strain distribution at the stationary jaw; **b** linear relationship between clamping force and strain with theoretical and FEA results

ical calculations and numerical simulations confirmed a linear relationship between strain and clamping force, as shown in Fig. 6.8b. Although differing boundary conditions resulted in slightly different slopes between the two models, both methods unequivocally demonstrated that gripping force can be accurately measured through strain detection.

Modal analysis simulations were conducted to evaluate the dynamic performance of the gripper. Figure 6.9 shows the first three mode shapes of the asymmetric microgripper. The first mode (944.26 Hz) featured parallel translation of the movable jaw within the gripper plane, perfectly aligning with the working mode direction. The second mode (1314.3 Hz) involved out-of-plane flipping of the jaw, while the third mode (2166.4 Hz) showed rotation of the bridge mechanism's movable side. These

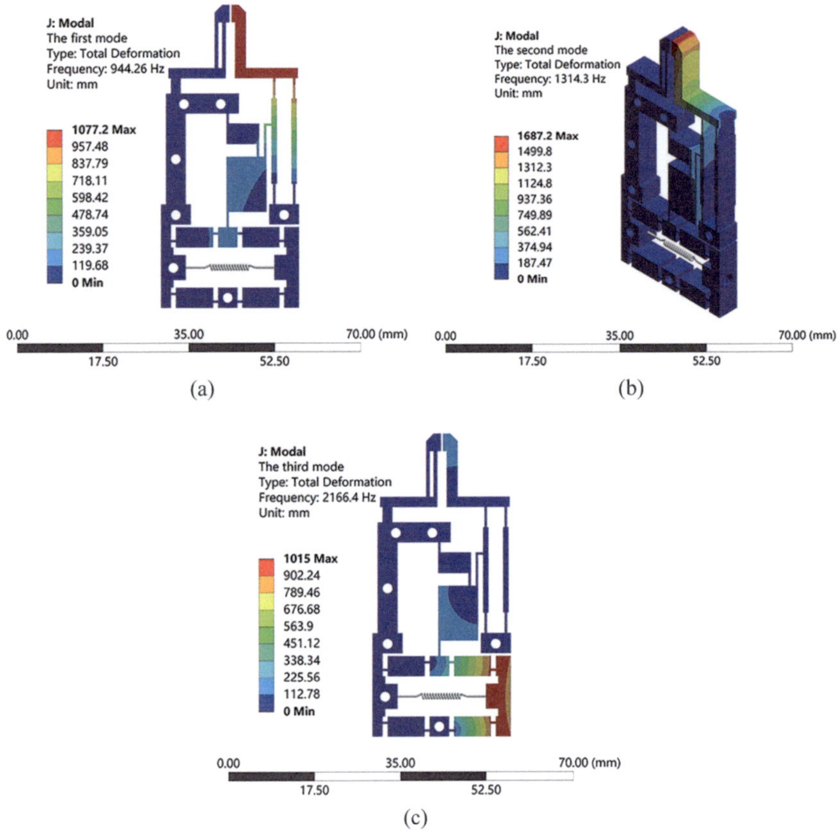

Fig. 6.9 Simulation results of the first-three resonant modal shapes of the asymmetrical gripper: **a** first mode (944.26 Hz)—in-plane parallel motion; **b** second mode (1314.3 Hz)—out-of-plane flipping; **c** third mode (2166.4 Hz)—rotational motion

results confirm two key advantages: (i) the first resonant frequency matches the working mode frequency, and (ii) the first three resonant frequencies are well-separated (944.26, 1314.3, and 2166.4 Hz), eliminating the dense mode problem that complicates control design. The high first resonant frequency of 944.26 Hz indicates excellent dynamic performance.

Comparative FEA analysis with a symmetric microgripper revealed significant differences. The symmetric gripper achieved a larger total output displacement of 239.38 μm under identical input conditions (Fig. 6.7b), but exhibited inferior dynamic characteristics. Its first three resonant frequencies (757.99, 803.82, and 1292.8 Hz) showed both mode misalignment and dense frequency spacing between the first two modes (45.83 Hz difference). As summarized in Table 6.3, the asymmetric design sacrifices 32.18% of working stroke but gains 186.27 Hz in first resonant frequency while eliminating mode alignment and density issues (Fig. 6.10). The

Table 6.3 Performance comparison between the designed symmetrical and asymmetrical grippers

Gripper type	Stroke (μm)	1st (Hz)	2nd (Hz)	3rd (Hz)	Working mode
Symmetrical gripper	239.38	757.99	803.82	1292.8	2nd
Asymmetrical gripper	162.33	944.26	1314.3	2166.4	1st

improved performance stems from two key design modifications: (i) implementation of an L-type gripper arm with aspect ratio ≈ 2, and (ii) addition of strategic constraints to the asymmetric structure. The stationary jaw was replaced with a more rigid structure, and additional fixing constraints were incorporated. These changes significantly enhanced the first resonant frequency while maintaining precise mode alignment, demonstrating the effectiveness of the asymmetric design approach for high-performance micromanipulation applications.

6.4 Experimental Tests and Results

6.4.1 Experimental Setup

The microgripper was fabricated from aluminum 7075 alloy using wire electrical discharge machining (WEDM) technology. The manufactured gripper is shown in Fig. 6.11a, while Fig. 6.11b displays the assembled microgripper mounted in its base for integration with other components. A real-time control system based on an NI cRIO-9022 controller was implemented on a vibration isolation platform (model: RS2000, Newport Corp.) to minimize environmental disturbances.

Figure 6.12a presents the block diagram of the experimental setup. The gripper was actuated by a SPCA (model: RP150/5 × 5/20) capable of providing 20 μm displacement at 150 V driving voltage. Actuation signals were generated by an NI-9263 analog output module and amplified through a power amplifier (model: EPA-104, PiezoSystems, Inc.). Displacement measurements at both input and output ends were obtained using laser displacement sensors (model: LK-H055, Keyence Corp.), with signals acquired through an NI-9215 analog input module.

For force sensing, a strain gauge (model: BFH 350-3AA, Chengtec Electronics Ltd.) with sensitivity factor 2.0 was bonded to the flexure beam at the stationary jaw. Strain signals were processed using NI 9945 and NI 9237 modules. Visual monitoring was provided by an industrial camera (model: MV-CE013-80UM, Hikvision Corp.) mounted above the gripper. The complete experimental setup is shown in Fig. 6.12b.

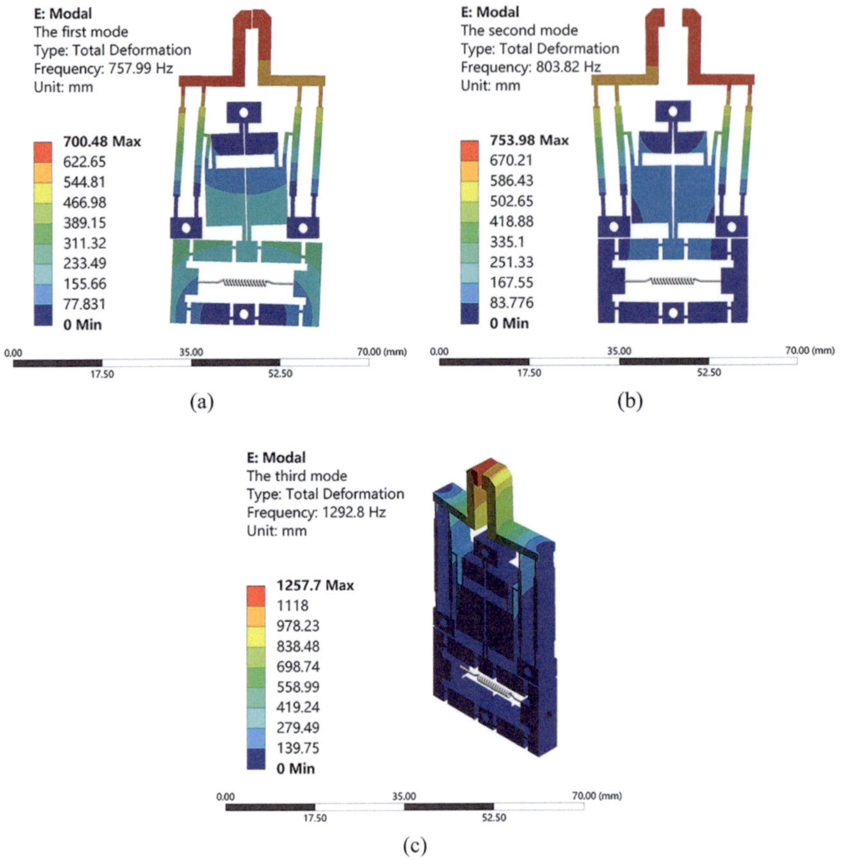

Fig. 6.10 Simulation results of the first-three resonant modal shapes of the symmetrical gripper: **a** first mode (757.99 Hz)—out-of-plane deformation; **b** second mode (803.82 Hz)—in-plane rotation; **c** third mode (1292.8 Hz)—complex hybrid motion

6.4.2 Experimental Results of Performance Test

6.4.2.1 Open-Loop Displacement Characterization

In open-loop testing, the SPCA was driven with forward and reverse voltage cycles ranging from 0 to 150 V. The output displacement of the gripper's movable jaw was measured using a laser displacement sensor. Figure 6.13a shows the relationship between input voltage and output displacement, revealing noticeable hysteresis behavior originating from the piezoelectric actuator's inherent nonlinearity. Signal chatter observed in the test data may be attributed to insufficient sensor warm-up time and environmental interference.

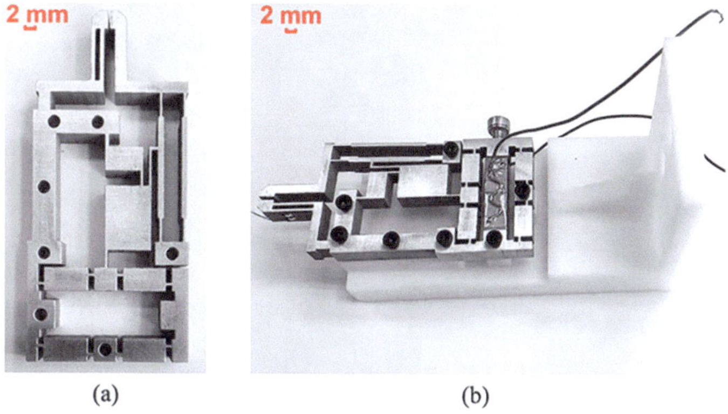

Fig. 6.11 Prototype of the developed microgripper: **a** fabricated microgripper showing monolithic aluminum 7075 structure; **b** assembled microgripper with integrated piezoelectric actuator and pre-loading mechanism

At maximum SPCA displacement (approximately 16 μm), the gripper achieved an output stroke of 131.2 μm. Figure 6.13b compares the input-output displacement relationships obtained from theoretical modeling, simulation, and experimental testing. Despite the hysteresis effects, the system maintains linearity between input and output displacements. The displacement amplification ratios were calculated as:

- Theoretical model: 9.62
- Simulation: 8.12
- Experimental: 8.10.

The close agreement between experimental and simulation results (1.5% difference) confirms the gripper's accuracy and stable performance. The measured system stiffness was 6.86 N/μm under 900 N driving force, which is higher than both theoretical (4.78 N/μm) and simulation (5.74 N/μm) predictions. This discrepancy is primarily due to the additional stiffness contribution from pre-tightening bolts in the physical system, which were not fully accounted for in the models.

6.4.2.2 Force Sensing and Calibration

A series of force sensing experiments were carried out to monitor the clamping force of the gripper for non-destructive operation. The strain gauge was calibrated before it was applied to measure the clamping force. The weights of different masses were suspended at the end of the stationary jaw. The calibration of the force sensor was realized by using the gravity of the weight as the clamping force. The relationship between the output voltage acquired by NI-9237 and the gravity of weight is shown in Fig. 6.14a. A good linear fitting relation can be obtained with the slope of 20,819, i.e., the force sensor gain is calculated as 20,819 mN/mV.

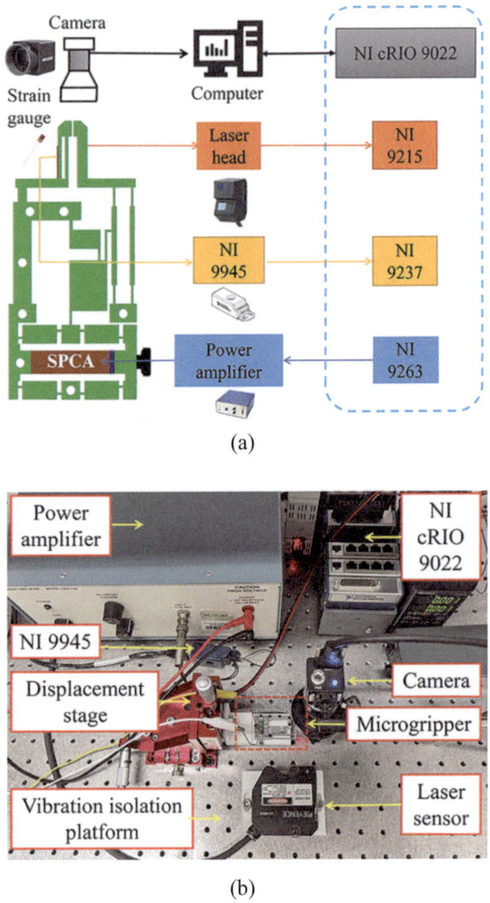

Fig. 6.12 Experimental setup: **a** block diagram showing signal flow and component connections; **b** photograph of the complete testing apparatus with vibration isolation

Afterward, the clamping force and output displacement of the gripper jaw have been measured when grasping a metal wire with a diameter of 300 μm. In the experiment, the input voltage was gradually increased from 0 to 150 V, and then gradually decreased to 0 V. The obtained experimental results are displayed in Fig. 6.14c. Before the gripper touches the small object, the displacement increases rapidly while the gripping force increases slowly. After the jaw contacts with the wire, the clamping force rises sharply and the displacement changes slowly. When the actuating voltage is 150 V, the maximum clamping force detected by the force sensor is about 337.35 mN.

In addition, the experimental data about input displacement and gripping force are extracted and presented in Fig. 6.14b. The experimental data are approximately consistent with the trend predicted by the simulation data. The ratio between the input displacement and grasping force is 26.23 mN/μm, obtained by linear least-

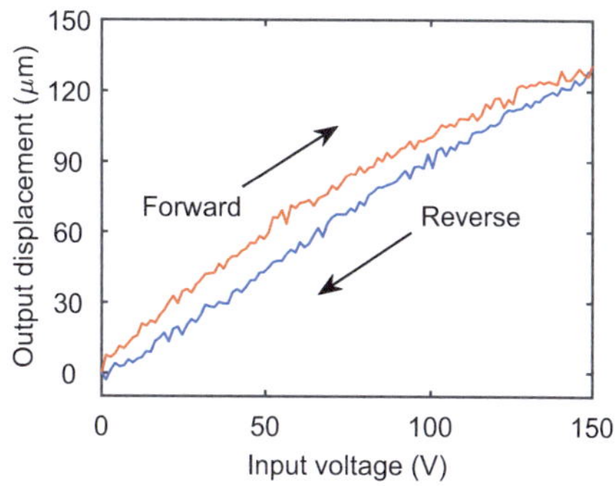

(a)

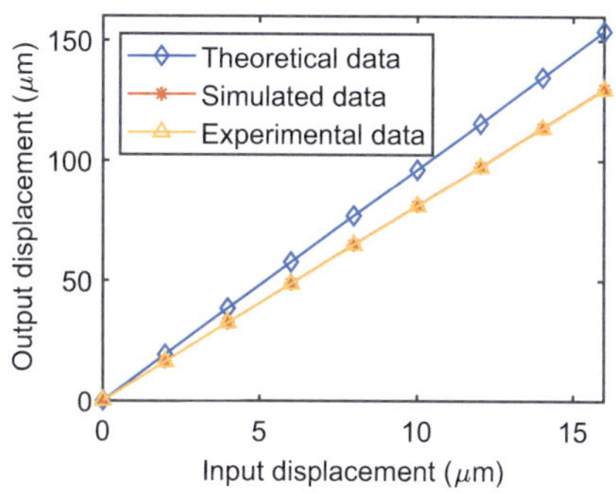

(b)

Fig. 6.13 Experimental results of gripping displacement test: **a** output displacement versus drive voltage showing hysteresis characteristics; **b** relationship between input displacement of SPCA and output displacement of gripper jaw with theoretical, simulation, and experimental results

squares fitting. The slope is determined by the stiffness of the wires adopted in the experiment.

6.4.2.3 Motion and Force Resolution

The motion resolution of the gripper was tested by performing a series of step responses under the implemented proportional-integral (PI) feedback controller. The amplitude of the step response is about 1 μm, and the experimental results are demonstrated in Fig. 6.15a. The error between the output signal and the reference signal conforms to the normal distribution with 2σ (σ is the standard deviation) of 0.6 μm. That is, 95% of sensor readings lie within the range of ± 0.6 μm near the reference. Thus, the motion resolution of the gripper is determined as ± 0.6 μm.

Similarly, the resolution measurement experiment of the gripping force was also carried out. The waveforms of the reference signal and the measured signal of the gripping force are shown in Fig. 6.15b. The error of measurement signal and reference signal also follows a normal distribution with 2σ of 2 mN. That is, 95% of sensor

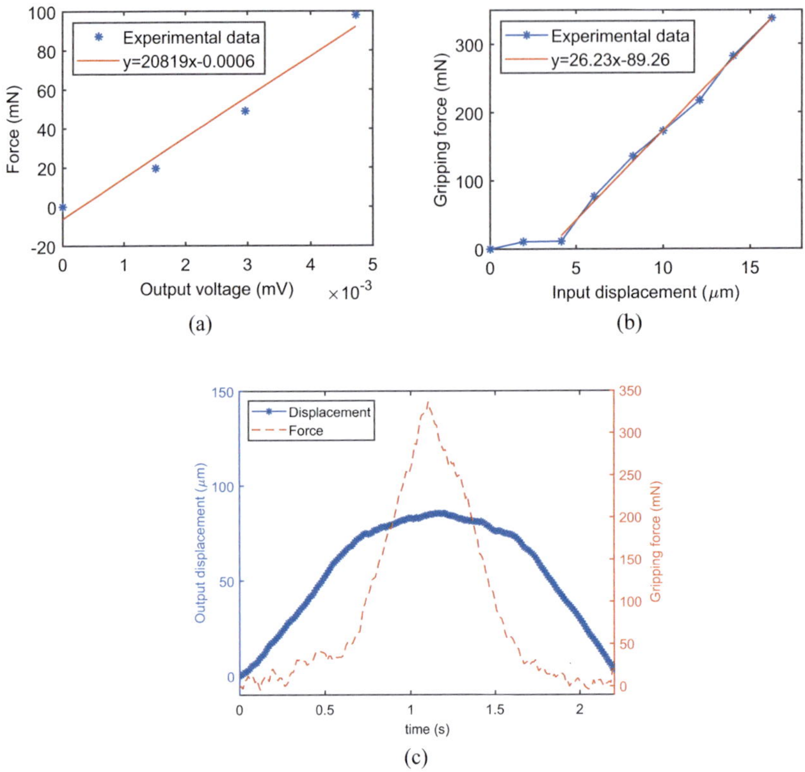

Fig. 6.14 Experimental results of gripping force test: **a** calibration result of the force sensor showing linear voltage-force relationship with gain of 20,819 mN/mV; **b** displacement and force signals during grasping of a 300 μm metal wire; **c** relationship between input displacement and gripping force with experimental and simulation data

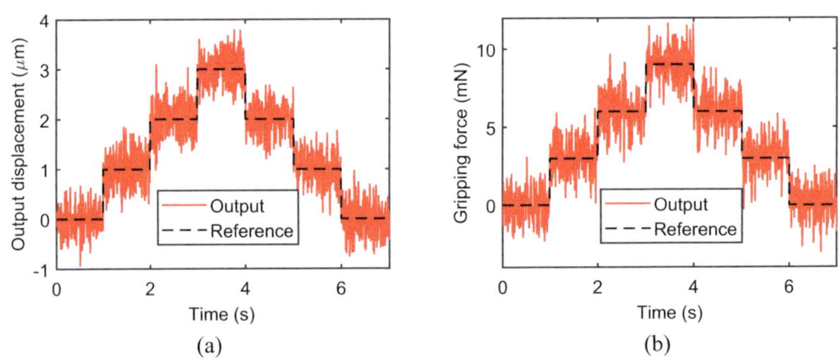

Fig. 6.15 Experimental results of resolution test: **a** displacement resolution of ±0.6 μm (2σ) under PI control; **b** force resolution of ±2 mN (2σ) at zero output

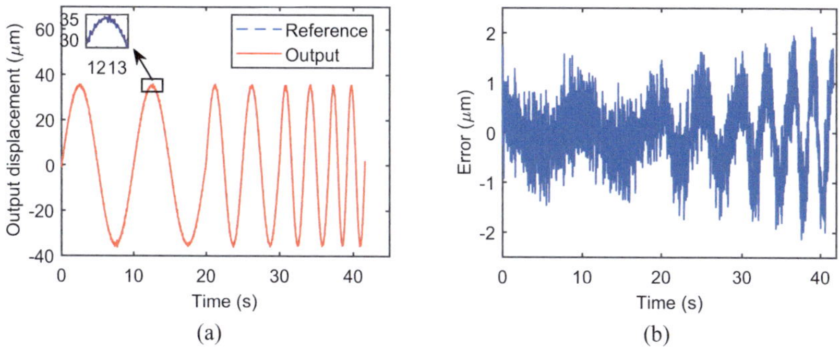

Fig. 6.16 Experimental results of sine-wave signal tracking test: **a** reference and actual displacement tracking results for 35 μm amplitude signals; **b** displacement tracking errors showing maximum errors of 1.58–2.13 μm across 0.1–0.4 Hz frequencies

readings lie within the range of ±2 mN at zero output, which is regarded as the resolution of the gripping force.

6.4.2.4 Tracking Performance

To verify the tracking ability of the gripper system, a sine-wave signal with varying frequencies was used. The amplitude of the sine-wave was set as 35 μm. The reference signal, tracking results, and tracking errors are displayed in Fig. 6.16. The maximum error reached 1.58 μm, 1.73 μm, 1.91 μm, and 2.13 μm corresponding to the signal frequency of 0.1 Hz, 0.2 Hz, 0.3 Hz, and 0.4 Hz, respectively. The percentage errors relative to the magnitude of the reference signal are 4.5%, 4.9%, 5.4%, and 6.1%, respectively. The root mean square error (RMSE) was calculated as 0.69 μm for the whole test process.

6.4.2.5 Dynamic Performance Evaluation

The dynamic performance of the asymmetrical microgripper was evaluated experimentally. In this work, the piezoelectric actuator was driven by applying a sweep-sine signal with an amplitude of 2 V and frequency ranging from 1 to 1000 Hz. The NI cRIO-9022 controller operated in FPGA mode at a sampling frequency of 4 kHz, satisfying Nyquist's sampling theorem. The displacement signal at the end of the gripper was captured by the laser sensor head. The frequency response relationship between the displacement signal and the excitation frequency is presented in Fig. 6.17.

The experimental results indicate that the first resonant mode of the gripper occurs at 705 Hz. This measured value is lower than both the simulation result (952.91 Hz) and the analytical model prediction (1159 Hz). The discrepancy arises from several factors: (i) the additional mass of assembly components including the SPCA, pre-

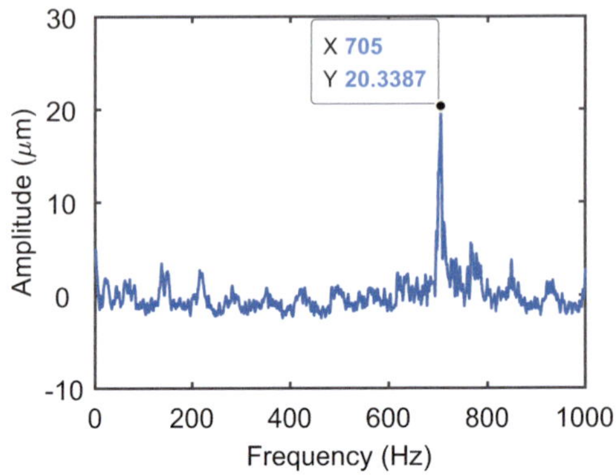

Fig. 6.17 Amplitude of frequency response test result of the microgripper showing first resonant frequency at 705 Hz with sweep-sine excitation (2 V amplitude, 1–1000 Hz range)

tightening bolts, and fixed bolts not accounted for in the models; (ii) unmodeled kinetic energy contributions from rigid connecting bars leading to an underestimated equivalent mass in theoretical calculations; and (iii) ideal boundary conditions in simulations differing from practical mounting constraints.

Despite these discrepancies, the experimentally obtained first resonant frequency of 705 Hz demonstrates excellent dynamic performance for the developed microgripper, indicating a high bandwidth suitable for precision micromanipulation applications.

6.4.3 Experimental Results of Object Grasping

A series of gripping experiments were conducted to validate the practical performance of the proposed microgripper. Three micro-targets of varying dimensions were selected for these operations. Two industrial cameras, positioned at different angles, provided multi-view monitoring of the grasping process. The captured images and target specimens are presented in Fig. 6.18.

Figure 6.18a, b demonstrate successful grasping of a metal ring with 110 μm thickness. Figure 6.18c, d show stable suspension of a rectangular metal sheet (20 mm × 5 mm × 200 μm). Figure 6.18e, f present secure handling of a wire specimen (40 mm length, 300 μm diameter). These experiments confirm the gripper's capability to manipulate objects of diverse geometries and sizes.

Table 6.4 provides a comparative analysis of piezoelectric-driven asymmetric microgrippers. For equitable comparison, all listed designs share the characteristics of first resonant mode operation and absence of mode density issues. The developed gripper demonstrates competitive advantages in multiple performance metrics:

- Suitable working stroke (131.2 μm)
- Highest first resonant frequency (705 Hz)
- Largest clamping force (337.35 mN).

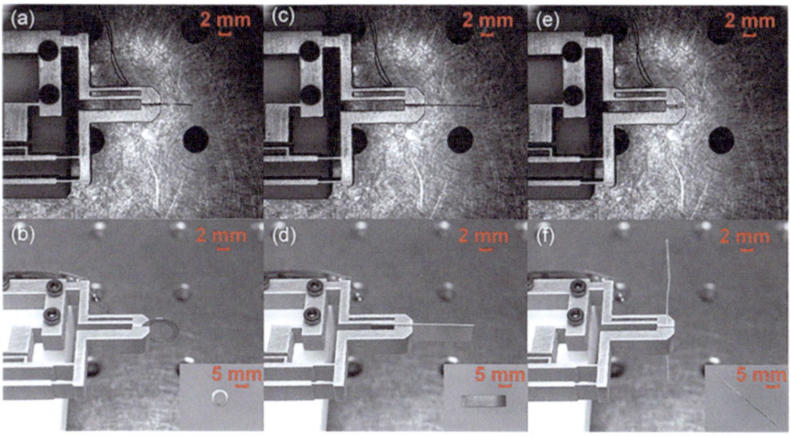

Fig. 6.18 Snapshots of clamping experiments for three micro-objects: **a–b** metal ring (110 μm thickness); **c–d** rectangular metal sheet (20 mm × 5 mm × 200 μm); **e–f** metal wire (40 mm length, 300 μm diameter)

Table 6.4 Performance comparison of piezoelectric-actuated asymmetrical microgrippers showing working stroke, resonant frequency, and gripping force characteristics

Reference	Stroke (μm)	1st resonance (Hz)	Gripping force (mN)	Mode dense	Working mode
[39]	180	–	300	–	–
[35]	249.39	–	280	–	–
[7]	632	–	270	–	–
[9]	483.31	207.72	50	No	1st
[16]	121.38	575	50	No	1st
This work	131.2	705	337.35	No	1st

These characteristics position this microgripper as a promising tool for advanced micromanipulation and microassembly applications, particularly where high dynamic response and precise force control are required. The monolithic asymmetric design effectively addresses common limitations of symmetrical configurations while maintaining practical manipulation capabilities.

6.5 Conclusion

This chapter proposes an asymmetrical microgripper based on SPCA with sufficient clamping stroke and high working frequency. The desired clamping stroke is achieved through a three-stage displacement amplification mechanism. The linear relationship between input displacement and output displacement is derived via kinematic modeling. The static model, dynamic model, and grasping force measurement model of the gripper are established based on PRBM. The structural parameters affecting the gripper's performance are optimized using response surface analysis.

FEA simulation results demonstrate that the first resonant mode aligns with the working mode for the asymmetrical gripper, and no mode density problem occurs, which enhances dynamic performance for high-speed operation. Additionally, the asymmetrical gripper exhibits advantages including a fixed positioning datum and force detection capability without affecting displacement amplification.

Experimental results align well with both theoretical predictions and FEA simulations, verifying the correctness and validity of the established PRBM-based models. The proposed microgripper demonstrates excellent working performance with a clamping stroke of 131.2 μm, first natural frequency of 705 Hz, and maximum gripping force of 337.35 mN. The displacement and force resolutions are measured as ±0.6 μm and ±2 mN, respectively. Successful grasping of three micro-objects with different shapes and dimensions demonstrates the adaptability and practical potential of the developed microgripper for micromanipulation applications.

References

1. Ai, W., Xu, Q.: New structural design of a compliant gripper based on the Scott-Russell mechanism. Int. J. Adv. Rob. Syst. **11**(12), 192 (2014)
2. Beyeler, F., Neild, A., Oberti, S., Bell, D.J., Sun, Y., Dual, J., Nelson, B.J.: Monolithically fabricated microgripper with integrated force sensor for manipulating microobjects and biological cells aligned in an ultrasonic field. J. Microelectromech. Syst. **16**(1), 7–15 (2007)
3. Cai, H., Wu, Z.: Control simulation of a quadruped robot using Matlab/Simulink module. In: Proceedings of 2021 36th Youth Academic Annual Conference of Chinese Association of Automation (YAC), pp. 322–326 (2021)
4. Chen, W., Qu, J., Chen, W., Zhang, J.: A compliant dual-axis gripper with integrated position and force sensing. Mechatronics **47**, 105–115 (2017)
5. Chen, W., Shi, X., Chen, W., Zhang, J.: A two degree of freedom micro-gripper with grasping and rotating functions for optical fibers assembling. Rev. Sci. Instrum. **84**(11), 115111 (2013)
6. Chen, X., Deng, Z., Hu, S., Gao, J., Gao, X.: Design of a flexible piezoelectric microgripper based on combined amplification principles. Nanotechnol. Precis. Eng. **2**(3), 138–143 (2019)
7. Chen, X., Deng, Z., Hu, S., Gao, J., Gao, X.: Design of a compliant mechanism based four-stage amplification piezoelectric-driven asymmetric microgripper. Micromachines **11**(1), 25 (2020)
8. Das, T.K., Shirinzadeh, B., Al-Jodah, A., Ghafarian, M., Pinskier, J.: A novel compliant piezoelectric actuated symmetric microgripper for the parasitic motion compensation. Mech. Mach. Theory **155**, 104069 (2021)
9. Das, T.K., Shirinzadeh, B., Ghafarian, M., Al-Jodah, A.: Design, analysis, and experimental investigation of a single-stage and low parasitic motion piezoelectric actuated microgripper. Smart Mater. Struct. **29**(4), 045028 (2020)
10. Das, T.K., Shirinzadeh, B., Ghafarian, M., Al-Jodah, A., Pinskier, J.: Characterization of a compact piezoelectric actuated microgripper based on double-stair bridge-type mechanism. J. Micro-Bio Robot. **16**(1), 79–92 (2020)
11. Das, T.K., Shirinzadeh, B., Ghafarian, M., Al-Jodah, A., Zhong, Y., Smith, J.: Design, analysis and experimental investigations of a high precision flexure-based microgripper for micro/nano manipulation. Mechatronics **69**, 102396 (2020)

12. Despa, V., Catangiu, A., Ivan, I.A., Gurgu, V., Ardeleanu, M.: Modeling and control of a microgripper based on electromagnetic actuation. Sci. Bull. Valahia Univ. **9**, 131–136 (2014)
13. Felix, A.A., Colón, D., Verona, B.M., Ramos, L.W., Cobas-Gomez, H., Gongora-Rubio, M.R.: Identification and robust controllers for an electrostatic microgripper. J. Vib. Eng. Technol. **9**, 389–397 (2021)
14. Ho, N.L., Dang, M.P., Dao, T.P.: Design and analysis of a displacement sensor-integrated compliant microgripper based on parallel structure. Vietnam J. Mech. **42**(4), 363–374 (2020)
15. Kim, K., Liu, X., Zhang, Y., Sun, Y.: Nanonewton force-controlled manipulation of biological cells using a monolithic mems microgripper with two-axis force feedback. J. Micromech. Microeng. **18**(5), 055013 (2008)
16. Liang, C., Wang, F., Shi, B., Huo, Z., Zhou, K., Tian, Y., Zhang, D.: Design and control of a novel asymmetrical piezoelectric actuated microgripper for micromanipulation. Sens. Actuators, A **269**, 227–237 (2018)
17. Liang, C., Wang, F., Tian, Y., Zhao, X., Zhang, H., Cui, L., Zhang, D., Ferreira, P.: A novel monolithic piezoelectric actuated flexure-mechanism based wire clamp for microelectronic device packaging. Rev. Sci. Instrum. **86**(4), 045106 (2015)
18. Liu, Y., Xu, Q.: Mechanical design, analysis and testing of a large-range compliant microgripper. Mech. Sci. **7**(1), 119–126 (2016)
19. Llewellyn-Evans, H., Griffiths, C., Fahmy, A.: An experimental study into displacement of a shape memory alloy actuated robotic microgripper. Eng. Res. Exp. **2**(1), 015027 (2020)
20. Llewellyn-Evans, H., Griffiths, C., Fahmy, A.: Microgripper design and evaluation for automated μ-wire assembly: a survey. Microsyst. Technol. **26**(6), 1745–1768 (2020)
21. Long, Z., Zhang, J., Liu, Y., Han, C., Li, Y., Li, Z.: Dynamics modeling and residual vibration control of a piezoelectric gripper during wire bonding. IEEE Trans. Compon. Packag. Manuf. Technol. **7**(12), 2045–2056 (2017)
22. Lyu, Z., Xu, Q.: Recent design and development of piezoelectric-actuated compliant microgrippers: a review. Sens. Actuators A: Phys. 113002 (2021)
23. Muntwyler, S., Kratochvil, B.E., Beyeler, F., Nelson, B.J.: Monolithically integrated two-axis microtensile tester for the mechanical characterization of microscopic samples. J. Microelectromech. Syst. **19**(5), 1223–1233 (2010)
24. Noveanu, S., Lobontiu, N., Lazaro, J., Mandru, D.: Substructure compliance matrix model of planar branched flexure-hinge mechanisms: design, testing and characterization of a gripper. Mech. Mach. Theory **91**, 1–20 (2015)
25. Ojalvo, I.: Efficient computation of modal sensitivities for systems with repeated frequencies. AIAA J. **26**(3), 361–366 (1988)
26. Piriyanont, B., Moheimani, S.R.: Mems rotary microgripper with integrated electrothermal force sensor. J. Microelectromech. Syst. **23**(6), 1249–1251 (2014)
27. Rong, W., Fan, Z., Wang, L., Xie, H., Sun, L.: A vacuum microgripping tool with integrated vibration releasing capability. Rev. Sci. Instrum. **85**(8), 085002 (2014)
28. Shi, B., Wang, F., Huo, Z., Tian, Y., Zhao, X., Zhang, D.: Design and characteristics of a novel compliant symmetric microgripper mechanism. In: Proceedings of 2018 IEEE International Conference on Manipulation, Manufacturing and Measurement on the Nanoscale (3M-NANO), pp. 65–69 (2018)
29. Shi, C., Dong, X., Yang, Z.: A microgripper with a large magnification ratio and high structural stiffness based on a flexure-enabled mechanism. IEEE/ASME Trans. Mechatron. (2021). https://doi.org/10.1109/TMECH.2021.3052806
30. Soon, R.H., Yu, L., Lim, C.T.: A soft sensorized microfluidic tubular actuating gripper. Adv. Mater. Technol. **5**(5), 2000150 (2020)
31. Verotti, M., Dochshanov, A., Belfiore, N.P.: A comprehensive survey on microgrippers design: mechanical structure. J. Mech. Des. **139**(6), 060801 (2017)
32. Wang, F., Fan, D.: Modeling and experimental study of a wire clamp for wire bonding. J. Electron. Packag. **137**(1), 011012 (2015)
33. Wang, F., Shi, B., Tian, Y., Huo, Z., Zhao, X., Zhang, D.: Design of a novel dual-axis micromanipulator with an asymmetric compliant structure. IEEE/ASME Trans. Mechatron. **24**(2), 656–665 (2019)

34. Xiao, S., Li, Y., Zhao, X.: Design and analysis of a novel micro-gripper with completely parallel movement of gripping arms. In: Proceedings of 2011 6th IEEE Conference on Industrial Electronics and Applications, pp. 2127–2132 (2011)
35. Xing, Q.: Design of asymmetric flexible micro-gripper mechanism based on flexure hinges. Adv. Mech. Eng. **7**(6), 1687814015590331 (2015)
36. Xing, Q., Ge, Y.: Parametric study of a novel asymmetric micro-gripper mechanism. J. Adv. Mech. Des. Syst. Manuf. **9**(5), JAMDSM0075 (2015)
37. Xu, Q.: Design and development of a novel compliant gripper with integrated position and grasping/interaction force sensing. IEEE Trans. Autom. Sci. Eng. **14**(3), 1415–1428 (2015)
38. Yoon, D.Y., Park, J.H.: Adaptive polynomials for the vibration analysis of an l-type beam structure with a free end. J. Mar. Sci. Eng. **9**(3), 300 (2021)
39. Zhang, D., Zhang, Z., Gao, Q., Xu, D., Liu, S.: Development of a monolithic compliant SPCA-driven micro-gripper. Mechatronics **25**, 37–43 (2015)

Chapter 7
Design of a Piezoelectrically Actuated Compliant Microgripper with High Area-Usage Efficiency

Abstract Microgrippers serve as crucial end-effectors in micromanipulation systems, performing essential pick-transport-release tasks. This chapter details the development of a novel piezoelectrically actuated compliant microgripper, covering its design, modeling, optimization, simulation, and experimental validation. To address space and cost constraints, the design emphasizes high area-usage efficiency and optimal piezoelectric actuator utilization. A three-stage amplification mechanism–combining bridge-type and leverage configurations–enables large jaw displacements. Analytical methods, including pseudo-rigid-body modeling and response surface optimization, were used to refine the compliant structure. Finite-element analysis and physical prototyping validated performance, demonstrating a gripping displacement of 548.42 μm, a natural frequency of 334 Hz, and a motion resolution of ±0.75 μm. The microgripper outperforms prior designs in displacement, compactness, and actuation efficiency, successfully manipulating objects of varying sizes and shapes.

Keywords Microgripper · Piezoelectric actuator · Compliant mechanism · PID controller · Micromanipulation

7.1 Introduction

Microelectromechanical systems (MEMS) offer unique advantages over traditional electromechanical systems, including compact size, lightweight construction, and high integration capability [25]. As MEMS research continues to expand, conventional 2D microstructures can no longer meet the growing demand for complex applications. Researchers are increasingly focusing on developing small-scale devices with sophisticated 3D structures assembled from diverse materials and components, such as minimally invasive medical instruments and micro-optical systems [3, 9, 13]. Consequently, microassembly technology suitable for micro/nano-scale operations has become a central focus in MEMS research [22]. The repetitive nature of microassembly tasks often leads to operator fatigue and reduced precision, highlighting the need for microassembly robots to automate these precision tasks.

Q. Xu, *Piezoelectric-Driven Flexible Robotic Microgrippers*,
https://doi.org/10.1007/978-981-95-4685-5_7

A microassembly robot is an automated system capable of performing operations with micron-, submicron-, or even nanometer-scale precision within confined workspaces. These systems typically comprise four key units: actuation, mechanical structure, sensing, and control [20, 22]. The actuation unit serves as the energy source for end-effector movement, with the choice of actuator significantly influencing the system's structure and performance. Common actuation methods include electrostatic, electromagnetic, piezoelectric, and shape memory alloy technologies [19]. Among these, piezoelectric actuator (PEA)-driven microgrippers are widely employed as end-effectors due to their rapid response and high resolution [4]. However, PEAs introduce challenges such as hysteresis effects and limited output displacement. These limitations can be mitigated through feedforward control strategies and appropriate feedback control methods [1, 10, 16]. Compliant displacement amplification and transmission mechanisms offer additional improvements, providing high accuracy, low cost, and friction-free operation [12, 21, 23, 30].

Current approaches to enhancing the working stroke of PEA-based compliant microgrippers fall into two main categories. First, using larger gripper structures or multi-stage amplification mechanisms increases displacement at the expense of larger volume and space requirements. However, microassembly robots typically operate in confined spaces with complex environments, necessitating compact designs with appropriate gripping ranges. To evaluate compactness, we define area-usage efficiency (η_a) as the ratio of stroke (d) to geometric area (A): $\eta_a = \frac{d}{A}$, representing displacement output per unit area. Note that stroke can be multiplied by unit length to obtain a dimensionless index. Previous designs have achieved varying efficiencies: Zubir et al. reached 66.67 nm/mm^2 [31], Sun et al. achieved 95.72 nm/mm^2 [24], Liang et al. obtained 129.3 nm/mm^2 [14], Wang et al. attained 140.74 nm/mm^2 [26], Liang et al. advanced to 299.4 nm/mm^2 [15], and Das et al. achieved 315.2 nm/mm^2 [7, 8]. While these microgrippers provide substantial displacement, their compactness remains insufficient for many applications.

Second, some researchers have employed multiple actuators to increase gripping displacement. Zhao et al. developed a dual-PEA microgripper with three-stage amplification providing 204.6 μm output [29], Yang et al. achieved 328.2 μm using dual PEAs with bridge-type mechanisms [28], and Chen et al. reached 632 μm with an asymmetric dual-PEA design [6]. While multiple actuators increase output, they also raise economic costs and energy consumption. To evaluate displacement efficiency, we introduce PEA utilization efficiency (η_p), defined as the ratio of stroke (d) to the number of PEAs (N): $\eta_p = \frac{d}{N}$, representing each actuator's contribution to output displacement. Higher η_p indicates better cost efficiency for achieving the same output.

This chapter presents a piezoelectrically actuated microgripper designed for micromanipulation tasks. The design features a large gripping stroke, exceptional area-usage efficiency, and superior PEA utilization efficiency, making it suitable for compact workspaces. A parameter-optimized symmetrical three-stage amplification mechanism ensures high performance. Finite element analysis (FEA) simulations evaluate maximum deformation and natural frequency, while experimental testing validates the design's open-loop characteristics (large stroke and high resonant frequency) and closed-loop performance (hysteresis suppression and improved resolution through PID control).

7.2 Mechanism Design and Analytical Modeling

This section presents the mechanical design and analytical modeling of the proposed piezoelectric-actuated compliant microgripper.

7.2.1 Mechanism Design

The computer-aided design (CAD) model of the proposed microgripper is shown in Fig. 7.1. The gripper features a compact symmetrical design with overall dimensions of 46 mm × 32 mm × 5 mm. This symmetrical configuration provides balanced operation and eliminates parasitic motions during gripping operations.

The microgripper comprises several key components: a stack-type piezoelectric actuator (PEA), a bridge-type amplification mechanism, a series of lever amplification mechanisms, a compliant gasket, and a preloading screw. The PEA is symmetrically mounted within the bridge-type mechanism, serving as the primary source of displacement and force generation.

To address the limited output displacement of the PEA, a three-stage displacement amplification system is incorporated, combining both bridge-type and lever

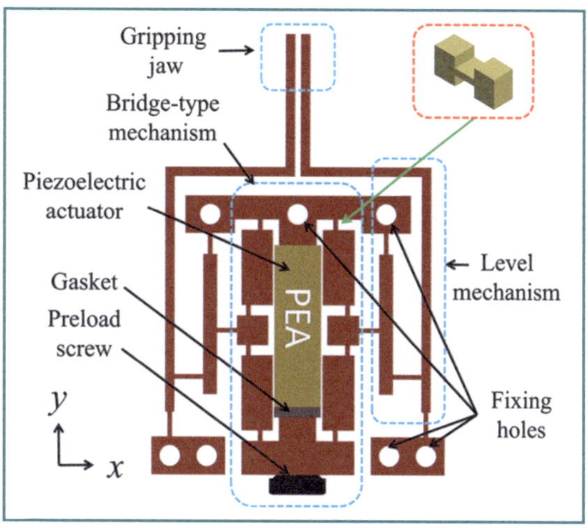

Fig. 7.1 CAD model of the designed compliant microgripper showing symmetrical three-stage amplification mechanism

mechanisms. The bridge-type mechanism was selected for its high structural rigidity, operational stability, and balanced force output characteristics. The cascaded lever mechanism provides additional displacement amplification while maintaining compact dimensions.

Rectangular flexure hinges are employed throughout the design to accommodate the required elastic deformations. These hinges offer several advantages including structural simplicity, manufacturing ease through wire electrical discharge machining (EDM), and a wide range of rotational motion.

A preloading screw mechanism enables precise adjustment of the initial jaw gap, allowing the gripper to adapt to various micro-object sizes. The monolithic symmetrical structure is specifically designed for manufacturing via wire EDM, ensuring high precision and structural integrity.

7.2.2 Analytical Modeling

The displacement amplification capability of the microgripper is analyzed using the pseudo-rigid-body model (PRBM) approach, as illustrated in Fig. 7.2. Due to the symmetrical structure, only one-half of the mechanism is modeled. In the PRBM, flexible hinges are replaced by revolute joints (indicated by circles) with associated torsional springs, while rigid links are represented by solid connecting lines. The model assumes that elastic deformation occurs exclusively at the flexure hinges.

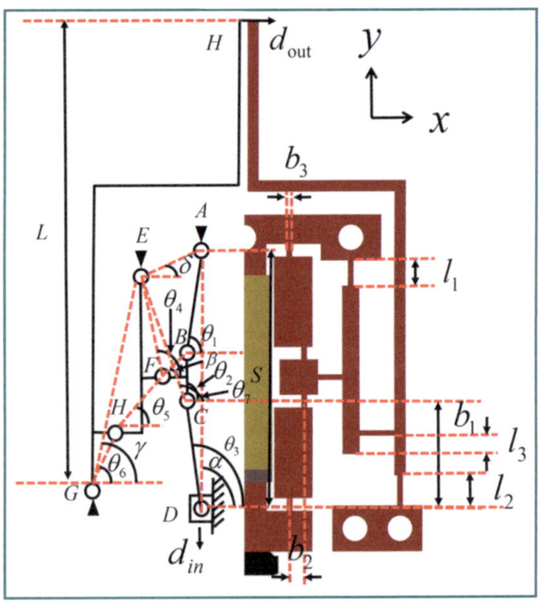

Fig. 7.2 Pseudo-rigid-body model (PRBM) of the microgripper showing flexure hinge replacements with torsional springs and critical geometric parameters for kinematic analysis

The initial angular positions of the linkages are defined as follows: θ_1 (BA), θ_2 (CB), θ_3 (DC), θ_4 (FE), θ_5 (HB), θ_6 (GE), and θ_7 (EA), with positive angles measured counterclockwise. Additional angular parameters include α (DC), β (FB), γ (GE), and δ (EA). The parameter S denotes the vertical distance between points A and D, representing the span between the driving ends of the bridge mechanism.

This chapter develops a simplified displacement model based on PRBM principles. For comprehensive details on PRBM theory and flexible mechanism modeling, readers are directed to foundational Refs. [2, 11, 17].

In the quadrilateral ABCD, the following kinematic relationships can be derived using the complex number method and geometric constraints:

$$l_{DC}e^{i\theta_3} + l_{CB}e^{i\theta_2} + l_{BA}e^{i\theta_1} = l_{DA}e^{i\alpha} \tag{7.1}$$

Using Euler's formula and the principle of equal terms, new scalar equations can be obtained below.

$$l_{DC}\cos\theta_3 + l_{CB}\cos\theta_2 + l_{BA}\cos\theta_1 = 0 \tag{7.2}$$

$$l_{DC}\sin\theta_3 + l_{CB}\sin\theta_2 + l_{BA}\sin\theta_1 = l_{DA}\sin\alpha \tag{7.3}$$

Differentiating Eqs. (7.2) and (7.3) with respect to time, gives

$$l_{DC}\omega_3\sin\theta_3 + l_{CB}\omega_2\sin\theta_2 + l_{BA}\omega_1\sin\theta_1 = 0 \tag{7.4}$$

$$l_{DC}\omega_3\cos\theta_3 + l_{CB}\omega_2\cos\theta_2 + l_{BA}\omega_1\cos\theta_1 = \dot{l}_{DA} = \dot{S} \tag{7.5}$$

Similarly, an equation can be acquired in quadrilateral ABEF as follows.

$$l_{FE}e^{i\theta_4} + l_{EA}e^{i\delta} = l_{FB}e^{i\beta} + l_{BA}e^{i\theta_1} \tag{7.6}$$

Using Euler's formula and the principle of equal terms, new scalar equations can be derived below.

$$l_{FE}\cos\theta_4 + l_{EA}\cos\delta = l_{FB}\cos\beta + l_{BA}\cos\theta_1 \tag{7.7}$$

$$l_{FE}\sin\theta_4 + l_{EA}\sin\delta = l_{FB}\sin\beta + l_{BA}\sin\theta_1 \tag{7.8}$$

Differentiating Eqs. (7.7) and (7.8) with respect to time, gives

$$l_{FE}\omega_4\sin\theta_4 = l_{FB}\omega_\beta\sin(\beta) + l_{BA}\omega_1\sin\theta_1 \tag{7.9}$$

$$l_{FE}\omega_4\cos\theta_4 = l_{FB}\omega_\beta\cos(\beta) + l_{BA}\omega_1\cos\theta_1 \tag{7.10}$$

Furthermore, the same equation can be drawn in quadrilateral EFGH as follows:

$$l_{GH}e^{i\theta_6} + l_{HF}e^{i\theta_5} + l_{FE}e^{i\theta_4} = l_{GE}e^{i\gamma} \tag{7.11}$$

The scalar equation can be rewritten below according to Euler's formula and the principle of equal terms:

$$l_{GH}\cos\theta_6 + l_{HF}\cos\theta_5 + l_{FE}\cos\theta_4 = l_{GE}\cos\gamma \tag{7.12}$$

$$l_{GH}\sin\theta_6 + l_{HF}\sin\theta_5 + l_{FE}\sin\theta_4 = l_{GE}\sin\gamma \tag{7.13}$$

Differentiating Eqs. (7.12) and (7.13) with respect to time, gives

$$l_{GH}\omega_6\sin\theta_6 + l_{HF}\omega_5\sin\theta_5 + l_{FE}\omega_4\sin\theta_4 = 0 \tag{7.14}$$

$$l_{GH}\omega_6\cos\theta_6 + l_{HF}\omega_5\cos\theta_5 + l_{FE}\omega_4\cos\theta_4 = 0 \tag{7.15}$$

Finally, a similar equation is computed in quadrilateral ABCE as follows:

$$l_{CB}e^{i\theta_2} + l_{BA}e^{i\theta_1} = l_{CE}e^{i\theta_7} + l_{EA}e^{i\delta} \tag{7.16}$$

According to the principles mentioned above, the following formula can be easily derived:

$$l_{CB}\cos\theta_2 + l_{BA}\cos\theta_1 = l_{CE}\cos\theta_7 + l_{EA}\cos\delta \tag{7.17}$$

$$l_{CB}\sin\theta_2 + l_{BA}\sin\theta_1 = l_{CE}\sin\theta_7 + l_{EA}\sin\delta \tag{7.18}$$

Differentiating Eqs. (7.17) and (7.18) with respect to time, yields

$$l_{CB}\omega_2\sin\theta_2 + l_{BA}\omega_1\sin\theta_1 = l_{CE}\omega_7\sin\theta_7 \tag{7.19}$$

$$l_{CB}\omega_2\cos\theta_2 + l_{BA}\omega_1\cos\theta_1 = l_{CE}\omega_7\cos\theta_7 \tag{7.20}$$

The velocity at the input end of the mechanism can be written as:

$$v_D = \frac{\partial d_{\text{in}}}{\partial t} = l'_{DA} = \dot{S} \tag{7.21}$$

The velocity at the output end of the mechanism is given as follows.

$$v_H = \frac{\partial d_{\text{out}}}{\partial t} = \omega_6 L \tag{7.22}$$

In addition, the displacement amplification ratio R_{amp} can be derived as follows.

$$R_{\text{amp}} = 2\frac{\partial d_{\text{out}}}{\partial d_{\text{in}}} = \frac{2r_5r_4L}{r_3l_{DC}\cos\theta_3 + r_2l_{CB}\cos\theta_2 + l_{BA}\cos\theta_1} \tag{7.23}$$

where $\omega_2 = r_2\omega_1, \omega_3 = r_3\omega_1, \omega_4 = r_4\omega_1$, and $\omega_6 = r_5 r_4 \omega_1$. Here, ω_i ($i = 2, 3, 4$, and 5) represents the scaling coefficients between different angle velocities, where

$$r_2 = \frac{l_{BA}\sin(\theta_7 - \theta_1)}{l_{CB}\sin(\theta_2 - \theta_7)}$$
$$r_3 = \frac{l_{BA}\sin(\theta_7 - \theta_1) + l_{BA}\sin\theta_1\sin(\theta_2 - \theta_7)}{-l_{DC}\sin\theta_3\sin(\theta_2 - \theta_7)}$$
$$r_4 = \frac{l_{BA}(\theta_1 - \beta)}{l_{EF}\sin(\theta_4 - \beta)}$$
$$r_5 = \frac{l_{FE}\sin(\theta_5 - \theta_4)}{l_{GH}\sin(\theta_6 - \theta_5)}$$

The geometric parameters that affect the gripper's displacement amplification capabilities are shown in Eq. (7.23).

The amplification ratio R serves as a critical performance metric and provides a foundational basis for selecting optimal structural parameters during mechanical optimization. Specifically, the geometric parameters directly influencing displacement amplification must be rigorously considered to maximize gripper performance.

7.3 Optimization and Simulation Study of the Microgripper

7.3.1 Response Surface Optimization

Structural optimization techniques are essential for enhancing the gripping performance of the preliminary microgripper design. An optimized physical structure significantly improves the overall performance and control system efficacy. Thus, systematic optimization of critical structural parameters is imperative.

The optimization problem is formulated through three key components: design variables, objective function, and constraint conditions. The design variables are selected as the parameter set $[b_1, b_2, b_3, l_1, l_2, l_3]$, where each parameter's physical significance is illustrated in Fig. 7.2. The subset $[b_1, b_2, b_3]$ governs the displacement amplification performance of the bridge-type mechanism, while $[l_1, l_2, l_3]$ determines the magnification characteristics of the series lever mechanism.

The objective function is defined as the maximization of the gripping stroke D of the microgripper. Constraints include the numerical bounds of each design variable and the requirement that the safety factor Q remains above a specified threshold to ensure structural integrity during operation. This optimization framework enables a balanced trade-off between performance enhancement and mechanical reliability.

Therefore, the following mathematical model represents the optimization problem:

$$\text{Maximize:} \quad D = f_1(b_1, b_2, b_3, l_1, l_2, l_3) \tag{7.24}$$

$$\text{Subject to:} \begin{cases} 7\,\text{mm} \le b_1 \le 9\,\text{mm} \\ 1\,\text{mm} \le b_2 \le 2\,\text{mm} \\ 0.5\,\text{mm} \le b_3 \le 0.7\,\text{mm} \\ 2\,\text{mm} \le l_1 \le 4\,\text{mm} \\ 2\,\text{mm} \le l_2 \le 4\,\text{mm} \\ 1\,\text{mm} \le l_3 \le 3\,\text{mm} \\ Q \ge 1 \end{cases} \tag{7.25}$$

The optimization study was conducted using ANSYS Workbench software. Figure 7.3 illustrates the flowchart for the structural parameter optimization of the microgripper design. The optimization process encompasses four main stages: parametric modeling, static structural analysis, response surface analysis, and

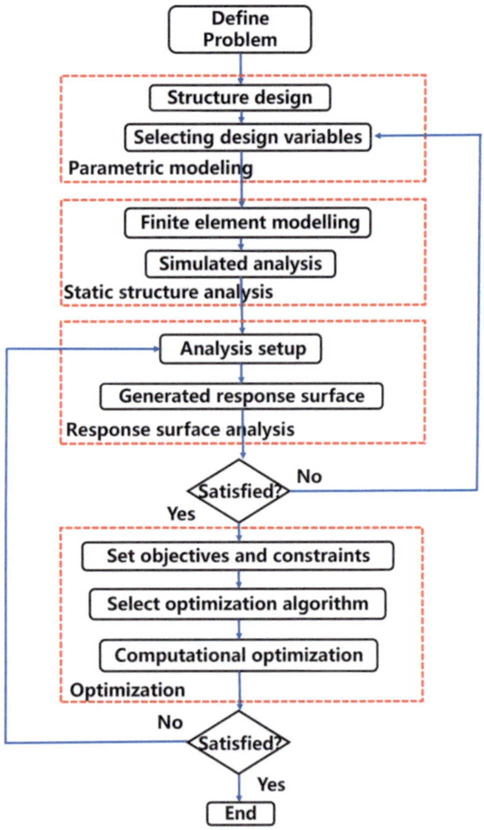

Fig. 7.3 Flowchart of the structural parameter optimization process for the microgripper

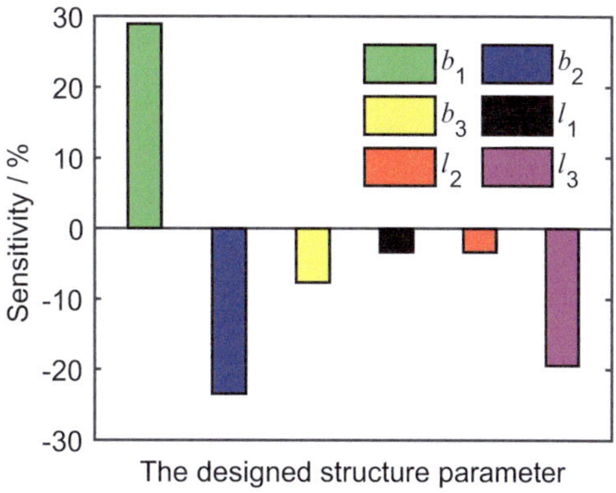

Fig. 7.4 Sensitivity results from response surface analysis showing the influence of geometric parameters

optimization. Response surface analysis enhances efficiency by filtering out insensitive parameters. The genetic algorithm (GA) was chosen as the optimization method due to its suitability for identifying global maxima and its independence from output parameter continuity. The number of initial samples was set to 300. Prior to optimization, response surface analysis was performed to assess the impact of selected design variables on the objective function.

From the response surface analysis, the sensitivity of structural parameters to the gripper's end displacement was quantified, with results shown in Fig. 7.4. Sensitivity values indicate the degree of influence each design variable has on the objective. The analysis reveals that design variables b_1, b_2, and l_3 significantly contribute to the gripping stroke, as evidenced by their high absolute sensitivity values. Meanwhile, parameters b_3, l_1, and l_2 also exhibit non-negligible effects, confirming the necessity of optimizing all selected variables. Additionally, response surfaces depicting the relationship between the optimization objective and design variables are plotted in Fig. 7.5, providing visual insights into parameter interactions and their collective impact on performance.

The response surface graphically illustrates the functional relationships between design variables and the output displacement. As shown in Fig. 7.5, the output displacement increases nonlinearly with b_1, exhibiting a gradually decreasing slope. In contrast, increases in b_2 and b_3 result in a sharp decline in displacement output, indicating that optimal performance requires these parameters to be minimized within feasible bounds.

For parameters l_1, l_2, and l_3, the displacement response follows a concave trend: initial increases enhance performance until an optimum is reached, beyond which further increases reduce displacement. This behavior confirms that their optimal values lie within intermediate ranges rather than at extremes.

The final optimized parameters, summarized in Table 7.1, align with these observations. The results validate the effectiveness of the genetic algorithm coupled with response surface analysis for achieving performance-driven parameter optimization. This approach not only enhances gripping stroke but also provides engineers with intuitive visual tools for design refinement.

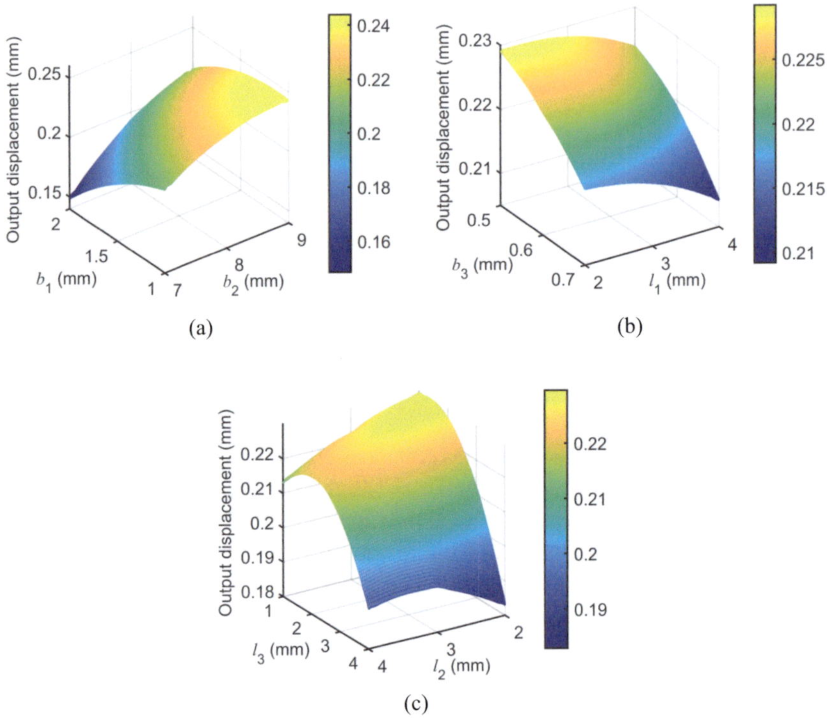

Fig. 7.5 Response surfaces illustrating the relationship between maximum output displacement and structural parameters: **a** b_1–b_2 interaction; **b** b_3–l_1 interaction; **c** l_3–l_2 interaction

Table 7.1 Optimization results of the microgripper parameters

Design variable	b_1	b_2	b_3	l_1	l_2	l_3
Value (mm)	9.0	1.3	0.5	2.2	2.6	1.6

7.3.2 FEA Simulation

A finite element analysis (FEA) simulation study was conducted using ANSYS Workbench software to validate the optimization results. In the simulation setup, a linear spring with equivalent stiffness was used to model the piezoelectric actuator (PEA). The total deformation state of the microgripper is shown in Fig. 7.6. The results demonstrate a maximum jaw displacement of 272.56 μm in the horizontal direction under an input displacement of 20 μm. Given the symmetrical dual-jaw design, the total gripping range is calculated as 545.12 μm. Under the maximum actuation force of 900 N, the safety factor of the microgripper is 4.34, confirming that the structure operates without yielding during full-range motion.

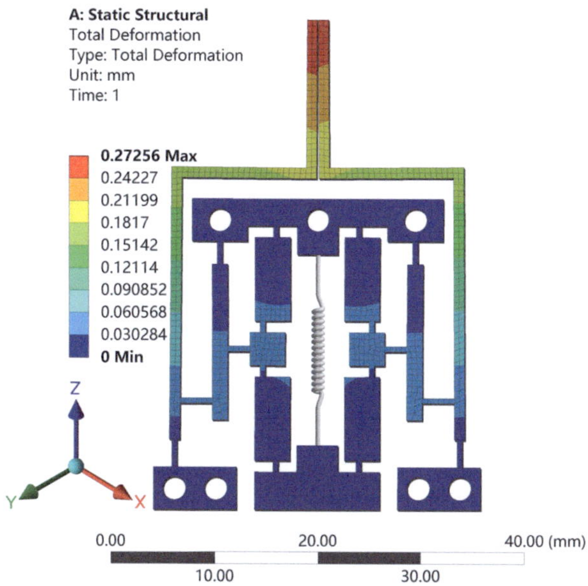

Fig. 7.6 Simulation result of the total deformation of the microgripper under maximum actuation, showing output displacement distribution and amplification mechanism performance

Modal analysis was performed to evaluate the dynamic characteristics of the gripper. The first three modal shapes are shown in Fig. 7.7a–c:

- The first mode (377.75 Hz) involves out-of-plane rotation of the jaws about the paper plane.
- The second mode (392.97 Hz) represents the working mode, with jaws moving symmetrically in the gripping direction.
- The third mode (872.79 Hz) corresponds to out-of-plane bending of the jaws.

Following engineering conventions, the control bandwidth of a compliant mechanism is typically one-third of its first natural frequency. The simulated natural frequencies indicate that the designed microgripper possesses suitable dynamic performance for high-speed micromanipulation tasks.

7.4 Experimental Setup and Results

The microgripper prototype was manufactured via a wire-EDM process. An experimental study was undertaken to evaluate its operational capabilities, validate the preceding performance analysis, and investigate its potential for precise movement in micromanipulation tasks.

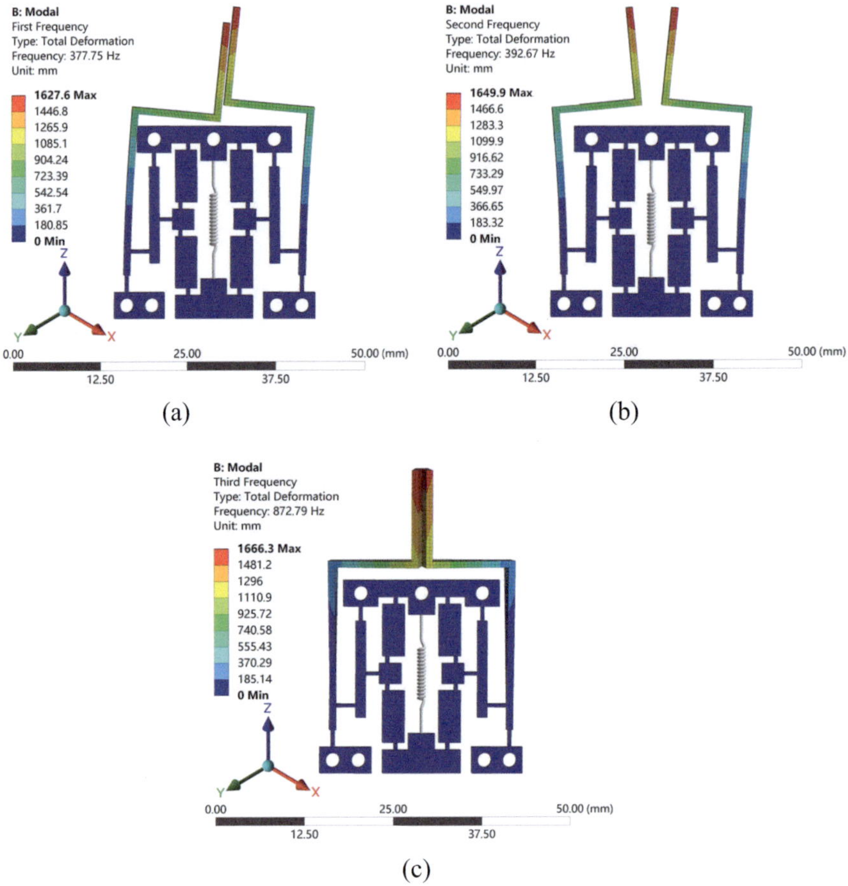

Fig. 7.7 Simulation results of the first three modal shapes of the microgripper

The experimental regimen consisted of two phases: open-loop and closed-loop testing. The open-loop phase was designed to characterize fundamental properties, including the working stroke and resonant frequency. The closed-loop phase aimed to improve motion precision by evaluating the system's hysteresis, step response, and resolution. Furthermore, the gripper's reliability was demonstrated through successful manipulation of various miniature objects.

7.4.1 Prototype Fabrication and Experimental Setup

To validate the optimization and simulation results, the designed microgripper was fabricated and subjected to experimental testing. The mechanical structure was machined from an Al-7075 alloy plate using wire electrical discharge machining

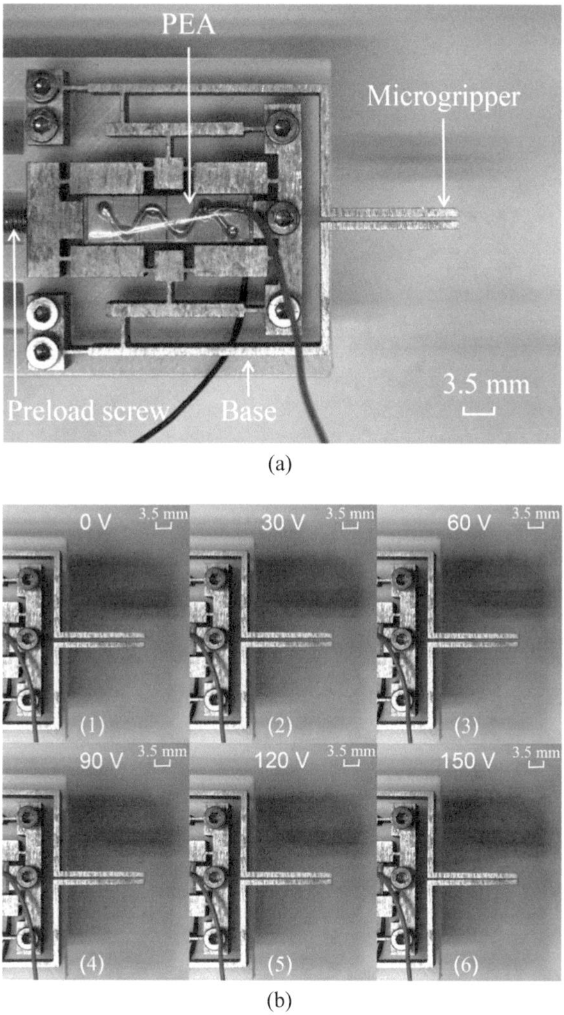

Fig. 7.8 Prototype fabrication and preliminary test result. **a** Assembled microgripper prototype; **b** snapshots of the gripper deformation under different driving voltages

(EDM). This material was selected for its favorable mechanical properties, including low density and high elasticity. Following fabrication, the mechanical body was assembled with other components–including the piezoelectric actuator (PEA) and preload bolts–to form the complete microgripper, as illustrated in Fig. 7.8a.

The selected PEA (model: RP150/5 $\times$ 5/20) has a nominal displacement output of 20 μm at a driving voltage of 150 V and a blocking force of 900 N. As the input voltage was increased linearly from 0 to 150 V, the PEA's displacement gradually increased to its maximum. This motion was transmitted through the mechanism, causing the gripper jaws to close smoothly for object manipulation. The entire flexible deformation process was captured by an industrial camera, as shown in Fig. 7.8b. The recorded sequence confirms that the jaw closure behavior matches the expected motion under a linearly increasing voltage excitation.

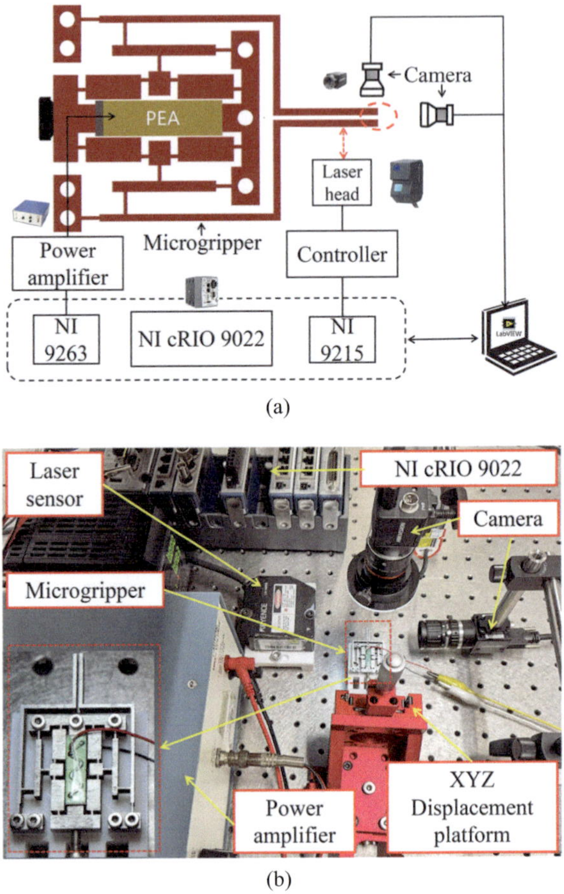

Fig. 7.9 Experimental setup for the microgripper system. **a** Schematic diagram of the hardware connection; **b** photo of the experimental apparatus

The schematic diagram and experimental setup for the performance test system are presented in Fig. 7.9. The entire apparatus is mounted on a vibration isolation table (Newport Corp., model: RS2000) to mitigate environmental disturbances. The microgripper, serving as the test target, is mounted via a fixing base to an XYZ motion stage (Newport Corp., model: HTCL25-X) to enhance positional flexibility. A voltage amplifier (Piezo Systems, Inc., model: EPA-104) amplifies the driving signal for the piezoelectric actuator. The gripper's output displacement is measured by a laser displacement sensor (Keyence Corp., model: LK-H055). This sensor has a 50 mm working distance and a high repeatability of 0.025 μm, enabling high-precision, non-contact measurement. Two industrial cameras (Hikvision Corp., model: MV-CE013-80UM) are positioned atop and beside the microgripper to monitor and record its gripping process. A real-time controller (National Instruments Corp., model: NI cRIO-9022), programmed in LabVIEW, is employed for system control, data acquisition, and waveform display.

7.4.2 Open-Loop Test of the Microgripper Performance

Open-loop experiments were conducted to characterize the fundamental working properties of the microgripper, specifically its output displacement and hysteresis. The input voltage applied to the piezoelectric actuator and the resulting displacement of a single gripper tip are presented in Fig. 7.10a, b, respectively.

A triangular waveform voltage signal, cycled three times between 0 and 150 V, was used as the input. At the maximum driving voltage of 150 V, the measured displacement for a single tip reached 274.21 μm. Owing to the gripper's symmetrical design, the total output displacement at the jaw is twice this value, amounting to 548.42 μm. This result aligns closely with the simulated performance described earlier.

The open-loop tests also characterized the system's hysteresis, as shown in Fig. 7.10c. The hysteresis errors for the three driving cycles are quantified in Fig. 7.10d. Using a low-frequency input (0.5 Hz) to minimize dynamic effects, the hysteresis error demonstrated good repeatability, varying within a range of 2.52–57.19 μm. This nonlinear behavior is primarily attributed to the inherent hysteresis

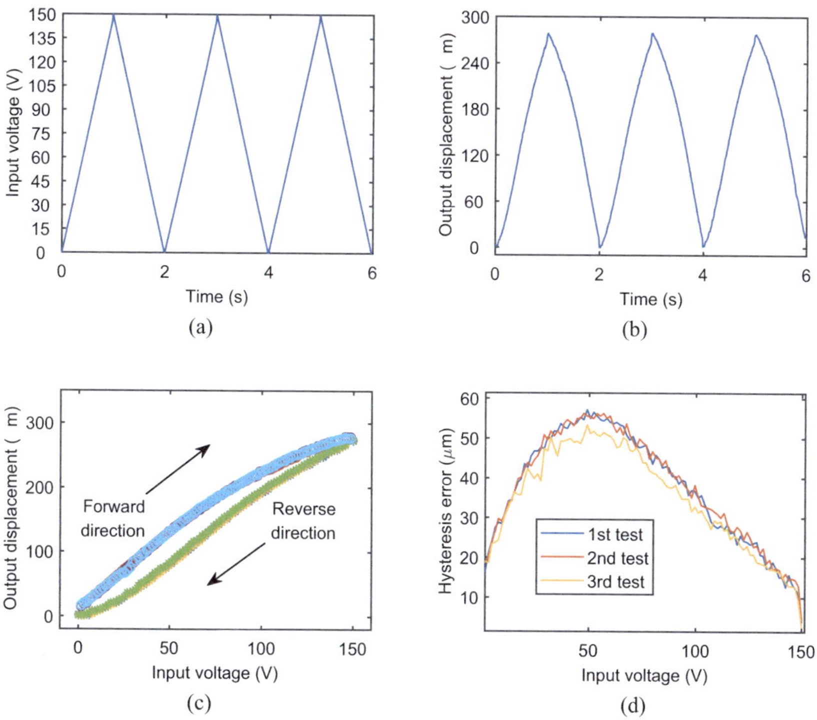

Fig. 7.10 Experimental results of gripping displacement. **a** Input voltage, **b** output displacement, **c** hysteresis effect, and **d** hysteresis error

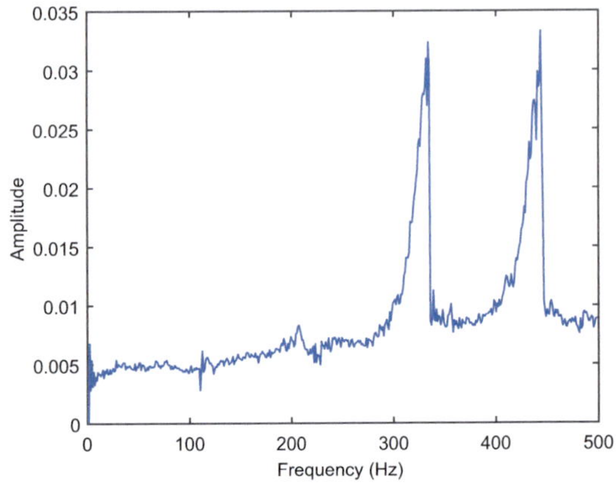

Fig. 7.11 Experimental results of the dynamic performance test

of the piezoelectric ceramic actuator. Additional contributing factors include electronic noise from the sensor, assembly errors, and external disturbances. Mitigating these nonlinearities is essential for achieving the precision required in micromanipulation applications.

Furthermore, the resonant frequency of the gripper was identified using a swept sinusoidal excitation signal. The system's sampling frequency was set to 2 kHz to comply with the Nyquist theorem. The sine wave amplitude was maintained at 1 V while its frequency was swept from 1 to 500 Hz. Resonance was indicated by a sharp increase in vibration amplitude accompanied by audible noise. The frequency response, shown in Fig. 7.11, exhibits pronounced peaks at approximately 334 and 444 Hz, identified as the first two natural frequencies of the system. These experimental values show reasonable agreement with the simulation results. The observed minor deviations are likely due to simplifications in the finite element model, such as neglecting the mass of assembled components including the PEA, gaskets, and preload bolts.

7.4.3 Closed-Loop Test of the Microgripper Performance

To enhance performance, a closed-loop control system using a PID strategy was implemented. The PID controller parameters were tuned empirically and set to $k_P = 35$, $k_I = 0.0002$, and $k_D = 0.00001$.

The system's dynamic response was evaluated using a step input with a 50 μm amplitude. As shown in Fig. 7.12, the gripper achieved a settling time of 0.05 s with an overshoot of 8.8%. This indicates a rapid and stable response, suitable for high-speed micromanipulation tasks.

Positioning accuracy and motion resolution were further characterized. A staircase reference signal was applied, consisting of 1 μm steps with a 2 s duration per step. The resulting positioning error, shown in Fig. 7.13, remained within ±0.75 μm. Statistical

Fig. 7.12 Experimental results of the step response test

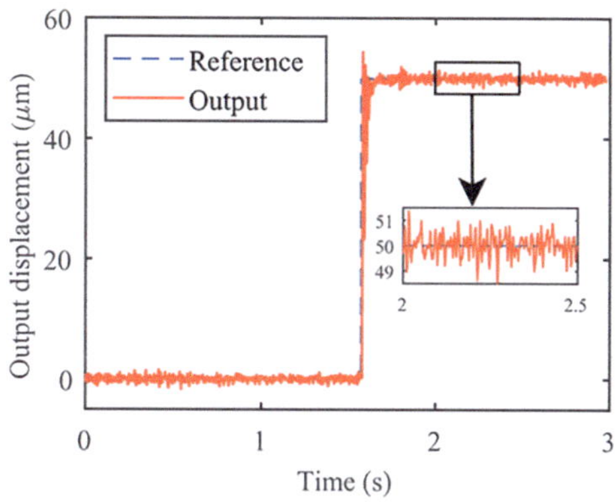

Fig. 7.13 Experimental results of the resolution test. **a** Motion resolution results; **b** error distribution

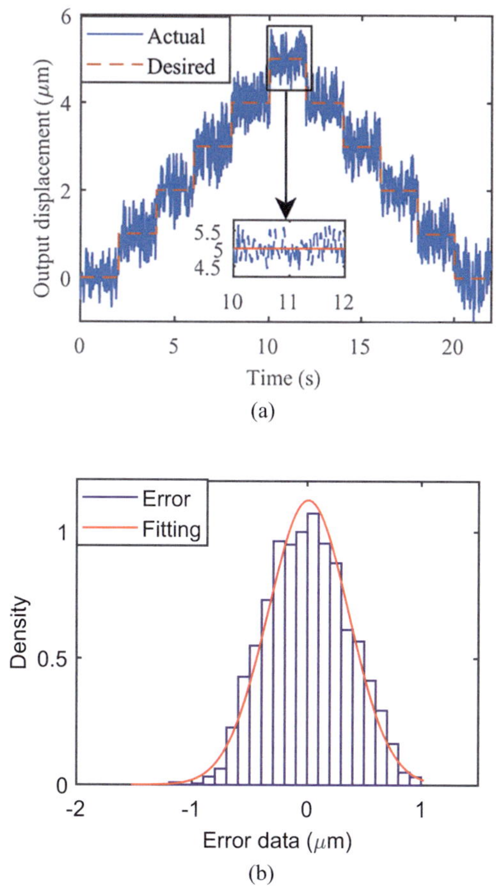

analysis revealed that the error follows a normal distribution with mean $\mu = 0.008$ and standard deviation $\sigma = 0.35$. Using the 2σ criterion, the motion resolution is determined to be 0.75 μm, indicating that 95% of the steady-state positioning errors fall within this bound under no-load conditions.

These results demonstrate that the PID controller provides stable and precise motion control. Further improvements in resolution may be achieved through the use of higher-precision displacement sensors and advanced control algorithms.

The hysteresis of the gripper's output was also evaluated under closed-loop control. A triangular wave signal commanding a 270 μm displacement was applied to the gripper. As shown in Fig. 7.14a, the hysteresis between the forward and reverse displacement paths was significantly suppressed compared to the open-loop case.

The experiment was repeated three times to assess repeatability. The resulting hysteresis error, plotted in Fig. 7.14b, c, remained within a narrow band of ±10 μm and exhibited a normal distribution. This represents a 63.4% reduction in the hysteresis error range compared to the open-loop experiment. The consistent results across multiple trials demonstrate the excellent repeatability of the closed-loop system and conclusively verify the effectiveness of the PID controller in mitigating nonlinearities, underscoring the necessity of closed-loop control for precision applications.

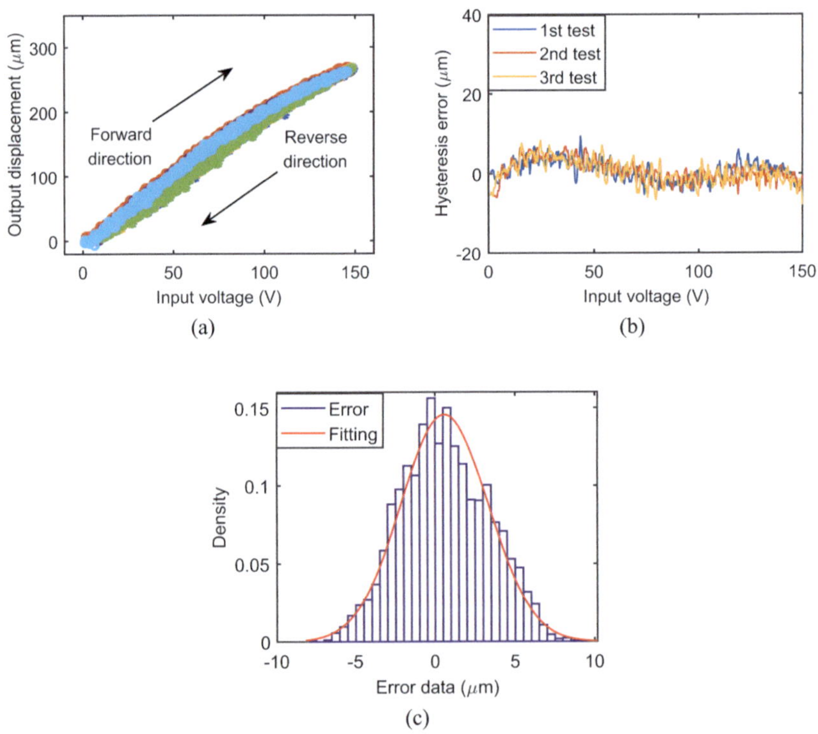

Fig. 7.14 Experimental results for the closed-loop test of the hysteresis effect. **a** Closed-loop hysteresis results; **b** closed-loop hysteresis error; **c** error distribution

7.4.4 Gripping Experiments and Performance Comparison

Gripping experiments were conducted to demonstrate the jaw's positional changes during object manipulation. The experiment involved grasping a metal microwire with a diameter of 200 μm. The input voltage, a triangular wave with an amplitude of 150 V applied to the PEA, and the resulting output displacement are shown in Fig. 7.15. As the voltage increased gradually, the jaw displacement initially increased until it reached the object, after which it remained constant despite further voltage increases. During the voltage decline phase, the displacement followed the opposite trend. This behavior is consistent with physical principles: the jaw moves freely until contact is established, after which the object's stiffness prevents further closure, resulting in a constant displacement reading.

To further evaluate the gripper's versatility, a series of experiments were performed on various miniature objects. The successful manipulation of three different targets is shown in Fig. 7.16. These include: (a) a metal ring with a thickness of 110 μm, grasped and suspended in air; (b) a microwire with a diameter of 300 μm; and (c) a rectangular sheet with a thickness of 500 μm. All objects were securely gripped without damage, demonstrating the microgripper's adaptability and reliability in handling diverse microscopic targets.

Key performance metrics–including gripper size, stroke, area-usage efficiency, PEA utilization efficiency, amplification ratio, and bandwidth–are compared with other piezoelectrically actuated microgrippers in Table 7.2. The comparative data demonstrate that the proposed microgripper offers a combination of a large gripping stroke, high area-usage efficiency, and excellent PEA utilization efficiency. Elevated values for the latter two metrics signify a more compact and economically viable mechanical design, which is highly advantageous for practical applications.

Furthermore, the parasitic motion in the microgripper's non-working direction is a non-negligible factor affecting performance. To enhance output stability, integrating a flexible guide mechanism at the input end of the bridge-type mechanism is a

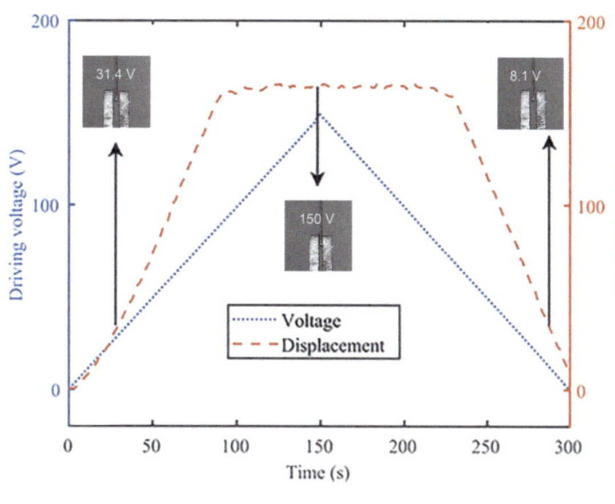

Fig. 7.15 Experimental results of gripping a microwire with a diameter of 200 μm

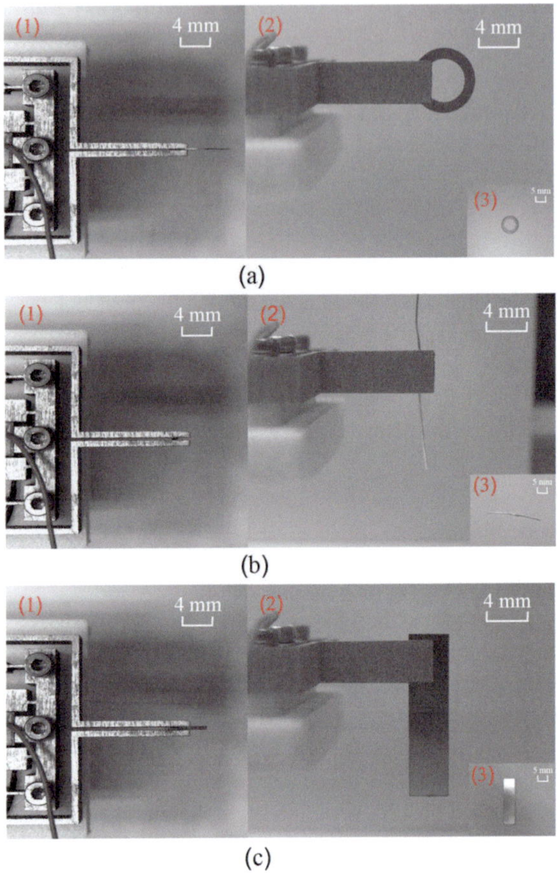

Fig. 7.16 Experimental results for gripping different objects. **a** A metal ring with a thickness of 110 μm; **b** a microwire with a diameter of 300 μm; **c** a rectangle with a thickness of 500 μm

promising approach. This strategy has been successfully employed in prior work, such as that by Ling et al. [18] and Chen et al. [5], to achieve precise and stable output displacement.

7.5 Conclusion

This chapter presented the design, optimization, and experimental validation of a novel compliant microgripper actuated by a piezoelectric actuator for microassembly and micromanipulation applications. The mechanism integrates a bridge-type amplifier with a leverage mechanism in a three-stage amplification structure to effectively transmit and amplify the PEA's displacement to the gripper jaws.

The pseudo-rigid-body model (PRBM) was employed to analyze the kinematic behavior and displacement amplification ratio of the gripper. Key structural

Table 7.2 Performance comparison of typical piezoelectric-actuated microgrippers

Reference	Size (mm × mm)	Stroke (μm)	Area usage efficiency (nm/mm^2)	Number of PEA	PEA utilization efficiency (μm)	Amplification ratio	Bandwidth (Hz)
[27]	110 × 58	213.9	33.53	1	213.9	21.4	125.5
[28]	88 × 67	328.2	55.66	2	164.1	16.4	157.5
[31]	50 × 30	100	66.67	1	100	5.7	–
[24]	40 × 35	134	95.72	1	134	15.5	–
[14]	49 × 22	139.4	129.31	1	129.31	13.94	531
[26]	50 × 27	190	140.74	1	190	22.8	953
[7]	66.2 × 22	338.4	232.35	1	338.4	23.96	1044
[15]	45 × 28	377.28	299.4	1	377.28	20.96	895
[8]	63.9 × 27	543.8	315.20	1	543.8	19.3	325.4
[29]	–	204.6	–	2	102.3	16.8	–
[6]	68.8 × 34.55	632	265.87	2	316	15.8	–
This work	46 × 32	545.12	370.32	1	545.12	27.26	334

parameters were systematically investigated and optimized using computational methods. Sensitivity analysis and response surface methodology confirmed that these parameters significantly influence the gripper's performance. Finite element analysis (FEA) was conducted to simulate and verify the mechanism's output displacement and natural frequencies, ensuring structural integrity and desired dynamic characteristics.

A prototype was fabricated and subjected to comprehensive experimental testing. Under closed-loop control with a PID controller, the hysteresis error was reduced by 63.4% compared to open-loop operation, demonstrating a substantial improvement in positioning accuracy. The microgripper successfully manipulated objects of varying shapes and sizes–including a 110 μm thick metal ring, a 300 μm diameter microwire, and a 500 μm thick rectangular sheet–confirming its versatility and practical applicability.

Performance benchmarks, including an area-usage efficiency of 372.56 nm/mm^2 and a PEA utilization efficiency reflected in a total jaw displacement of 548.42 μm, were found to be competitive with or superior to existing designs reported in the literature. Future work will focus on the development of advanced hybrid control strategies integrating both displacement and force feedback to enable precise and damage-free manipulation of delicate micro-objects.

References

1. Bhagat, U., Shirinzadeh, B., Clark, L., Qin, Y., Tian, Y., Zhang, D.: Experimental system identification, feed-forward control, and hysteresis compensation of a 2-DOF mechanism. Int. J. Intell. Mechatron. Robot. **3**(3), 1–21 (2013)
2. Bilancia, P., Berselli, G.: An overview of procedures and tools for designing nonstandard beam-based compliant mechanisms. Comput. Aided Des. 103001 (2021)
3. Chai, J., Zhang, K., Xue, Y., Liu, W., Chen, T., Lu, Y., Zhao, G.: Review of MEMS based Fourier transform spectrometers. Micromachines **11**(2), 214 (2020)
4. Chen, G., Ma, Y., Li, J.: A tensural displacement amplifier employing elliptic-arc flexure hinges. Sens. Actuators, A **247**, 307–315 (2016)
5. Chen, W., Kong, C., Lu, Q., Liang, Y., Luo, L., Wei, H.: Nonlinear analysis, optimization, and testing of the bridge-type compliant displacement amplification mechanism with a single input force for microgrippers. Precis. Eng. **73**, 166–182 (2022)
6. Chen, X., Deng, Z., Hu, S., Gao, J., Gao, X.: Design of a compliant mechanism based four-stage amplification piezoelectric-driven asymmetric microgripper. Micromachines **11**(1), 25 (2020)
7. Das, T.K., Shirinzadeh, B., Al-Jodah, A., Ghafarian, M., Pinskier, J.: Computational parametric analysis and experimental investigations of a compact flexure-based microgripper. Precis. Eng. **66**, 363–373 (2020)
8. Das, T.K., Shirinzadeh, B., Ghafarian, M., Al-Jodah, A., Pinskier, J.: Characterization of a compact piezoelectric actuated microgripper based on double-stair bridge-type mechanism. J. Micro-Bio Robot. **16**(1), 79–92 (2020)
9. Ghazali, F.A.M., Hasan, M.N., Rehman, T., Nafea, M., Ali, M.S.M., Takahata, K.: MEMS actuators for biomedical applications: a review. J. Micromech. Microeng. **30**(7), 073001 (2020)
10. Gu, G.Y., Zhu, L.M., Su, C.Y., Ding, H.: Motion control of piezoelectric positioning stages: modeling, controller design, and experimental evaluation. IEEE/ASME Trans. Mechatron. **18**(5), 1459–1471 (2012)

11. Howell, L.L., Magleby, S.P., Olsen, B.M.: Handbook of Compliant Mechanisms
12. Iqbal, S., Malik, A.: A review on MEMS based micro displacement amplification mechanisms. Sens. Actuators, A **300**, 111666 (2019)
13. Johal, Z.Z., Crippen, J.W., Forsman, A.C., Lundgren, E.H., Moreno, K.A., Nikroo, A.: Robust capsule and fill tube assemblies for the national ignition campaign. Fusion Sci. Technol. **55**(3), 331–336 (2009)
14. Liang, C., Wang, F., Shi, B., Huo, Z., Zhou, K., Tian, Y., Zhang, D.: Design and control of a novel asymmetrical piezoelectric actuated microgripper for micromanipulation. Sens. Actuators, A **269**, 227–237 (2018)
15. Liang, C., Wang, F., Tian, Y., Zhao, X., Zhang, H., Cui, L., Zhang, D., Ferreira, P.: A novel monolithic piezoelectric actuated flexure-mechanism based wire clamp for microelectronic device packaging. Rev. Sci. Instrum. **86**(4), 045106 (2015)
16. Liaw, H.C., Shirinzadeh, B., Smith, J.: Robust neural network motion tracking control of piezoelectric actuation systems for micro/nanomanipulation. IEEE Trans. Neural Netw. **20**(2), 356–367 (2009)
17. Ling, M., Howell, L.L., Cao, J., Chen, G.: Kinetostatic and dynamic modeling of flexure-based compliant mechanisms: a survey. Appl. Mech. Rev. **72**(3), 030802 (2020)
18. Ling, M., Wang, J., Wu, M., Cao, L., Fu, B.: Design and modeling of an improved bridge-type compliant mechanism with its application for hydraulic piezo-valves. Sens. Actuators, A **324**, 112687 (2021)
19. Lyu, Z., Xu, Q.: Recent design and development of piezoelectric-actuated compliant micro-grippers: a review. Sens. Actuators A: Phys. 113002 (2021)
20. Popa, D.O., Stephanou, H.E.: Micro and mesoscale robotic assembly. J. Manuf. Process. **6**(1), 52–71 (2004)
21. Reissman, T., Garcia, E., Lobontiu, N., Nam, Y.: Transfer function characterization of thermally-actuated displacement-amplified micro-electromechanical systems (MEMS). In: Varadan, V.K. (ed.) Smart Structures and Materials 2006: Smart Electronics, MEMS, BioMEMS, and Nanotechnology, vol. 6172, pp. 350–361. SPIE (2006)
22. Ruggeri, S., Fontana, G., Fassi, I.: Micro-assembly. In: Micro-Manufacturing Technologies and Their Applications, pp. 223–259. Springer (2017)
23. Shi, C., Dong, X., Yang, Z.: A microgripper with a large magnification ratio and high structural stiffness based on a flexure-enabled mechanism. IEEE/ASME Trans. Mechatron. (2021)
24. Sun, X., Chen, W., Tian, Y., Fatikow, S., Zhou, R., Zhang, J., Mikczinski, M.: A novel flexure-based microgripper with double amplification mechanisms for micro/nano manipulation. Rev. Sci. Instrum. **84**(8), 085002 (2013)
25. Varadan, V.K., Varadan, V.V.: Microsensors, microelectromechanical systems (MEMS), and electronics for smart structures and systems. Smart Mater. Struct. **9**(6), 953 (2000)
26. Wang, F., Liang, C., Tian, Y., Zhao, X., Zhang, D.: Design of a piezoelectric-actuated micro-gripper with a three-stage flexure-based amplification. IEEE/ASME Trans. Mechatron. **20**(5), 2205–2213 (2014)
27. Yang, Y.L., Wei, Y.D., Lou, J.Q., Tian, G., Zhao, X.W., Fu, L.: A new piezo-driven microgripper based on the double-rocker mechanism. Smart Mater. Struct. **24**(7), 075031 (2015)
28. Yang, Y.L., Wei, Y.D., Lou, J.Q., Xie, F.R., Fu, L.: Development and precision position/force control of a new flexure-based microgripper. J. Micromech. Microeng. **26**(1), 015005 (2015)
29. Zhao, Y., Huang, X., Liu, Y., Wang, G., Hong, K.: Design and control of a piezoelectric-driven microgripper perceiving displacement and gripping force. Micromachines **11**(2), 121 (2020)
30. Zhu, J., Hao, G.: Design and test of a compact compliant gripper using the Scott-Russell mechanism. Arch. Civil Mech. Eng. **20**(3), 1–12 (2020)
31. Zubir, M.N.M., Shirinzadeh, B.: Development of a high precision flexure-based microgripper. Precis. Eng. **33**(4), 362–370 (2009)

Chapter 8
Design of a Piezoelectric-Actuated Biaxial Compliant Microgripper With Long Stroke

Abstract This chapter presents a piezoelectric-actuated biaxial compliant microgripper capable of long-stroke automated gripping and rolling of micro-scale objects. To enhance the working stroke while maintaining a compact footprint, a counter-side distributed two-stage lever amplifier with parallelogram mechanisms is introduced. Using the pseudo-rigid-body model, analytical formulations for the displacement amplification ratio, input stiffness, and natural frequency are established for both gripper arms. The design is optimized and simulated via finite element analysis. Experimental validation demonstrates that the microgripper achieves gripping and rolling of a 200 μm optical fiber by over 45°, showcasing its potential for precision applications such as optical fiber alignment.

Keywords Microgripper · Compliant mechanism · Fiber alignment · Gripping and rubbing · Micromanipulation · Motion control

8.1 Introduction

As the end-effector of micromanipulation systems, the microgripper plays an essential role in both scientific research and industrial production [10, 11, 23, 28]. It operates by driving the jaws to open and close via various actuators, catering to diverse application scenarios. For instance, electrothermal microgrippers are capable of handling biomedical samples such as cyanobacteria cells and zebrafish embryos [33, 34], while electrostatic microgrippers are suitable for micro/nano-scale operations including hair clamping and mechanical characterization of plant cells [15, 27].

Beyond the biomedical field, there is also significant industrial demand for microgrippers. Owing to their compact footprint and substantial stroke, piezoelectrically actuated microgrippers have been widely applied in microelectronics, optical engineering, and precision assembly [12, 14, 19, 20]. A variety of single-degree-of-freedom (DOF) piezoelectric microgrippers have been developed and investigated for micromanipulation tasks, benefiting from their large working stroke, controllable clamping force, and high resolution [2, 6, 8, 16–18, 21, 26, 35].

Q. Xu, *Piezoelectric-Driven Flexible Robotic Microgrippers*,
https://doi.org/10.1007/978-981-95-4685-5_8

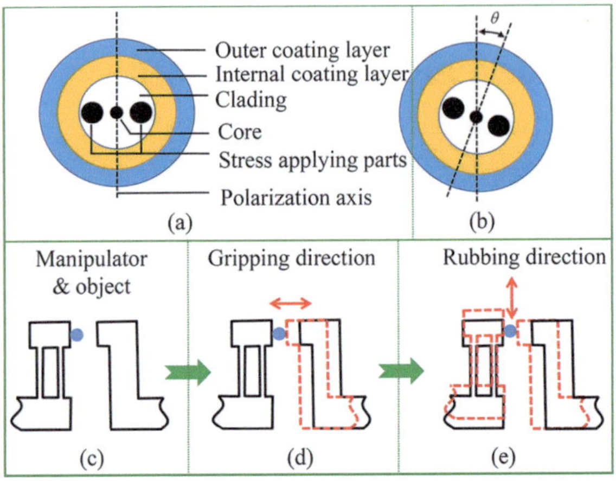

Fig. 8.1 Structure of a PANDA polarization-maintaining optical fiber (PMOF) and the operational concept of a 2-DOF microgripper. **a** Schematic of the PANDA PMOF structure, **b** illustration of possible polarization axis misalignment, **c** overview of the 2-DOF microgripper design, **d** gripping motion demonstration, and **e** rubbing motion demonstration

However, the limited functionality of single-DOF microgrippers makes them insufficient for certain complex operations. For example, in the assembly process of MEMS chip-based optical switches, the microgripper must provide rotational capability [5]. Conventional single-DOF microgrippers require rotation as a whole to achieve optical fiber alignment, which is constrained by the accuracy of the overall rotation and the dimensions of the micromanipulation workspace. Therefore, a 2-DOF microgripper is highly desirable, as it can provide not only gripping but also longitudinal rubbing functionality [22, 30].

A particularly critical application is the rolling operation of polarization-maintaining optical fiber (PMOF) to ensure rotational alignment [1, 13, 29]. Misalignment can lead to transmission loss due to changes in light mode. Figure 8.1a, b illustrate the structure of a typical PANDA PMOF and the potential deflection of its polarization axis. Figure 8.1c–e demonstrate the operational process of a biaxial microgripper capable of both gripping and rubbing for optical fiber alignment.

To integrate these two functions into a monolithic architecture, numerous symmetric and asymmetric biaxial microgrippers have been proposed [3, 4, 7, 9, 24, 25, 31, 32]. For example, Xu designed a symmetric 2-DOF microgripper using a modular and decoupled approach, which offers advantages such as low cost and ease of maintenance [25]. However, its displacement amplification ratio is relatively small. Jia et al. developed a dual-axis gripper based on a flexure mechanism and genetic algorithm, achieving 2-DOF motion in each jaw [9], but it suffers from limited working stroke and complex control requirements. Chen et al. proposed an asymmetric 2-DOF microgripper integrating an L-lever mechanism and a Roberts mechanism for optical fiber assembly [4], though its small rolling stroke complicates and prolongs the alignment process. Subsequent improvements by Chen et al. [3] and Royson et al. [7] incorporated parallel motion and force sensing, enhancing the safety and reliability of micromanipulation. Zhang et al. significantly improved the rubbing capability using a differential lever mechanism, achieving a rotating stroke of up to 223.4 μm [32], but the magnification along the clamping direction was weak, limiting its operational range. Wang et al. demonstrated a compliant microgripper with fine gripping and rubbing capabilities, enabling switching control between force and displacement [24], yet it still exhibits insufficient stroke in both functions.

Motivated by these advancements, this work aims to develop a biaxial microgripper that combines a large working stroke with a compact mechanical structure for practical applications in complex operations. To this end, a biaxial piezoelectric-actuated microgripper is designed and developed for complex tasks such as PMOF alignment. To achieve a large working stroke within a compact footprint, a counter-side distributed two-stage lever amplifier with a parallelogram mechanism is introduced in the flexure-based compliant microgripper design. Compared to prior work, the proposed dual-axis microgripper offers both a large working stroke and a high resonant frequency, meeting the demands for fast and precise fiber manipulation. The right-side mechanism handles object clamping, while the left-side mechanism performs dexterous rubbing tasks. The gripper's performance is ensured through rational mechanism design and parameter optimization. Analytical modeling, simulation studies, and experimental investigations are conducted to validate the capabilities of the proposed microgripper.

8.2 Mechanism Design of the Microgripper

Compared to traditional rigid-body structures, compliant mechanisms offer several advantages, including no clearance, no friction, no need for assembly or lubrication, monolithic design, and vacuum compatibility. Therefore, they are a preferred choice for developing a dual-axis microgripper utilizing flexible displacement amplification and transmission mechanisms. Commonly used displacement amplification mechanisms include lever, rhombus-type, bridge, Scott–Russell (SR), double-rocker mechanisms, among others. Among these, the bridge amplifier stands out due to its high stiffness, excellent stability, high amplification ratio, and compact structure.

8.2.1 Design of Displacement Amplification Mechanism

To achieve a large amplification ratio within a compact structure, a displacement amplification mechanism is designed first. The parallel-type flexible bridge structure features two output modes: single-end output and double-end output, as illustrated in Fig. 8.2a, b, respectively. Both modes share the common characteristic of amplifying the actuator's input displacement and converting it into an orthogonal output displacement. The double-end output bridge mechanism is often employed in symmetrical microgrippers to enhance the working stroke. However, under constraints of workspace and displacement amplification requirements, this configuration tends to compromise structural compactness. Therefore, a bridge mechanism with a sin-

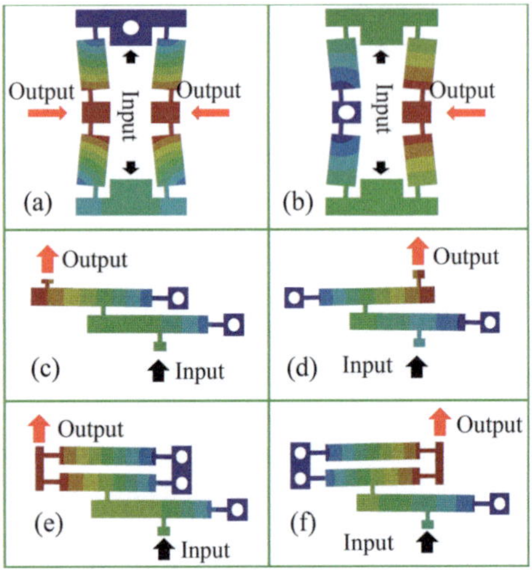

Fig. 8.2 Compliant displacement amplification mechanisms. **a** Double-output bridge-type mechanism, **b** single-output bridge-type mechanism, **c** two-stage lever mechanism with ipsilateral configuration, **d** two-stage lever mechanism with contralateral configuration, **e** two-stage lever mechanism with ipsilateral configuration and integrated parallelogram flexures, and **f** two-stage lever mechanism with contralateral configuration and integrated parallelogram flexures

gle output terminal is adopted as the first stage of the displacement amplification mechanism in the gripper design. Specifically, horizontal bridge mechanisms are positioned on the left side, while vertical bridge mechanisms are placed on the right side, aligning with their respective functional orientations.

In addition to the bridge mechanism, the lever mechanism is another widely used displacement amplification method due to its simple design and high amplification efficiency. To enhance displacement magnification, lever structures can be configured in series, parallel, or differential arrangements. Generally, parallel lever configurations effectively reduce parasitic displacement at the output end, while differential levers offer remarkable displacement amplification. However, these designs often require considerable space, making them unsuitable for asymmetric microgripper applications.

A two-stage lever mechanism connected in series is illustrated in Fig. 8.2c. An alternative design is shown in Fig. 8.2d, where the fixed ends are positioned on either side of the moving end. Finite element analysis indicates that the latter configuration achieves a higher displacement amplification ratio under identical input displacements and structural parameters, albeit with increased parasitic displacement. To mitigate parasitic displacement, a flexible parallelogram mechanism can be incorporated, as demonstrated in Fig. 8.2e, f.

A performance comparison of the four mechanisms under the same input displacement is provided in Table 8.1. Coupling the lever mechanism with the parallelogram

Table 8.1 Simulation results for different lever mechanisms as illustrated in Fig. 8.2c–f

Type	Input displacement (mm)	Output displacement (mm)	Parasitic displacement (mm)	Magnification	Parasitic ratio (%)
Figure 8.2c	0.1	0.2073	0.0289	2.073	13.94
Figure 8.2d	0.1	0.3111	0.0629	3.111	20.22
Figure 8.2e	0.1	0.1834	0.0020	1.834	1.09
Figure 8.2f	0.1	0.2634	0.0038	2.634	1.44

mechanism results in a slight reduction in displacement magnification but significantly reduces parasitic displacement. Notably, the lever mechanism integrated with a flexible parallelogram mechanism, as depicted in Fig. 8.2d, achieves substantial displacement amplification with acceptable levels of parasitic motion. Consequently, this cascaded flexible amplification mechanism is adopted for the rubbing operation of the microgripper. Additionally, the flexible parallelogram mechanism is incorporated into the clamping mechanism (on the right side) to enable parallel gripping of small objects.

8.2.2 *Design of the Microgripper Mechanism*

Based on the previously described displacement amplification mechanisms, a 2-DOF microgripper is designed as shown in Fig. 8.3a. The monolithic gripper assembly consists of a pair of piezoelectric actuators (PZTs), a set of gaskets, and two preload screws. The mechanical body is functionally divided into left and right sections, connected by a Z-shaped rigid structure. The right-side gripper is designed to grasp tiny objects when actuated by PZT 1, while the left-side gripper, driven by PZT 2, performs rubbing motions with its gripping jaw.

Flexure hinges are essential components for motion transmission within the gripper. Rectangular flexure hinges are used due to their advantages in manufacturability and low stiffness. The left and right sections are connected only through rigid elements, ensuring they operate independently without interference.

A double cantilever beam is integrated between the parallelogram mechanisms on the left side and the rubbing jaws. This beam functions as a force sensing element; by mounting a strain gauge sensor on it, clamping force or rotational friction can be detected. The double-cantilever design provides passive compliance, allowing it to deform elastically under applied forces.

By attaching strain gauges and performing calibration experiments, the gripping force can be accurately measured for safe and precise operation. The initial clearance between the two jaws can be adjusted using the preloading device (preload bolts and gaskets) to suit various applications. A through-hole is incorporated to mount the

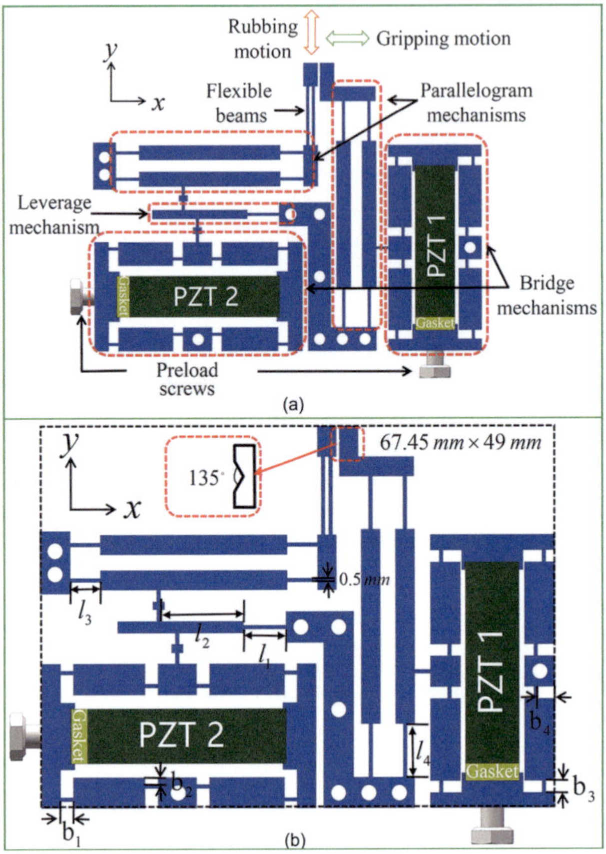

Fig. 8.3 Designed biaxial compliant microgripper. **a** Overall mechanism design, and **b** key geometric parameters with dimensional annotations

entire gripper onto a micropositioning platform, enabling delicate and dexterous operations.

8.3 Analytical Modeling of the Microgripper

In this work, the pseudo-rigid-body model (PRBM) method is adopted to analyze the displacement amplification performance of the gripper. PRBM offers advantages in simplicity and computational efficiency. In this model, the motion characteristics of flexure hinges are simulated using rigid linkages with revolute joints, while the stiffness characteristics are represented by additional torsional springs.

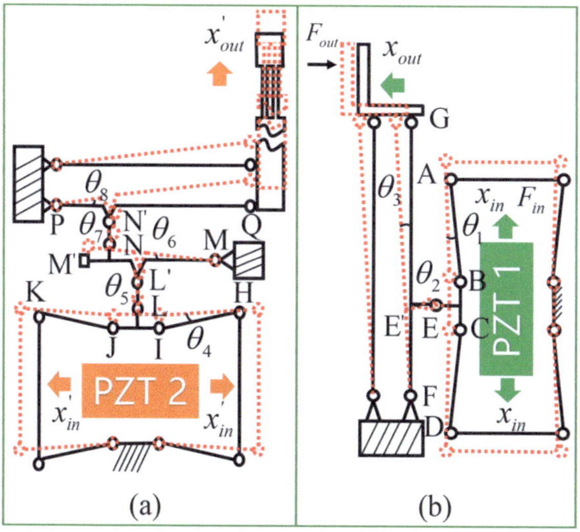

Fig. 8.4 Deformation analysis of the microgripper mechanism. **a** Left-side deformation under actuation, and **b** right-side deformation under actuation

The deformation diagram of the proposed microgripper is illustrated in Fig. 8.4. The solid black lines indicate the initial position of the gripper mechanism according to the PRBM, while the red dotted lines represent the final position after actuation by PZT 1 and PZT 2. The angle increments of linkages AB, EE′, FG, HI, LL′, MN, NN′, and PQ are denoted by θ_i $(i = 1, 2, \ldots, 8)$. The input and output displacements of the right-side mechanism are represented by x_{in} and x_{out}, respectively. Similarly, x'_{in} and x'_{out} denote the input and output displacements of the left-side structure. It can be observed that the ends of the left and right grippers undergo small parallel motions in different directions.

Figure 8.5a illustrates the kinematics model and velocity vector of the right-side gripper, while Fig. 8.5b shows the PRBM and kinematics model of the left-side gripper.

8.3.1 Kinematic Modeling

The displacement magnification ratio (R_{amp}) of the gripper is defined as the ratio of the output displacement (x_{out} or x'_{out}) to the input displacement (x_{in} or x'_{in}). Alternatively, R_{amp} can be interpreted as the ratio of the velocity at the input end to the velocity at the output terminal of the mechanism. The displacement amplification ratios for the left-side and right-side mechanisms are denoted as R_l and R_r, respectively, and are calculated as follows.

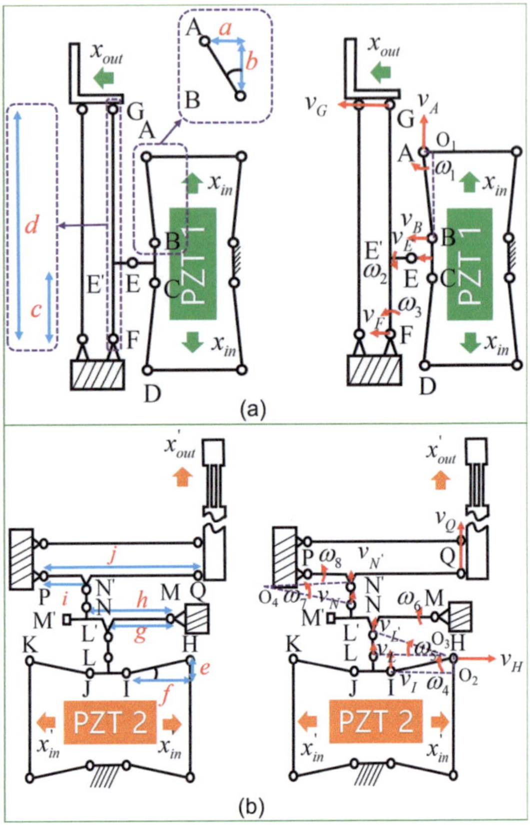

Fig. 8.5 Analytical modeling of the microgripper. **a** Pseudo-rigid-body model (PRBM) and kinematic diagram for the right-side mechanism, and **b** PRBM and kinematic diagram for the left-side mechanism

The instantaneous center method is an effective approach for velocity analysis in planar mechanisms. For a rigid body, the instantaneous center can be determined once the absolute velocities at two points are known. Specifically, the instantaneous center is located at the intersection point of two lines perpendicular to the velocity directions at those points.

For the right-side gripper, the displacement magnification ratio R_r is derived from the instantaneous velocities at points A and G. The approximate instantaneous centers for linkages AB and GF are point O_1 and point F, respectively. Assuming that linkage BCE is a rigid structure with uniform horizontal velocity across its right part, the instantaneous velocities at points A, B, E, F, and G can be determined. Their relationships are expressed as follows:

$$\begin{aligned}
v_A &= \omega_1 \cdot O_1 A \\
v_B &= \omega_1 \cdot O_1 B \\
v_E &= v_B \\
v_E &\approx \omega_3 \cdot FE \\
v_H &= \omega_3 \cdot FG
\end{aligned}$$

where ω_1 and ω_3 are the instantaneous angular velocities of the linkages AB and EF, respectively.

Based on the above kinematic and geometrical relationships, the displacement magnification ratio can be derived as:

$$R_r = \frac{x_{\text{out}}}{2x_{\text{in}}} \approx \frac{v_G}{2v_A} = \frac{bd}{2ac} \tag{8.1}$$

Similarly, the displacement amplification ratio R_l for the left-side mechanism can be derived from the instantaneous velocities at points H and Q. The approximate instantaneous centers of linkages HI, MN, and PQ are points O_2, M, and P, respectively. The relationships of the instantaneous velocities at points H, I, L, N, and Q are described as follows:

$$\begin{aligned}
v_H &= \omega_4 \cdot O_2 H \\
v_I &= \omega_4 \cdot O_2 I \\
v_L &= v_I \\
v_L &\approx \omega_6 \cdot ML^{'} \\
v_N &= \omega_6 \cdot MN \\
v_{N'} &\approx v_N = \omega_8 \cdot PN^{'} \\
v_Q &= \omega_8 \cdot PQ
\end{aligned}$$

where ω_4, ω_6, and ω_8 are the instantaneous angular velocities of the linkages HI, MN, and PQ, respectively.

According to the preceding mathematical relationships, R_l can be easily obtained below.

$$R_l = \frac{x^{'}_{\text{out}}}{2x^{'}_{\text{in}}} \approx \frac{v_Q}{2v_H} = \frac{fhj}{2egi} \tag{8.2}$$

Based on the above analysis, a linear relationship between the end displacement of the gripper and the actuation displacement of the piezoelectric actuator can be observed. This linear input-output relationship simplifies the design of control schemes, facilitating precise operational performance.

It is important to note that certain approximations–such as the substitution of linkage velocities in the kinematic model–may introduce deviations between the analytical displacement amplification model and actual values. These approximations, while necessary for theoretical modeling, highlight the need for empirical validation.

To address this, subsequent simulations and experiments will be conducted to verify the accuracy of the derived analytical model and refine it as necessary.

8.3.2 Static Modeling

The static modeling begins with the analysis of the right-side gripper's clamping mechanism. The angle increments for the kinetic linkages AB, EE′, and FG are illustrated in Fig. 8.4b. Under the assumption that the input displacement from the piezoelectric actuator is sufficiently small, the angular increments θ_i (where $i = 1, 2, 3$) of the linkages can be treated as the time integrals of their corresponding angular velocities ω_i (where $i = 1, 2, 3$).

The rotational angles φ_i (where $i = \text{A}, \text{B}, \ldots, \text{G}$) of the compliant hinges A through G are derived as follows.

$$\varphi_A = \varphi_B = \varphi_C = \varphi_D = \frac{bx_{\text{in}}}{2a\sqrt{a^2 + b^2}}$$

$$\varphi_E = \varphi_F = \varphi_G = \theta_2 = \theta_3 = \frac{bx_{\text{in}}}{ac}$$

According to PRBM, all of the compliant hinges are replaced with equivalent revolute joints coupled with torsional springs. The spring stiffness for the short rectangular flexure hinge can be derived below.

$$K_r^i = \frac{Eb_t t_i^3}{12l_i} = \eta_i K_r^A \tag{8.3}$$

where $i = \text{A} \sim \text{G}$, and E represents the Young's modulus of the material. The parameters b_t, t_i, and l_i denote the thickness, width, and length of the flexure hinges, respectively. The stiffness K_r^A corresponds to the torsional stiffness of flexure hinge A, and η_i represents the stiffness ratio between the hinge A and other hinges.

For the right-side gripper, the work contributed by the external force F_{in} of PZT 1 is expressed as:

$$W_r = \frac{1}{2} F_{\text{in}} x_{\text{in}}. \tag{8.4}$$

Certainly, the work done by the external force is primarily converted into elastic potential energy stored in the flexible hinges. If the jaw of the right-side gripper contacts a small target, a portion of the mechanical energy will be dissipated. To simplify the model, the output force F_{out} is not considered in the theoretical analysis. Therefore, the strain energy of the gripper under no-load conditions can be expressed as:

$$E_p^r = 2\sum_{i=A}^{B} K_r^i \varphi_i^2 + \frac{1}{2} K_r^E \varphi_E^2 + \sum_{i=F}^{G} K_r^i \varphi_i^2. \tag{8.5}$$

Furthermore, based on the conservation of energy, the following equation can be derived.

$$F_{\text{in}} = \frac{2b^2 K_r^A x_{\text{in}}}{a^2} \left(\frac{1}{a^2 + b^2} + \frac{\eta_E + 4\eta_F}{2c^2} \right). \tag{8.6}$$

Hence, the equivalent stiffness of the right-side mechanism can be obtained as:

$$K_{\text{in}}^r = \frac{F_{\text{in}}}{x_{\text{in}}} = \frac{2b^2 K_r^A}{a^2} \left(\frac{1}{a^2 + b^2} + \frac{\eta_E + 4\eta_F}{2c^2} \right) \tag{8.7}$$

where K_{in}^r is the input stiffness of the right-side gripper. Analysis reveals that the input stiffness is influenced by the linkage dimensions–denoted by lengths a, b, and c–as well as the rotational stiffness K_r of the short rectangular flexure hinge.

For the left-side gripper, the input stiffness can be derived as follows:

$$K_{\text{in}}^l = \frac{2f^2 K_r^H}{e^2} \left(\frac{1}{e^2 + f^2} + \frac{\eta_{L'} + \eta_M + \eta_N}{8g^2} + \frac{\eta_N (2h - i)^2 + 16\eta_P h^2}{8g^2 i^2} \right) \tag{8.8}$$

where K_r^H is the stiffness of the flexure hinge H, and η_i denotes the stiffness ratio between the hinge H and other hinges.

8.3.3 Dynamic Modeling

Using Lagrange's equation, the dynamic model of the right-side mechanical structure can be expressed as:

$$\frac{d}{dt}\left(\frac{\partial T}{\partial \dot{q}_j} \right) - \frac{\partial T}{\partial q_j} + \frac{\partial V}{\partial q_j} = F_j \ \ (\text{for } j = 1, 2, \ldots, n) \tag{8.9}$$

where T and V represent the total kinetic and potential energies of the mechanical system, respectively. Here, q_j is the generalized coordinate, n is the number of generalized coordinates, and F_j denotes the generalized nonconservative force. The proposed right-side gripper has one degree of freedom (DOF).

In the dynamic analysis, the kinetic energy of the gripper is assumed to be the sum of the translational and rotational kinetic energies of its rigid linkages. The kinetic energy of the right-side gripper can be expressed as:

$$\begin{aligned} T_r = 2I_{AB}\omega_1^2 + I_{FG}\omega_3^2 + \frac{1}{2}I_{EE'}\omega_2^2 + m_{\text{in}}\dot{x}_{\text{in}}^2 \\ + \frac{1}{2}m_{BCE}\omega_1^2(a^2+b^2) + \frac{1}{2}m_{\text{out}}\dot{x}_{\text{out}}^2 \end{aligned} \tag{8.10}$$

where $\dot{x}_{\text{in}}$ is the velocity of the input terminal; m_{in}, m_{out}, and m_{BCE} denotes the mass of the input end, linkage BCE, and output end of the right-side gripper; I_{AB}, $I_{EE'}$, and I_{FG} represent the moment of inertia of the linkages AB, EE$'$, and FG, respectively.

The total potential energy of the right-side mechanism is equal to its strain energy, denoted as V_r:

$$V_r = E_p^r. \tag{8.11}$$

Further simplification yields the following second-order dynamic equation:

$$M_r\ddot{x}_i + K_r x_i = F_r \tag{8.12}$$

where M_r, K_r, and F_{in} represent the equivalent mass, equivalent stiffness, and generalized force of the right-side gripper, respectively. Their expressions are derived as follows.

$$\begin{aligned} M_r = \frac{b^2 I_{AB}}{a^2(a^2+b^2)} + \frac{2I_{FG}b^2}{a^2c^2} + \frac{b^2 I_{EE'}}{a^2c^2} \\ + 2m_{\text{in}} + \frac{b^2 m_{BCE}}{4a^2} + \frac{b^2d^2}{a^2c^2}m_{\text{out}} \end{aligned} \tag{8.13}$$

$$K_r = K_{\text{in}}^r \tag{8.14}$$

Finally, the natural frequency of the right-side gripper can be expressed as

$$f_r = \frac{1}{2\pi}\sqrt{\frac{M_r}{K_r}}. \tag{8.15}$$

In the same way, the natural frequency of the left-side gripper cab be derived as

$$f_l = \frac{1}{2\pi}\sqrt{\frac{M_l}{K_l}}. \tag{8.16}$$

where the equivalent mass M_l and equivalent stiffness K_l for the left-side mechanism are obtained as follows.

$$\begin{aligned} M_l = \frac{I_{HI}f^2}{e^2(e^2+f^2)} + \frac{(I_{LL'}+I_{ML'})f^2}{4e^2g^2} + I_{NN'}\left(\frac{2fh-fi}{2egi}\right)^2 \\ + 2m_{\text{in}}' + \frac{2I_{PQ}f^2h^2 + f^2h^2j^2m_{\text{out}}'}{e^2g^2i^2} + \frac{(m_{L'N'N}f^2h^2) + m_{ILJ}f^2g^2}{4e^2g^2} \end{aligned} \tag{8.17}$$

$$K_l = K^l_{\text{in}}. \tag{8.18}$$

Indeed, to increase the natural frequency of the mechanism, the equivalent stiffness of the structure can be increased, or its equivalent mass can be reduced. This relationship is evident from the natural frequency equation.

8.4 Structure Optimization and FEA Verification

8.4.1 Structure Optimization

To achieve a large displacement magnification, the structural parameters of the gripper must be appropriately adjusted. The goal of structural optimization is to determine the optimal design parameters under given constraints, such as minimizing weight or maximizing stiffness. In this work, the optimization objective is to maximize the end strokes of both sides of the gripper. The geometric parameters referenced in Eqs. (8.1) and (8.2) are denoted in the pseudo-rigid-body model (PRBM) as illustrated in Fig. 8.5. These parameters depend on the dimensions of the associated flexure hinges and rigid linkages. The actual structural parameters, which govern these variables, are marked in Fig. 8.3b.

The parameter set $[b_1, b_2, b_3, b_4]$ influences the magnification of the bridge mechanisms, while the set $[l_1, l_2, l_3, l_4]$ affects the deformability of the lever mechanisms. The final values of these physical parameters were determined using the direct optimization module in ANSYS software. The Multi-Objective Genetic Algorithm (MOGA) was employed, and the optimization process converged after 2492 evaluations. The optimized structural parameters of the gripper are summarized in Table 8.2.

In addition to these key variables, other architectural parameters were also defined. For instance, the thickness of the rectangular flexure hinges was set to 0.5 mm due to manufacturing constraints. The jaw of the right gripper features a V-shaped groove with an opening angle of 135° to accommodate tiny targets such as optical fibers. This design ensures effective rolling contact between the fiber and the gripper, as derived in previous work [4]. The overall dimensions of the gripper are approximately 67.45 mm × 49 mm × 5 mm, resulting in a compact and efficient structure.

Table 8.2 Optimum values of the structural design parameters

Design variable	Optimum value	Design variable	Optimum value
b_1 (mm)	1.7	l_1 (mm)	5.5
b_2 (mm)	0.9	l_2 (mm)	10.8
b_3 (mm)	1.5	l_3 (mm)	4
b_4 (mm)	2.3	l_4 (mm)	6.8

8.4.2 FEA Simulation Results

The performance of the designed gripper was verified through finite element analysis (FEA) simulations conducted using ANSYS Workbench software. The piezoelectric actuator (PZT) adopted in the design provides a nominal displacement of 30 μm under a driving force of 900 N. In the simulation, the PZT was modeled as a spring with equivalent stiffness to approximate the physical system. The mounting holes on the gripper were assigned as fixed supports.

To ensure optimal deformation performance of the flexure hinges, aluminum 7075 was selected as the material due to its high yield-strength-to-elastic-modulus ratio. The material properties include a density of 2810 kg/m^3, Young's modulus of 71.7 GPa, Poisson's ratio of 0.33, and a yield strength of 503 MPa.

The simulation results for the clamping and rubbing strokes are shown in Fig. 8.6a. The maximum deformations occur at the tip ends of the left and right sides of the gripper. Under an input displacement of 30 μm, the clamping stroke reaches 357.94 μm,

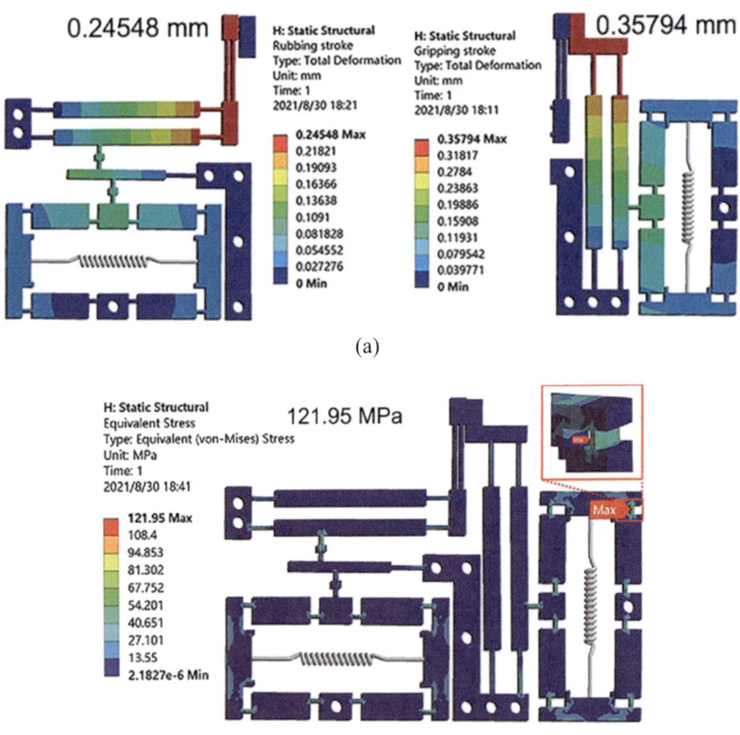

Fig. 8.6 Finite element simulation results of the microgripper. **a** Output displacement analysis showing rubbing stroke and clamping stroke, and **b** equivalent stress distribution under maximum actuation

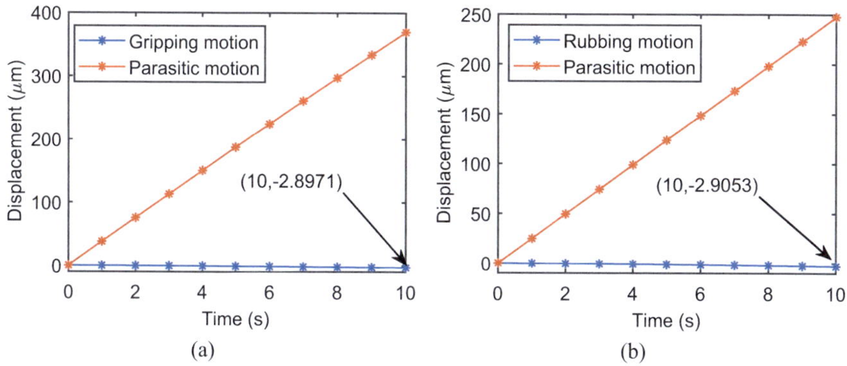

Fig. 8.7 Simulation results of output displacements. **a** Gripping displacement and parasitic displacement versus time for the right-side mechanism, and **b** rubbing displacement and parasitic displacement versus time for the left-side mechanism

and the rubbing stroke reaches 245.48 μm, corresponding to displacement amplification ratios of 11.52 and 8.18, respectively.

The maximum equivalent stress during deformation occurs at the flexure hinge of the right-side bridge mechanism, as shown in Fig. 8.6b. The peak stress is 121.95 MPa, well below the material's yield strength of 503 MPa. The resulting safety factor at maximum driving force is 4.69, confirming that the gripper operates within safe structural limits. Additionally, the working and parasitic displacements of both mechanisms were simulated (Fig. 8.7). The parasitic displacements remain below 3 μm throughout the working stroke, ensuring linear output motion at the end-effector.

Modal analysis was performed to evaluate the natural frequencies of the microgripper. The first three mode shapes are illustrated in Fig. 8.8:

- The first mode involves translational motion of the right gripping arm along its working direction.
- The second mode corresponds to vertical translational motion of the left gripper, aligned with its functional mode.
- The third mode represents in-plane rotation of the left bridge mechanism.

The natural frequencies for these modes are 459.58 Hz, 634.16 Hz, and 952.66 Hz, respectively. These results indicate a high operational bandwidth, suitable for dynamic and precision applications.

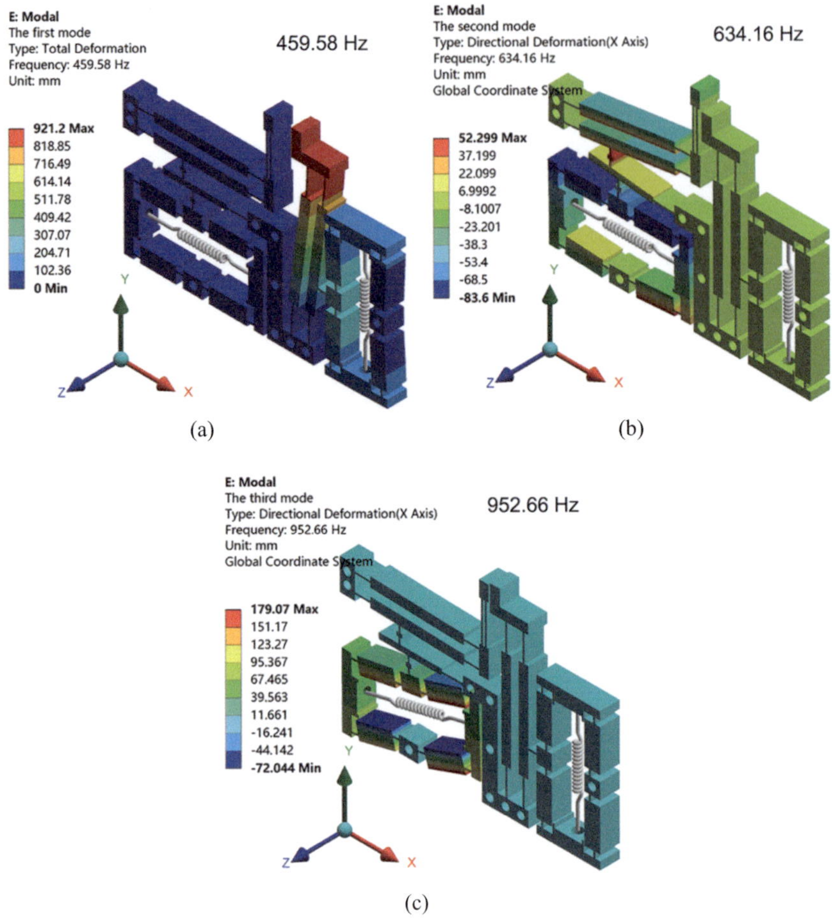

Fig. 8.8 Finite element simulation results of the first three modal shapes of the microgripper. **a** First mode shape (459.58 Hz), **b** second mode shape (634.16 Hz), and **c** third mode shape (952.66 Hz)

8.5 Experimental Investigations and Discussion

8.5.1 Experimental Setup

The performance of the microgripper, having been enhanced and preliminarily validated through optimization and simulation, was further evaluated via physical experimentation. The microgripper was fabricated from aluminum alloy 7075 using wire electro-discharge machining (WEDM) technology. The assembled device is mounted on a connecting base, as illustrated in Fig. 8.9a. The entire experimental setup is placed on a vibration isolation table (model: RS2000, Newport Corp.) to minimize environmental disturbances, as shown in Fig. 8.9c.

Fig. 8.9 Experimental setup for microgripper performance evaluation. **a** Fabricated monolithic prototype, **b** block diagram of the closed-loop control system architecture, and **c** complete experimental test configuration

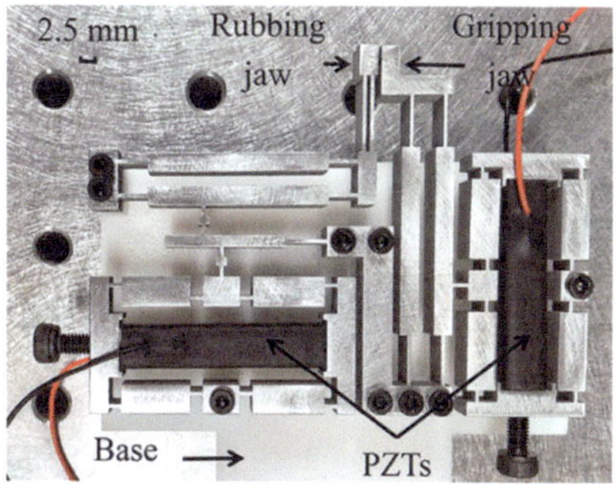

(a)

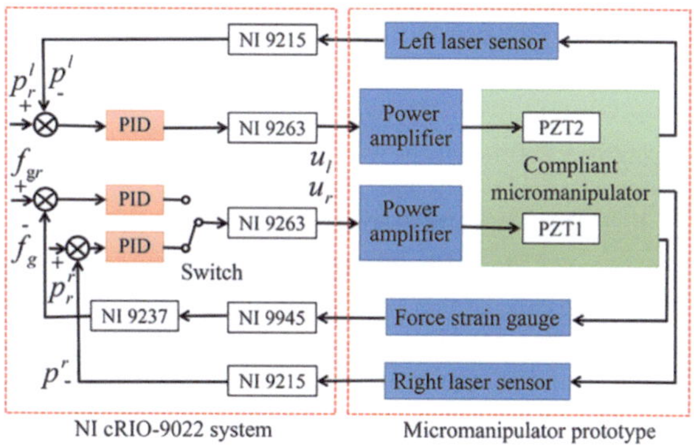

(b)

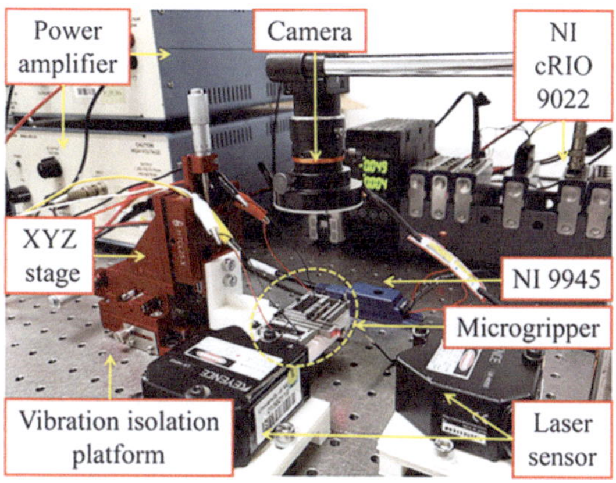

(c)

Key components of the experimental system include:

- The designed microgripper,
- An XYZ micropositioning stage (model: HTCL25-X, Newport Corp.),
- Two voltage amplifiers (model: EPA-104, PiezoSystems, Inc.),
- Two laser displacement sensors (model: LK-H055, Keyence Corp.),
- An NI cRIO-9022 real-time control system,
- An industrial camera (model: MV-CE013-80UM, Hikvision Corp.),
- A personal computer.

Two piezoelectric actuators (model: RP150/5×5/30) provide driving displacements for the clamping and rotating directions, respectively. A calibrated strain gauge (model: BFH 350-3AA, Chengtec Electronics Ltd.) with a sensitivity factor of 2.0 is attached to the double flexible beam on the left-side microgripper to monitor force feedback during micromanipulation.

To comprehensively assess the microgripper's performance, open-loop, closed-loop, and micromanipulation experiments were conducted. A proportional-integral-derivative (PID) controller was implemented in the closed-loop experiments to evaluate the gripper's controllability. The control block diagram of the closed-loop system is depicted in Fig. 8.9b, integrating the microgripper prototype with the NI cRIO-9022 control system. Three PID controllers were utilized to regulate gripping displacement, rubbing displacement, and clamping force, respectively. The control parameters for each PID controller were tuned empirically using a trial-and-error approach.

8.5.2 Open-Loop Experimental Results

The open-loop experiments were conducted to characterize the working stroke, displacement magnification, and resonant frequency of the microgripper. The output displacement was measured using a laser sensor (with sub-micrometer resolution over a 20 mm range). Due to the challenge of directly measuring the input displacement of the bridge-type mechanism–which features two input terminals–the nominal output displacement of the actuator was used as a proxy. This substitution, being slightly larger than the actual structural input, partially accounts for the observed reduction in displacement magnification.

Figure 8.10a, b present scatter plots and fitted lines relating the output stroke to the input displacement for both sides of the gripper. The slopes of these fitted lines indicate displacement amplification ratios of 8.63 and 7.80 for the gripping and rubbing mechanisms, respectively. The output displacements exhibit a strong linear relationship with the input displacements under discrete voltage drives.

Hysteresis characteristics were also examined under a 0.5 Hz triangular voltage excitation. Figure 8.10c, d show the hysteresis loops for the gripping and rubbing displacements at the end effector. The total clamping stroke reached 251.2 μm, while the maximum rubbing stroke was 225.0 μm.

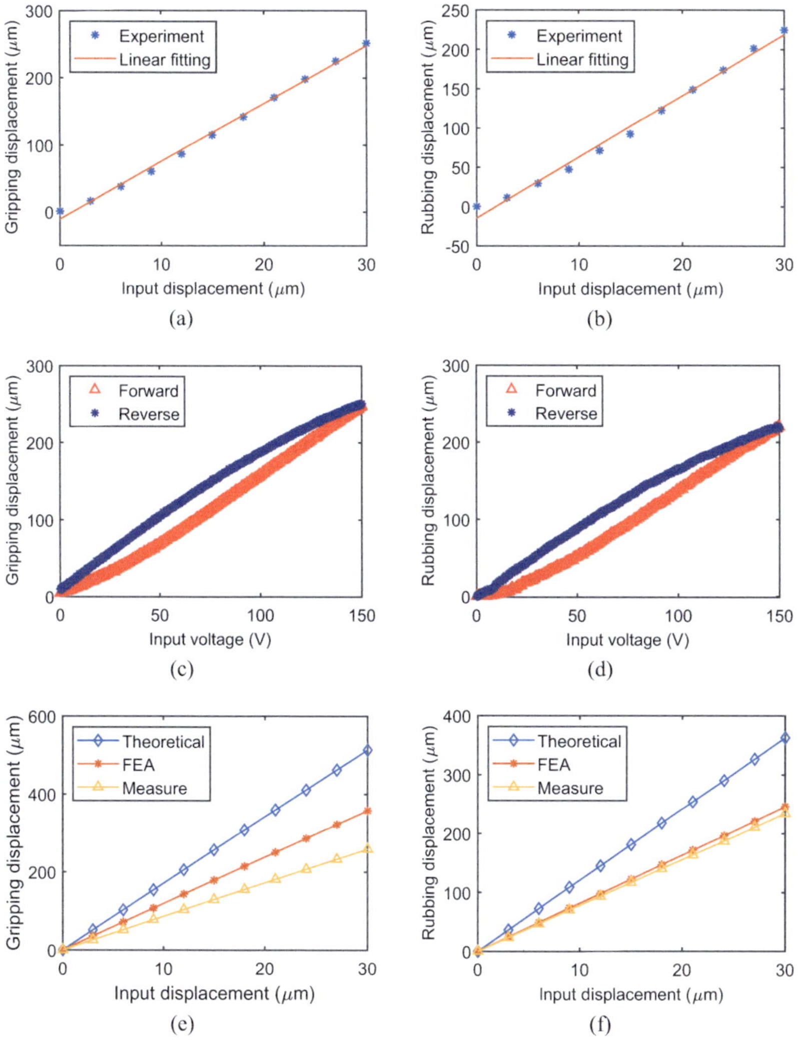

Fig. 8.10 Open-loop experimental characterization of output displacements. **a** Gripping displacement versus input displacement, **b** rubbing displacement versus input displacement, **c** hysteresis behavior in gripping motion, **d** hysteresis behavior in rubbing motion, **e** comparative analysis of gripping displacement amplification ratios, and **f** comparative analysis of rubbing displacement amplification ratios

Figure 8.10e, f compare the linear relationships between output and input displacements from theoretical analysis, FEA simulations, and experimental results. The FEA simulations align closely with experimental data, whereas the theoretical analysis overestimates the amplification ratios (17.12 and 12.09). Several factors contribute to this discrepancy:

- The theoretical model neglects the thickness of the gripper.
- Deformation modes beyond flexure hinge bending (e.g., axial drift) are not accounted for.
- Theoretical rigid linkages exhibit practical deformations.
- Idealized assumptions–such as perfectly horizontal motion of linkage BCE–deviate from real behavior.

Additional sources of error include simulation approximations and manufacturing tolerances. These results underscore the importance of integrated modeling and experimental validation in refining compliant mechanism design.

To calibrate the force–electric conversion coefficient of the strain gauge sensor, multiple weights of known mass were suspended to apply precise forces. The gravitational force of each weight was premeasured using a precision balance (model: HZK-FA110, Huazhi Scientific Instrument Co. Ltd.). The experimental setup and calibration results are shown in Fig. 8.11a, b, respectively. The calibration yielded a conversion coefficient of 6761.8 mN/mV between the strain gauge output voltage and the clamping force. The resolution of the strain-gauge sensor was determined to be approximately 8 mN under zero input conditions.

The dynamic performance of the gripper was evaluated through frequency-sweeping experiments. Given the distinct resonant modes associated with the left and right sides of the microgripper, separate frequency sweeps were conducted for each side. As illustrated in Fig. 8.11c, d, the first two resonant frequencies were identified as 350.63 Hz and 603.37 Hz, corresponding to the operational modes of the right and left sides, respectively.

These experimental results align reasonably with the theoretical predictions (539 and 762 Hz) and FEA simulations (459 and 634 Hz). The slight reduction in experimentally measured frequencies is attributed to additional mass from actuators, preload bolts, and other components not fully accounted for in the idealized models. Table 8.3 summarizes the displacement amplification ratios and resonant frequencies obtained from theoretical, simulation, and experimental approaches, demonstrating consistent trends across all three methods.

8.5.3 Closed-Loop Experimental Results

To perform dexterous and precise micro/nano operations–such as clamping, rubbing, and releasing–the microgripper requires accurate displacement output and controllable force. A PI controller (i.e., a PID controller without derivative action) was

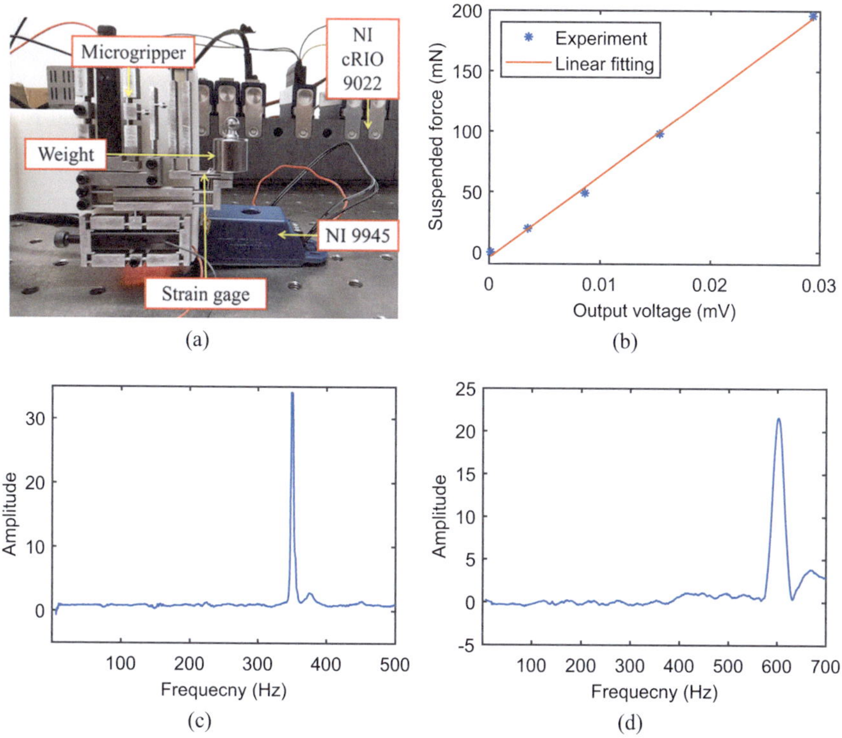

Fig. 8.11 Open-loop experimental characterization of output forces. **a** Force calibration setup for clamping force measurement, **b** results of clamping force calibration experiment, **c** frequency response of the right-side gripper mechanism, and **d** frequency response of the left-side gripper mechanism

Table 8.3 Performance comparison of displacement amplification ratios and resonant frequencies between the right and left sides of the microgripper

Parameters	Theory model	Simulation	Experiment
R_r	17.12	11.52	8.63
R_l	12.09	8.18	7.8
f_r (Hz)	539	459	350
f_l (Hz)	762	634	603

implemented to regulate both output displacement and clamping force. The control parameters were tuned empirically to achieve rapid response and minimal steady-state error.

Step signals with an amplitude of 50 μm were applied to both the left and right grippers to evaluate system response. The results are shown in Fig. 8.12a, b:

- For the right-side gripper, the overshoot was 2.97%, settling time 140 ms, and steady-state error ±1.3 μm.
- For the left-side gripper, the overshoot was 3.57%, settling time 60 ms, and steady-state error ±1.5 μm.

Motion resolution was assessed using consecutive step signals. As depicted in Fig. 8.12c, d, the displacement errors (relative to the reference signals) follow a normal distribution. Within a 95% confidence interval:

- The gripping motion resolution is 1.12 μm,
- The rubbing motion resolution is 1.26 μm.

Step response and resolution tests were also conducted for the clamping force sensor. Under a 50 mN step input (Fig. 8.12e, f):

- The overshoot was 6%, settling time 67 ms, and steady-state error ±2.5 mN.
- The clamping force resolution is approximately 3 mN.

These results demonstrate the microgripper's capability for high-precision, closed-loop control in dynamic micromanipulation tasks.

8.5.4 Grasping and Rubbing Experimental Results

To validate the performance of the designed microgripper, clamping and rubbing experiments were conducted on tiny objects. A 200 μm-diameter optical fiber was successfully clamped in free space, as shown in Fig. 8.13a. Figure 8.14a shows the displacement and force sensor signals during the clamping process. As the actuation voltage increased, the gripper displacement rose rapidly initially and then slowed, while the gripping force remained constant before increasing sharply and linearly. The reverse behavior was observed when decreasing the voltage. The maximum clamping force measured was 484 mN at 50 V. The experiment was repeated three times with consistent results, confirming the microgripper's repeatability and reliability.

A custom-made small flag attached to a 200 μm-diameter metal wire was used to evaluate the rotation performance. The initial and final states are shown in Fig. 8.13b, c. The flag rotated over 45° as measured by image processing. The experimental rotation angle was smaller than theoretically predicted due to three main factors: energy dissipation to overcome static friction, slippage during rubbing that reduced effective stroke, and minor friction between the flag and gripper surface.

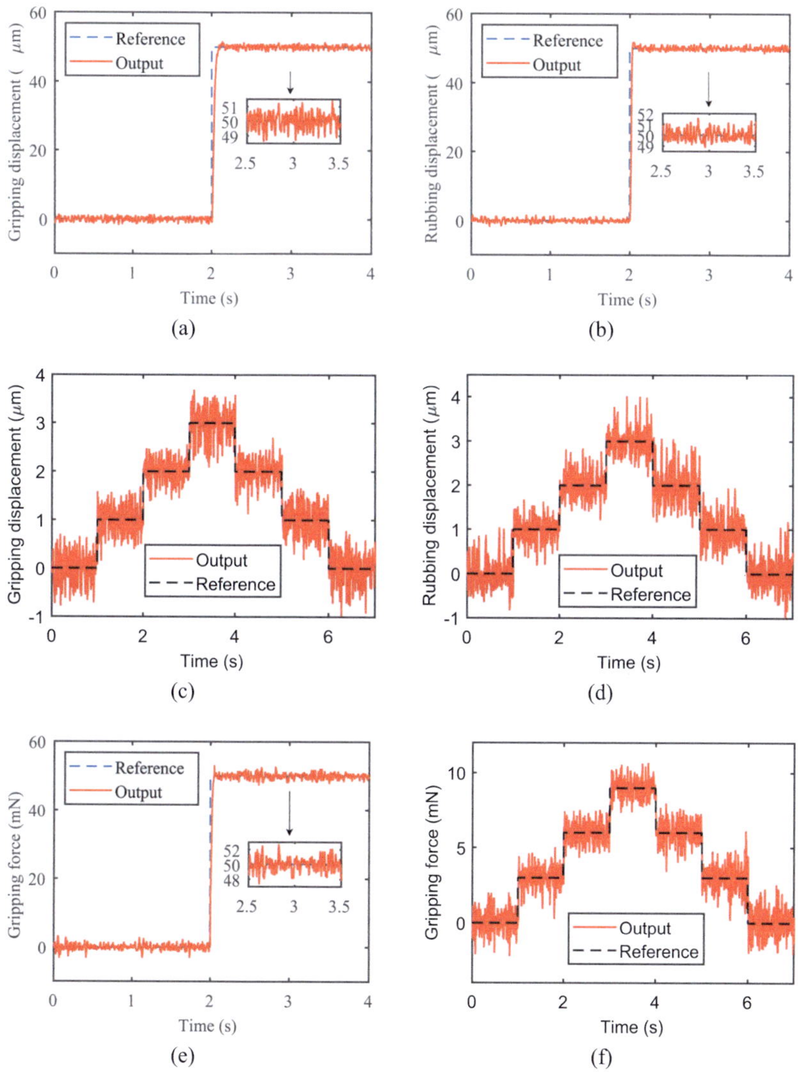

Fig. 8.12 Closed-loop experimental performance evaluation. **a** Step response characterization of gripping displacement, **b** step response characterization of rubbing displacement, **c** resolution analysis of gripping displacement, **d** resolution analysis of rubbing displacement, **e** step response characterization of gripping force, and **f** resolution analysis of gripping force

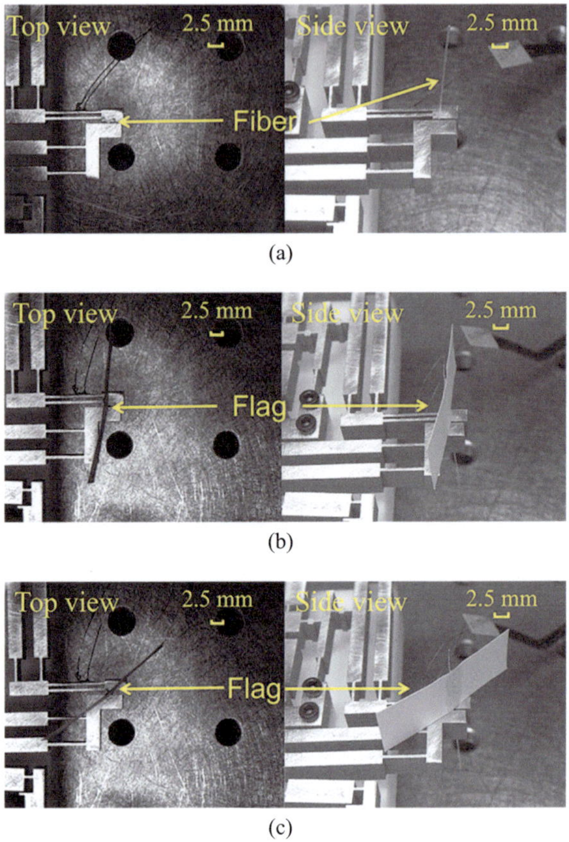

Fig. 8.13 Experimental demonstration of micromanipulation capabilities. **a** Optical fiber gripping operation, **b** initial position of calibration flag before rotation, and **c** final position of calibration flag after rotation

Figure 8.14b shows the rubbing displacement and clamping force signals during rotation. Three repeated tests confirmed process repeatability. The first rubbing stroke was notably larger than subsequent strokes, with a corresponding force slip peak, likely due to initial jaw posture adjustment and static friction overcoming. Subsequent strokes were smaller than no-load values due to opposing friction torque. Force sensor output showed periodic fluctuations within safe limits [4]. These results demonstrate the microgripper's effectiveness in clamping and rotating wire-type targets, confirming its potential for optical fiber alignment applications.

8.5.5 Further Discussion

Table 8.4 presents a performance comparison of typical 2-DOF compliant microgrippers, where the acronyms IS, GS, RS, TS, DR, and FR denote input stroke, gripping stroke, rubbing stroke, total stroke, displacement resolution, and force resolution,

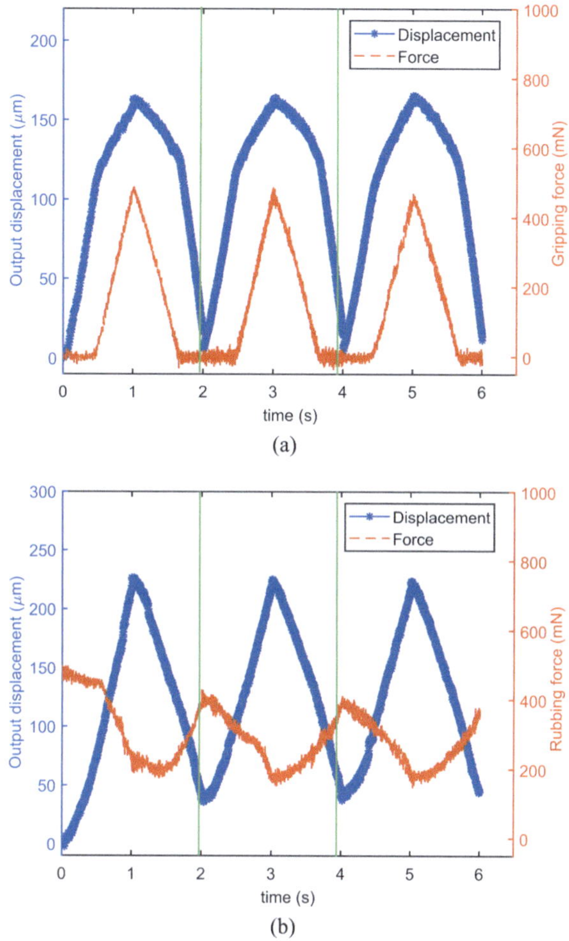

Fig. 8.14 Experimental results of micromanipulation performance. **a** Gripping displacement and force profiles during fiber gripping experiment, and **b** rubbing displacement and gripping force profiles during flag rotation experiment

respectively. Compared to prior designs, the proposed microgripper demonstrates advantages in large working strokes for both grasping and rotating operations, high first natural frequency, and high gripping force resolution. Additionally, the monolithic integration of the actuator and compliant mechanism contributes to a compact structure and reduced operational space occupancy.

The current displacement resolution of 1.2 μm requires further improvement. Enhancements can be pursued through advancements in actuator performance, sensor technology, and control strategies. While the piezoelectric ceramic actuator remains a suitable choice, upgrading the sensor system–for instance, replacing the triangulation-based laser sensor with a higher-resolution laser interferometer–could significantly improve resolution. Furthermore, advanced control techniques such as sliding mode control should be investigated to compensate for hysteresis in the piezo-

Table 8.4 Comparative performance analysis of 2-DOF compliant microgrippers

Reference	IS (μm)	GS (μm)	RS (μm)	TS (μm)	1st Freq. (Hz)	2nd Freq. (Hz)	DR (μm)	FR (mN)	Size (mm × mm)
[25]	15	36.45	34.2	70.65	236.29	302.37	–	–	>70 × 70 (with PZT)
[4]	20	190	85	275	221	1003	–	–	≈58 × 34
[7]	30	240.53	133.34	373.87	–	–	–	–	≈55 × 50
[3]	20	163.3	83.1	246.4	326.8	599.5	0.40	4.72	≈50 × 35
[32]	20	108	223.4	331.4	–	–	0.5	–	≈62 × 50 (with PZT)
[24]	20	222.4	92.8	315.2	–	–	0.15	4	≈68 × 58 (with PZT)
This work	30	251.2	225.0	476.2	350.63	603.37	1.2	3	67.45 × 49 (with PZT)

electric actuator and enhance positioning accuracy. Given the precision demands of micro/nano-manipulation, these improvements will be prioritized in future work.

8.6 Conclusion

This chapter has presented the mechanism design, theoretical analysis, structural optimization, simulation, and experimental validation of a biaxial microgripper. A cascaded configuration combining a flexible single-output bridge mechanism and a parallelogram mechanism was employed to achieve large displacement output with minimal parasitic motion. Theoretical models were developed to characterize the displacement amplification ratio, input stiffness, and natural frequency for both sides of the microgripper. Key structural parameters influencing performance were optimized and verified through finite element analysis.

A prototype was fabricated to experimentally evaluate the microgripper's capabilities. Results demonstrated strong agreement between theoretical predictions, simulations, and experimental measurements. The microgripper successfully clamped a 200 μm optical fiber and performed precise rubbing operations on a miniature flag, confirming its dexterous and stable bilateral functionality. Compared to existing designs, the proposed microgripper offers simultaneously large working strokes and high resonant frequencies, enhancing both the reliability and speed of micromanipulation tasks. Future work will focus on implementing advanced control strategies to further improve precision and conducting targeted experiments for optical fiber alignment applications.

References

1. Aalto, T.T., Harjanne, M., Kapulainen, M.: Method for the rotational alignment of polarization-maintaining optical fibers and waveguides. Opt. Eng. **42**(10), 2861–2867 (2003)
2. Chen, F., Gao, Y., Dong, W., Du, Z.: Design and control of a passive compliant piezo-actuated micro-gripper with hybrid flexure hinges. IEEE Trans. Industr. Electron. **68**(11), 11168–11177 (2021)
3. Chen, W., Qu, J., Chen, W., Zhang, J.: A compliant dual-axis gripper with integrated position and force sensing. Mechatronics **47**, 105–115 (2017)
4. Chen, W., Shi, X., Chen, W., Zhang, J.: A two degree of freedom micro-gripper with grasping and rotating functions for optical fibers assembling. Rev. Sci. Instrum. **84**(11), 115111 (2013)
5. Chen, W.J., Lin, W.: Fiber assembly of mems optical switches with u-groove channels. IEEE Trans. Autom. Sci. Eng. **5**(2), 207–215 (2008)
6. Das, T.K., Shirinzadeh, B., Al-Jodah, A., Ghafarian, M., Pinskier, J.: Computational parametric analysis and experimental investigations of a compact flexure-based microgripper. Precis. Eng. **66**, 363–373 (2020)
7. Dsouza, R.D., Navin, K.P., Theodoridis, T., Sharma, P.: Design, fabrication and testing of a 2 DOF compliant flexural microgripper. Microsyst. Technol. **24**(9), 3867–3883 (2018)
8. Grossard, M., Rotinat-Libersa, C., Chaillet, N., Boukallel, M.: Mechanical and control-oriented design of a monolithic piezoelectric microgripper using a new topological optimization method. IEEE/ASME Trans. Mechatron. **14**(1), 32–45 (2009)
9. Jia, Y., Zhang, X., Xu, Q.: Design and optimization of a dual-axis PZT actuation gripper. In: Proceedings of 2014 IEEE International Conference on Robotics and Biomimetics (ROBIO 2014), pp. 321–325. Bali, Indonesia (2014)
10. Komati, B., Clévy, C., Lutz, P.: High bandwidth microgripper with integrated force sensors and position estimation for the grasp of multistiffness microcomponents. IEEE/ASME Trans. Mechatron. **21**(4), 2039–2049 (2016)
11. Liu, S., Li, Y.F., Wang, X.W.: A novel dual-probe-based micrograsping system allowing dexterous 3-d orientation adjustment. IEEE Trans. Autom. Sci. Eng. **17**(4), 2048–2062 (2020). https://doi.org/10.1109/TASE.2020.2986992
12. Long, Z., Zhang, J., Liu, Y., Han, C., Li, Y., Li, Z.: Dynamics modeling and residual vibration control of a piezoelectric gripper during wire bonding. IEEE Trans. Compon. Packag. Manuf. Technol. **7**(12), 2045–2056 (2017)
13. Lu, K., Zhang, J., Chen, W., Jiang, J., Chen, W.: A monolithic microgripper with high efficiency and high accuracy for optical fiber assembly. In: Proceedings of 2014 9th IEEE Conference on Industrial Electronics and Applications, pp. 1942–1947. Hangzhou, China (2014)
14. Lyu, Z., Xu, Q.: Recent design and development of piezoelectric-actuated compliant micro-grippers: a review. Sens. Actuators, A **331**, 113002 (2021)
15. Muntwyler, S., Kratochvil, B.E., Beyeler, F., Nelson, B.J.: Monolithically integrated two-axis microtensile tester for the mechanical characterization of microscopic samples. J. Microelectromech. Syst. **19**(5), 1223–1233 (2010)
16. Qian, J., Yan, P., Liu, P.: Position/force modeling and analysis of a piezo-driven compliant micro-gripper considering the dynamic impacts of gripping objects. Smart Mater. Struct. **30**, 075036 (2021)
17. Shi, C., Dong, X., Yang, Z.: A microgripper with a large magnification ratio and high structural stiffness based on a flexure-enabled mechanism. IEEE/ASME Trans. Mechatron. **26**(6), 3076–3086 (2021)
18. Shi, Q., Yu, Z., Wang, H., Sun, T., Huang, Q., Fukuda, T.: Development of a highly compact microgripper capable of online calibration for multisized microobject manipulation. IEEE Trans. Nanotechnol. **17**(4), 657–661 (2018)
19. Shi, X., Chen, W., Zhang, J., Chen, W.: Design, modeling, and simulation of a 2-DOF micro-gripper for grasping and rotating of optical fibers. In: Proceedings of 2013 IEEE/ASME International Conference on Advanced Intelligent Mechatronics, pp. 1597–1602. Wollongong, NSW, Australia (2013)

20. Wang, D., Yang, Q., Dong, H.: A monolithic compliant piezoelectric-driven microgripper: design, modeling, and testing. IEEE/ASME Trans. Mechatron. **18**(1), 138–147 (2011)
21. Wang, F., Liang, C., Tian, Y., Zhao, X., Zhang, D.: Design of a piezoelectric-actuated microgripper with a three-stage flexure-based amplification. IEEE/ASME Trans. Mechatron. **20**(5), 2205–2213 (2014)
22. Wang, F., Liang, C., Tian, Y., Zhao, X., Zhang, D.: Design and control of a compliant microgripper with a large amplification ratio for high-speed micro manipulation. IEEE/ASME Trans. Mechatron. **21**(3), 1262–1271 (2016)
23. Wang, F., Shi, B., Huo, Z., Tian, Y., Zhang, D.: Design and control of a spatial micromanipulator inspired by deployable structure. IEEE Trans. Industr. Electron. **69**(1), 971–979 (2022)
24. Wang, F., Shi, B., Tian, Y., Huo, Z., Zhao, X., Zhang, D.: Design of a novel dual-axis micromanipulator with an asymmetric compliant structure. IEEE/ASME Trans. Mechatron. **24**(2), 656–665 (2019)
25. Xu, Q.: Mechanism design and analysis of a novel 2-DOF compliant modular microgripper. In: Proceedings of 2012 7th IEEE Conference on Industrial Electronics and Applications (ICIEA), pp. 1966–1971. Singapore (2012)
26. Xu, Q.: Design and smooth position/force switching control of a miniature gripper for automated microhandling. IEEE Trans. Industr. Inf. **10**(2), 1023–1032 (2013)
27. Xu, Q.: Design, fabrication, and testing of an mems microgripper with dual-axis force sensor. IEEE Sens. J. **15**(10), 6017–6026 (2015)
28. Xu, Q.: Design and development of a novel compliant gripper with integrated position and grasping/interaction force sensing. IEEE Trans. Autom. Sci. Eng. **14**(3), 1415–1428 (2017)
29. Yan, J., Miao, L., Huang, T., Che, S., Shu, X.: Development of method for polarization alignment of panda polarization maintaining fiber. Opt. Fiber Technol. **53**, 101999 (2019)
30. Yang, Y.L., Wei, Y.D., Lou, J.Q., Fu, L., Fang, S.: Design and control of a multi-DOF micromanipulator dedicated to multiscale micromanipulation. Smart Mater. Struct. **26**(11), 115016 (2017)
31. Ye, T., Ling, J., Yao, T., Xiao, X.: Design of a 2-DOF constant force compliant microgripper for optical switch assembly. In: Proceedings of the 46th Annual Conference of the IEEE Industrial Electronics Society (IECON 2020), pp. 4403–4408. Singapore (2020)
32. Zhang, J., Lu, K., Chen, W., Jiang, J., Chen, W.: Monolithically integrated two-axis microgripper for polarization maintaining in optical fiber assembly. Rev. Sci. Instrum. **86**(2), 025105 (2015)
33. Zhang, R., Chu, J., Wang, H., Chen, Z.: A multipurpose electrothermal microgripper for biological micro-manipulation. Microsyst. Technol. **19**(1), 89–97 (2013)
34. Zhang, Z., Yu, Y., Song, P., Zhang, Y., Tian, D., Zhang, H., Wei, H., Cui, M., Si, G., Zhang, X.: Automated manipulation of zebrafish embryos using an electrothermal microgripper. Microsyst. Technol. **26**(6), 1823–1834 (2020)
35. Zubir, M.N.M., Shirinzadeh, B., Tian, Y.: A new design of piezoelectric driven compliant-based microgripper for micromanipulation. Mech. Mach. Theory **44**(12), 2248–2264 (2009)

Chapter 9
Design of a Large-Workspace XY Compliant Manipulator with Triple-Stage Parallelogram Flexure

Abstract This chapter presents an XY compliant parallel manipulator designed for large workspace and minimal motion coupling, achieved through a planar multistage parallelogram mechanism. The design utilizes a triple-stage parallelogram flexure for motion transmission and decoupling, with fixed and moving ends employing planar and spatial connections, respectively. This configuration enables the construction of diverse multistage parallelogram flexures. The driving stiffness and resonant frequency are derived using theoretical statics and Lagrangian dynamics. Analytical model accuracy is verified via finite element analysis, and performance is compared across three manipulator variants. A prototype was fabricated and tested, demonstrating a workspace of 1.51 cm $\times$ 1.48 cm and cross-axis coupling ratios below 1.2%. A volume ratio metric is introduced to evaluate compactness against existing designs. The centimeter-scale workspace enhances the suitability of this manipulator for broader industrial applications.

Keywords Compliant mechanism · XY compliant manipulator · Microgripper · Large stroke · Leaf flexure

9.1 Introduction

A compliant mechanism is a type of mechanism that utilizes elastic deformation to transmit or convert motion and force. Compared to conventional rigid-body mechanisms (e.g., gears and lead screws), compliant mechanisms eliminate backlash, friction, wear, and the need for lubrication [1, 4, 6, 15, 20]. As a typical application, the XY compliant manipulator has been extensively employed in precision manufacturing, micro/nano-manipulation, semiconductor packaging, and biomedical engineering [3, 7, 24, 32, 35]. For example, Moheimani et al. designed a flexure-guided nanopositioner to observe dynamic biological processes using high-speed atomic force microscopy [23]. Zhang et al. developed a millimeter-travel compliant nanomanipulator to eliminate image stitching errors in electron beam lithography [18]. The working stroke is a critical performance index for XY compliant manipulators, gov-

Q. Xu, *Piezoelectric-Driven Flexible Robotic Micrograppers*,
https://doi.org/10.1007/978-981-95-4685-5_9

erning their application fields. Improving the motion stroke requires consideration of two key aspects: actuator selection and structural design.

In the literature, typical actuators for precise motion drives include shape memory alloys, giant magnetostrictive actuators, normal-stressed electromagnetic actuators, piezoelectric actuators (PEAs), and voice coil motors (VCMs) [26, 28, 34]. Among these, PEAs and VCMs are the most widely used [2, 9, 14, 25]. Although PEAs are compact and have a fast response, they suffer from disadvantages such as high driving voltage and limited stroke. Conversely, VCMs are more desirable for motion travels at the millimeter level or above due to their long stroke, high precision, and good stability [22, 30]. For instance, Lee et al. fabricated an XY nanopositioner using stereolithography and investigated its motion cross-coupling, achieving a range of ± 1 mm per axis with cross-coupling of 3.4 and 8.1% [14]. Herpe et al. developed a VCM-actuated XY manipulator where a symmetrical flexible beam serves as the guiding mechanism and a parallelogram structure acts as the decoupling mechanism [9]. This stage, designed for a hybrid miniaturized product assembly system, provides a travel range of ± 2.3 mm with a maximum cross-coupling error of 2.5% per axis. Moreover, Liu et al. adopted a VCM to drive an XY nanomanipulator, achieving a 2 mm $\times$ 2 mm workspace with active suppression of parasitic rotation [17]. Jadhav et al. [11] reported a modular XY positioning stage with a translational motion range of ± 5 mm. For application scenarios such as femtosecond laser direct writing, automated microinjection, and soft contact lithography, large strokes (millimeter or centimeter level) are essential [3, 5, 10]. For example, microinjection of zebrafish embryos requires a precise XY positioning stage with a stroke exceeding 1 cm [10]. Therefore, to achieve large-stroke precision motion, this work employs a VCM to drive the XY compliant manipulator.

Regarding structural design, existing XY compliant manipulators can be classified into planar and spatial architectures. Planar manipulators consist of a single-layer flexure hinge with the transmission mechanism primarily distributed in-plane. In contrast, spatial XY manipulators typically feature a multi-layer mechanism in the vertical direction, addressing motion transmission and decoupling through a layered design [21, 33]. The latter can achieve a larger stroke within a compact configuration than the former. For example, Xu introduced an XY precision positioning system with a two-layer flexible mechanism that delivers an 11 mm $\times$ 11 mm workspace [29], where each layer is responsible for motion transmission and decoupling in one direction. Zhu et al. adopted a layered design and mirror symmetry to optimize the XY compliant parallel manipulator proposed by Awtar et al., achieving a large stroke output (5.19 mm $\times$ 5.19 mm) and low parasitic coupling motion [36]. However, layered spatial XY manipulators have drawbacks. The layered architecture increases production cost, introduces assembly errors, and raises the system's fault rate due to the increased number of compliant parts. Consequently, this chapter aims to achieve a large displacement output (>1 cm) using a single-layer architecture.

The selection and arrangement of compliant are also crucial aspects of structural design. Flexible hinges for motion transmission in XY manipulators are primarily of two types: notch hinges and leaf hinges [13, 16]. Although notch hinges offer excellent rotational accuracy, their performance is limited by stress concentration. Leaf hinges, alternatively, provide high compliance in the working direction and high stiffness in off-axis directions, which is critical for motion transmission and decoupling. Current transmission structures based on leaf hinges are predominantly single-stage or double-stage. Multi-stage series leaf hinges offer advantages in withstanding large deformation and distributing stress. However, the application of cascaded leaf hinges beyond two stages is rarely reported, and it remains uncertain whether a multi-stage parallelogram flexure can achieve the desired motion transmission and decoupling.

To address this, this chapter presents a concatenation method for a multi-stage compliant parallelogram mechanism. The movable ends of the multi-stage parallelogram flexure are connected spatially, while the fixed ends are connected in-plane. Specifically, a triple-stage parallelogram flexure (TSPF) is adopted as the basic unit for motion transmission to construct an XY compliant manipulator capable of two-dimensional large-stroke (>1 cm) motion. Compared to a traditional manipulator based on a double-stage parallelogram flexure (DSPF), this design expands the movement stroke without increasing the planar footprint. Except for the spatial connection bars, the manipulator's primary deformation mechanisms are designed and fabricated in a single layer. This offers advantages of low vertical occupancy, reduced manufacturing cost, and fewer assembly steps compared to hierarchically designed XY manipulators. The performance of the designed XY compliant manipulator is evaluated and verified through theoretical analysis, finite element analysis (FEA) simulation, and experimental tests.

9.2 Structural Improvement and Mechanism Design

9.2.1 Compound Parallelogram Flexures

To achieve larger strokes while maintaining precision, most existing XY manipulators employ a combination of flexible guiding mechanisms. Among these, the basic parallelogram mechanism (BPM)–composed of leaf flexures–and its various evolved forms are widely adopted due to their large stroke, low stiffness, ease of fabrication, and effective stress dispersion. Figure 9.1a depicts the basic parallelogram flexure (BPF) alongside its double-stage (DSPF) and triple-stage (TSPF) compound forms. Figure 9.1b compares the performance of these three mechanisms in terms of stiffness and strength. The mechanism's stiffness decreases as the number of compound stages increases, and the equivalent stress under an identical motion stroke is also reduced.

However, compound parallelogram mechanisms with three or more stages are rarely discussed in the literature. This is primarily because their moving platform is concealed within the center of the structure, complicating actuation and integration.

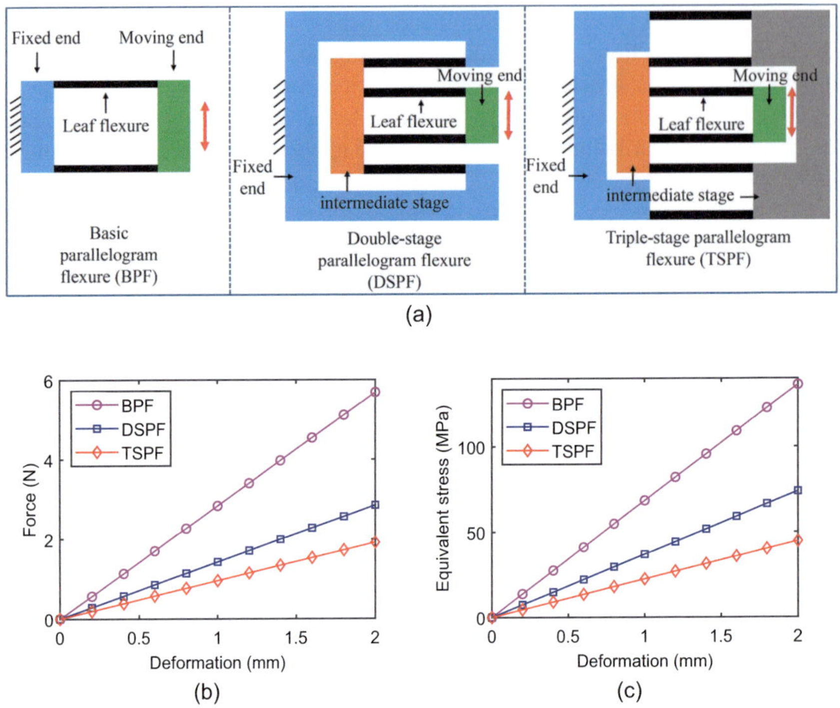

Fig. 9.1 Basic parallelogram mechanism and its compound forms: **a** structure diagrams of the basic (BPF), double-stage (DSPF), and triple-stage (TSPF) flexures; **b** stiffness comparison; and **c** stress comparison

9.2.2 4-PP Parallel-Kinematic Architecture

A prevalent configuration for XY manipulators is the 4-PP (P: prismatic) parallel-kinematic model. This architecture employs symmetrical branches to provide redundant constraints, which helps mitigate undesirable deformations. As shown in Fig. 9.2a, a 4-PP model contains four PP joints, each composed of two orthogonally combined P joints. Within a PP joint, one P joint acts as the active guider, while the other serves as the passive decoupler. This arrangement is the crucial element for the transmission and decoupling of linear motion.

Following this 4-PP configuration, two XY manipulators were constructed using the BPF and the DSPF, as illustrated in Fig. 9.2b, c, respectively. Although the planar footprint of the DSPF-based design ($232 \times 232\,\text{mm}^2$) is larger than that of the BPF-based design ($110 \times 110\,\text{mm}^2$), the DSPF-based manipulator delivers a significantly larger linear motion stroke, as later quantified in Table 9.3. In other words, the latter design offers superior structural compactness, as evaluated by the ratio of the motion stroke to the mechanism's occupancy area. The stage structure shown in Fig. 9.2c, however, can be further improved.

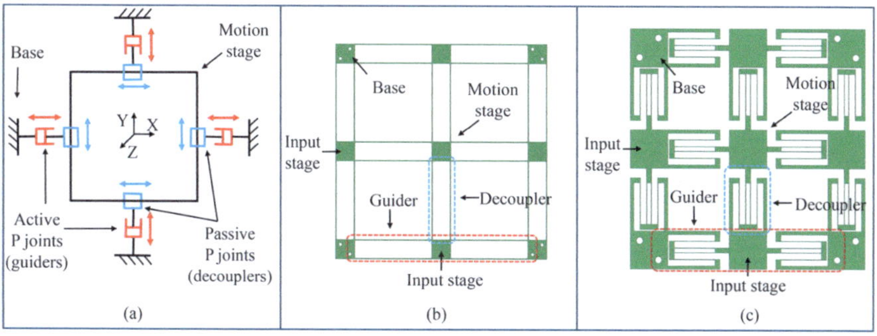

Fig. 9.2 Design evolution of the XY compliant manipulator: **a** the 4-PP kinematic model, **b** an implementation using basic parallelogram flexures (BPF), and **c** an implementation using double-stage parallelogram flexures (DSPF)

9.2.3 Design Improvement Using Spatial Connection Bars

Figure 9.3a shows a feasible design for an improved XY manipulator, while Fig. 9.3b depicts the corresponding simulation results under identical displacement actuation. In the basic parallelogram mechanism, the two sides are rigid rods that connect the leaf flexure to the fixed end. These rods do not participate in deformation but occupy a significant amount of planar space. A potential improvement is to replace these rigid rods with leaf flexures to utilize their deformation capability, as shown in Fig. 9.3a. However, as illustrated in Fig. 9.3b, simply thinning the rigid rods into flexures degrades the mechanism's movability. This occurs because the two front-end connecting blocks of the basic parallelogram mechanism (which act as the decoupler) undergo complex deformations, including stretching and bending. Ideally, the decoupler should deform primarily through tension or compression when subjected to an axial load, rather than bending.

To address this, a spatial connection bar is applied to fasten these two rigid blocks together, as shown in Fig. 9.3a. These bars stiffen the assembly, reducing bending deformation in the decoupler's components and ensuring reliable motion transmission. To further mitigate undesired deformation, spatial connection bars are also inserted at both ends of the guiders. The simulation results in Fig. 9.3b demonstrate that with these additions, the stage mechanism achieves the intended motion transmission equivalent to the previous design.

In essence, this guiding mechanism with a spatial connection bar is functionally equivalent to the TSPF discussed earlier and shown in Fig. 9.1a. Figure 9.4a illustrates the conversion between these two comparable mechanisms. The modified design effectively transforms the right-side planar rigid connection bar of the TSPF into a spatial form. This transformation allows the moving end–which was originally concealed within the structure–to be extended outward, facilitating interaction with the external environment.

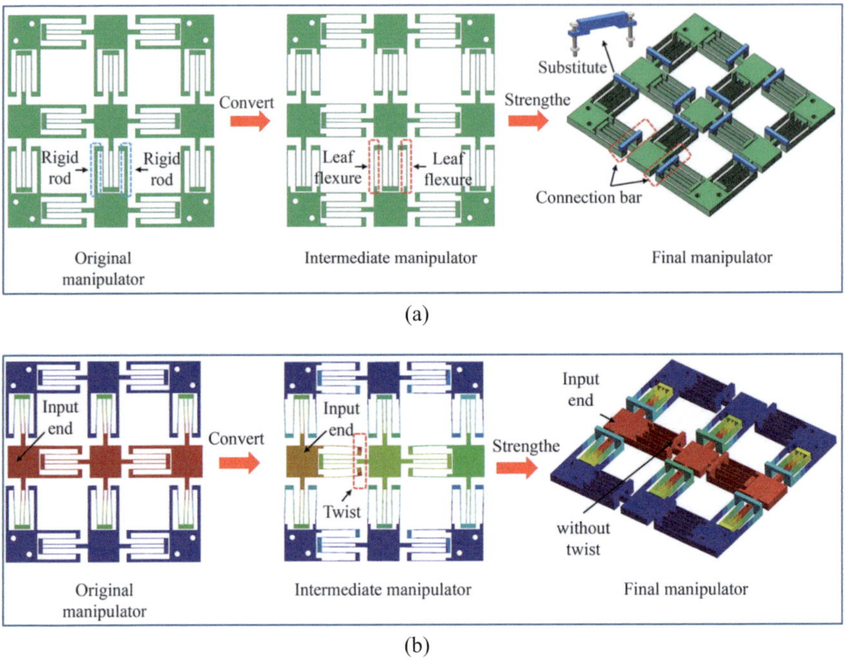

Fig. 9.3 Structural improvement of the XY manipulator. **a** Proposed design modification using spatial connection bars to enhance performance. **b** Corresponding simulation results comparing the deformation and stress of the original and modified designs under identical linear displacement actuation

A performance comparison of the two mechanisms, based on simulation results, is shown in Fig. 9.4b, c. The stiffness of the two designs is approximately identical. However, under identical end displacement, the modified mechanism exhibits lower equivalent stress. This indicates that the modified triple-stage guiding mechanism has a greater potential for larger deformation before yielding. Therefore, the method of utilizing a spatial connection bar is highly beneficial.

It is important to note that these connecting bars do not need to be manufactured monolithically with the platform mechanism. Instead, they can be fabricated independently and subsequently attached to the stage using bolts and nuts. This approach aims to reduce manufacturing complexity and maintenance costs. Furthermore, the strategy of using a spatial connection bar on the moving end and planar connection bars on the fixed side can be extended to other multi-stage parallelogram mechanisms, as demonstrated in Fig. 9.5.

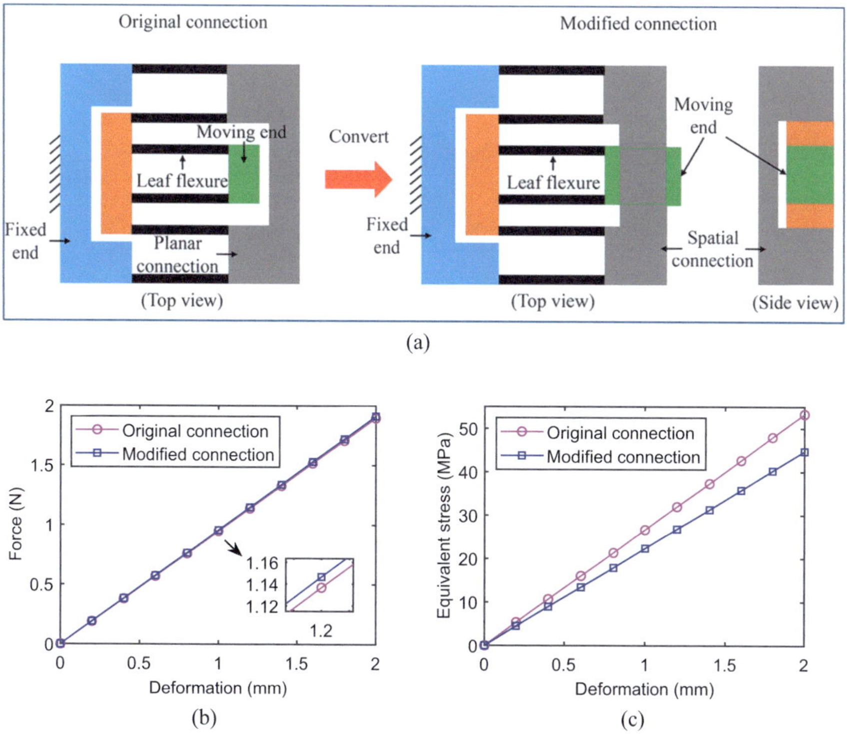

Fig. 9.4 Comparison of triple-stage parallelogram flexure (TSPF) connection schemes. **a** Structural diagrams of the planar and spatially-connected TSPF variants, **b** output stiffness comparison, and **c** maximum equivalent stress (von Mises) comparison under load

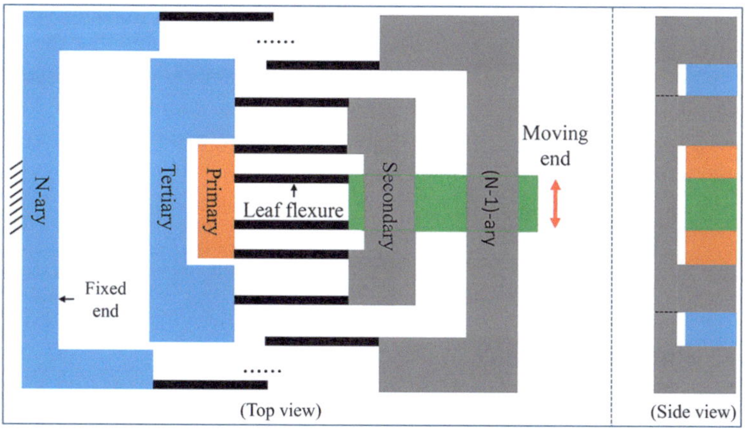

Fig. 9.5 Generalized design concept for multi-stage parallelogram mechanisms. The fixed ends employ planar connections, while the moving ends utilize spatial connections to externalize the motion platform

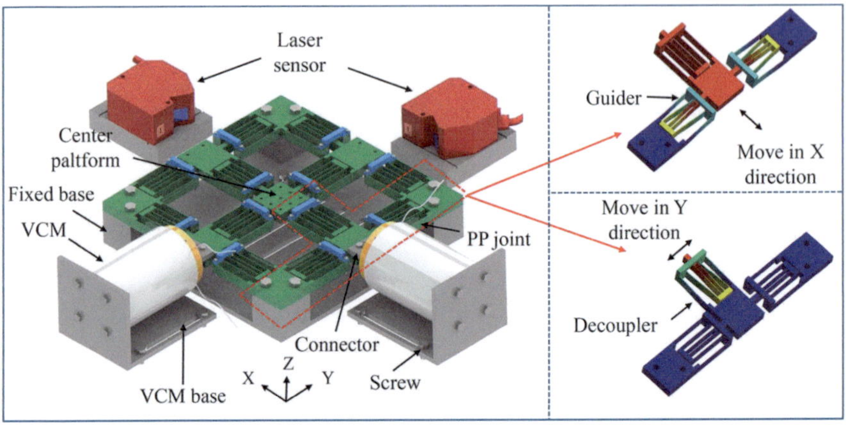

Fig. 9.6 Prototype of the single-layer, large-stroke XY compliant manipulator. The design utilizes modified triple-stage parallelogram flexures to achieve centimeter-scale travel without a multi-layer (hierarchical) architecture, reducing vertical occupancy and manufacturing complexity

9.2.4 Final Manipulator Assembly

Figure 9.6 depicts the complete architecture of the proposed compliant manipulator, which is constructed based on the modified triple-stage parallelogram mechanisms. The entire mechanism is bolted to a fixed base, a design choice that also accommodates the height of the actuator. As previously mentioned, the manipulator adopts a 4-PP parallel-kinematic configuration. A single PP joint, representative of those used in the design, is highlighted within the red dotted line in Fig. 9.6. The right part of Fig. 9.6 illustrates the mechanism's motion deformations under actuation displacements in two orthogonal directions. The simulation demonstrates that the guiders and decouplers independently and effectively accomplish motion transmission.

In addition to the mechanical structure, two VCMs are adopted to provide long-stroke actuation, and two laser displacement sensors are equipped to measure the stage's motion.

9.3 Theoretical Modeling Analysis

9.3.1 Stiffness Analysis

Stiffness is a crucial parameter for investigating the kinematic characteristics of the micropositioning stage. It directly influences static performance traits, such as the selection of an appropriate actuator, and also contributes to determining the resonant frequency of the stage's working mode.

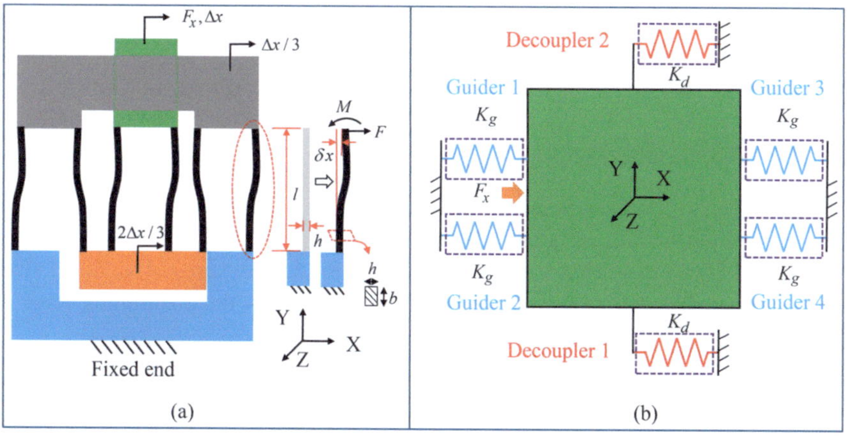

Fig. 9.7 Theoretical model for stiffness analysis. **a** Kinematic diagram of the modified triple-stage parallelogram mechanism under an external load, and **b** equivalent stiffness model of the entire stage in one working direction, accounting for guiders and decouplers

Figure 9.7a displays the kinematic analysis of the modified triple-stage parallelogram mechanism. When an external force is applied to the moving end, all leaf flexures undergo an identical magnitude of displacement. Considering the motion superposition of the cascaded mechanism, the total displacement of the moving end is three times the deformation of a single flexure.

Since the leaf flexures act as fixed-guided beams, their guided ends are constrained by a moment M. Applying knowledge of theoretical mechanics and the boundary conditions at the end of the flexure, the following equations are derived:

$$0 = \frac{Fl^2}{2EI} - \frac{Ml}{EI}, \quad \delta x = \frac{Fl^3}{3EI} - \frac{Ml^2}{2EI} \tag{9.1}$$

where δx is the displacement of the guided end of a single beam, E is the elastic modulus, and I is the area moment of inertia, expressed as $I = bh^3/12$. The variables l, b, and h denote the length, width (in-plane), and thickness (out-of-plane) of the flexure, respectively.

Solving the above equations yields the relationship between the end force and the displacement:

$$F = \frac{2M}{l}, \quad \delta x = \frac{Fl^3}{12EI}. \tag{9.2}$$

Considering the mechanism's symmetry and the principle of displacement superposition, the stiffness of the modified triple-stage parallelogram mechanism is expressed as follows:

$$K_{3s} = \frac{F_x}{\Delta x} = \frac{2F}{3\delta x} = \frac{2Ebh^3}{3l^3}. \tag{9.3}$$

Furthermore, the stiffness for a modified n-stage parallelogram mechanism can be generalized by the equation:

$$K_{ns} = \frac{2Ebh^3}{nl^3}. \tag{9.4}$$

The stiffness model of the entire stage in one working direction is illustrated in Fig. 9.7b, which includes the contributions from both the guiders and the decouplers. Since both the guiders and decouplers employ leaf flexures of identical size, they possess equal stiffness. Consequently, the stiffness of the positioning stage is computed as:

$$K_{\text{stage}} = 4K_g + 2K_d = 6K_{3s} = \frac{4Ebh^3}{l^3} \tag{9.5}$$

where K_g and K_d denote the stiffness of a single guider and decoupler, respectively.

Based on theoretical mechanics, the driving stiffness of the proposed XY manipulator is determined and remains constant for a given set of structural parameters. This indicates a linear force-deflection relationship, a benefit derived from the designed TSPF. For general engineering design, a linear model is simpler, more intuitive, and more time-efficient. From this perspective, the model presented here is sufficient for practical engineering applications.

To validate the model's accuracy, finite element analysis (FEA) and experimental tests are conducted in the following sections. It is acknowledged that nonlinear behavior is unavoidable in beams undergoing large deflections. Numerous nonlinear modeling methods have been reported to precisely predict the trajectory of the beam's end. For instance, Herpe et al. proposed a method combining a linear model with a simplified nonlinear model to achieve accurate prediction of large deformations [9]. Zhu and Wu employed the beam constraint model for nonlinear modeling in their compliant mechanism, obtaining acceptable results [27, 36]. Alternatively, Shang and Liu et al. selected the compliance matrix method, which is suitable for compliant mechanisms with complex configurations, as their theoretical modeling approach [18, 22]. A comprehensive survey of large-deflection nonlinear modeling methods for compliant manipulators can be found in [15]. In future work, we plan to integrate these nonlinear factors to develop a more comprehensive and accurate predictive model.

The resonant frequency is another critical index for analyzing the dynamic characteristics of the stage. The Lagrangian equations are typically employed to derive the stage's dynamic model. For an XY stage with two degrees of freedom, the equations are:

$$\frac{d}{dt}\left(\frac{\partial L}{\partial \dot{x}}\right) - \frac{\partial L}{\partial x} = F_1, \quad \frac{d}{dt}\left(\frac{\partial L}{\partial \dot{y}}\right) - \frac{\partial L}{\partial y} = F_2 \tag{9.6}$$

where $L = T - U$ is the Lagrangian function, T and U are the total kinetic and potential energy of the system, respectively, x and y are the stage's displacements in the two working directions, and F_i $(i = 1, 2)$ is the generalized force excluding conservative forces.

The dynamic model of the XY manipulator under free motion $(F_i = 0)$ can be derived as:

$$M_x\ddot{x} + K_x x = 0, \quad M_y\ddot{y} + K_y y = 0 \tag{9.7}$$

where M_x and M_y are the equivalent masses, K_x and K_y are the equivalent stiffnesses of the stage in the two working directions, and x (y) denotes the displacement of the stage in the X (Y)-axis direction.

The characteristic equation for the system is given by:

$$(K_x - \omega_1^2 M_x)\Phi_x = 0, \quad (K_y - \omega_2^2 M_y)\Phi_y = 0 \tag{9.8}$$

where the eigenvector Φ_i $(i = 1, 2)$ denotes the mode shape and ω_i $(i = 1, 2)$ represents the corresponding natural circular frequency. The natural frequencies can be obtained by solving the following determinant equations:

$$|K_x - \omega_1^2 M_x| = 0, \quad |K_y - \omega_2^2 M_y| = 0. \tag{9.9}$$

Finally, the resonant frequencies of the first two modes (in Hertz) for the XY manipulator are expressed as:

$$f_1 = \frac{1}{2\pi}\sqrt{\frac{K_x}{M_x}}, \quad f_2 = \frac{1}{2\pi}\sqrt{\frac{K_y}{M_y}}. \tag{9.10}$$

Given the stage's symmetrical structure and identical dimensions in both directions, it can be predicted that the resonant frequencies of the first two modes are identical, with a calculated value of 28.32 Hz.

9.4 Parameter Determination and Simulation Study

9.4.1 Parameter Determination

Prior to the simulation study, the structural and material parameters of the positioning stage must be determined. The primary structural parameters were specified based on a trade-off among stroke requirements, spatial constraints, and manufacturing considerations (Table 9.1). The leaf flexure dimensions are set to a length of 40 mm, a width of 0.5 mm, and a height of 10 mm. The overall planar dimension of the mechanism is $232 \times 232\,\text{mm}^2$, and the central platform measures $37 \times 37\,\text{mm}^2$.

The material selected for stage fabrication is Al-7075, chosen for its excellent mechanical properties which make it suitable for compliant mechanisms such as nano-manipulators, microgrippers, and XY-θ micro-motion stages [17, 19, 31]. Its

Table 9.1 Key geometric dimensions of the proposed XY compliant manipulator

Parameter	Leaf flexure	Whole stage
Dimension	$40 \times 0.5 \times 10\,\text{mm}^3$	$232 \times 232\,\text{mm}^2$

key material parameters are as follows: Young's modulus of 71.7 GPa, yield strength of 503 MPa, Poisson's ratio of 0.33, and density of $2810\,\text{kg m}^{-3}$.

9.4.2 FEA Simulation Results

9.4.2.1 FEA Setup and Mesh Convergence

Finite element analysis (FEA) was conducted using the static structural and modal analysis modules in ANSYS Workbench software. The simulation settings were configured as follows. The grid partition utilized the software's default method. The relevance was set to 100 and the relevance centre was set to "fine" to refine the mesh density. The element type was "tet10", resulting in 63,888 elements and 134,217 nodes, with an average element quality of 0.52. Fixed constraints were applied to the mounting through-holes of the XY manipulator as the boundary condition. The meshing and boundary conditions are shown in Fig. 9.9a, b, respectively.

9.4.2.2 Static Structural Analysis

The driving stiffness is defined as the ratio of the driving force to the resulting motion displacement. By applying a series of displacements to the input end and monitoring the reaction force, the stiffness was determined. The driving stiffness values obtained from theoretical calculation and simulation analysis are 5.60 N/mm and 6.42 N/mm, respectively, as shown in Fig. 9.8a. Both methods yield force-displacement data with a strong linear relationship. The discrepancy arises because the theoretical model simplifies the computation by considering only the bending of the leaf flexures, whereas the simulation software comprehensively accounts for multiple deformation modes. This likely explains why the theoretical stiffness is lower than the simulated value.

Figure 9.8b displays the total deformation and equivalent stress of the stage when both input ends are actuated simultaneously. The corresponding deformation and equivalent stress distribution nephograms are shown in Fig. 9.9. At a driving displacement of 9 mm, the maximum equivalent stress on the mechanism is 280 MPa, which is below the material's yield strength of 503 MPa. Therefore, considering the bidirectional driving capability of the VCM, the developed stage can achieve centimeter-scale travel without material failure.

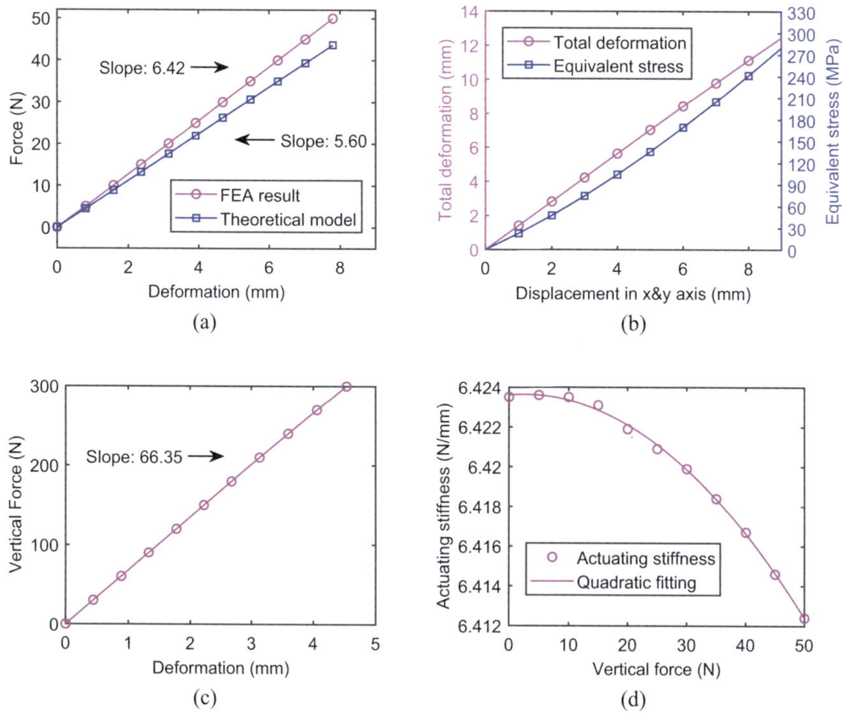

Fig. 9.8 Finite element analysis (FEA) results for the XY compliant manipulator. **a** Driving stiffness comparison between theoretical and simulation results. **b** Total deformation and equivalent von Mises stress distribution under full actuation. **c** Out-of-plane (Z-axis) stiffness characterization. **d** Variation of in-plane driving stiffness under different vertical loads

The driving stiffness of the platform in the vertical direction was also simulated, as shown in Fig. 9.8c. The vertical stiffness is approximately ten times the in-plane driving stiffness, indicating superior resistance to Z-axis loading. The stage's in-plane driving stiffness varies under different vertical loads, as shown in Fig. 9.8d. This occurs because vertical loads induce axial deformation (stretching) in the leaf flexures, which in turn influences their bending stiffness. Curve fitting results demonstrate that the variation in the mechanism's stiffness follows a quadratic polynomial distribution. When the vertical load is less than 1 kg, the driving stiffness changes minimally and remains approximately constant. Thus, the stiffness model derived earlier remains applicable under moderate vertical loads.

9.4.2.3 Modal Analysis

The first six modal analysis results of the stage are shown in Fig. 9.10. The first two modes correspond to translational motion in the two working directions, with

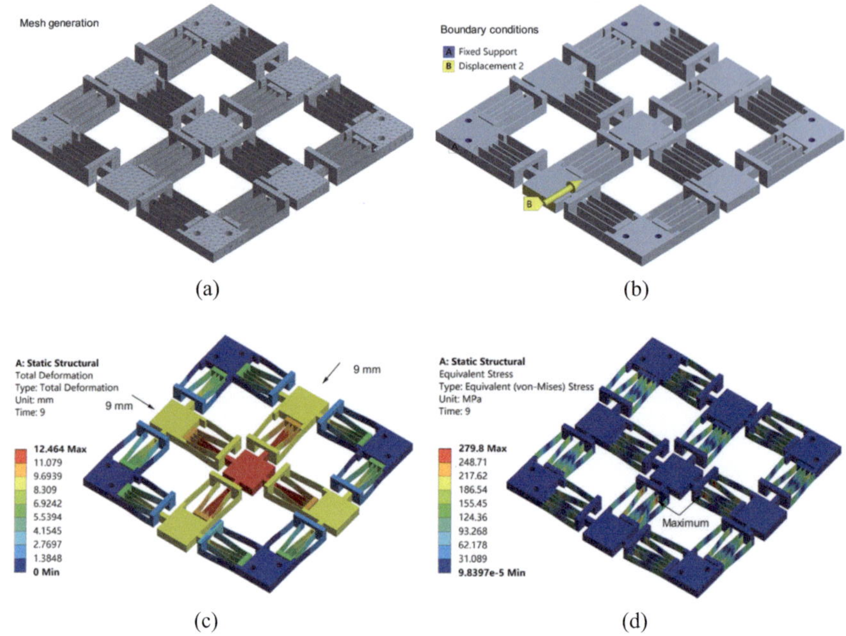

Fig. 9.9 Finite element analysis setup and results. **a** Mesh configuration and statistics, **b** applied boundary conditions with fixed supports, **c** total deformation nephogram under actuation, and **d** von Mises stress distribution nephogram

natural frequencies of 30.01 Hz and 30.03 Hz, respectively. Their nearly identical values confirm that the mechanism exhibits consistent dynamic behavior in both driving axes. The subsequent four modes correspond to in-plane vibrations of the leaf flexures, such as swinging and rotation. The working mode is well isolated from other modes, as evidenced by the third-mode natural frequency of 116.04 Hz, which is nearly four times that of the first two modes. To further suppress these higher-order mode shapes, spatially connecting parts with similar motion forms is recommended [8].

9.4.2.4 Fatigue Analysis

Fatigue analysis is pivotal for compliant manipulators intended for high-cycle applications. Due to the destructive nature of experimental testing, FEA was utilized for this analysis. The fatigue tool parameters were set as follows: a fatigue strength factor of 0.8, a fully reversed load type, and the stress-life method. Table 9.2 presents the minimum life cycles of the mechanism for various stroke settings. It can be observed that the lifecycle decreases as the output motion increases, which is attributable to the flexible beams bearing higher equivalent stress under larger deformations. When the

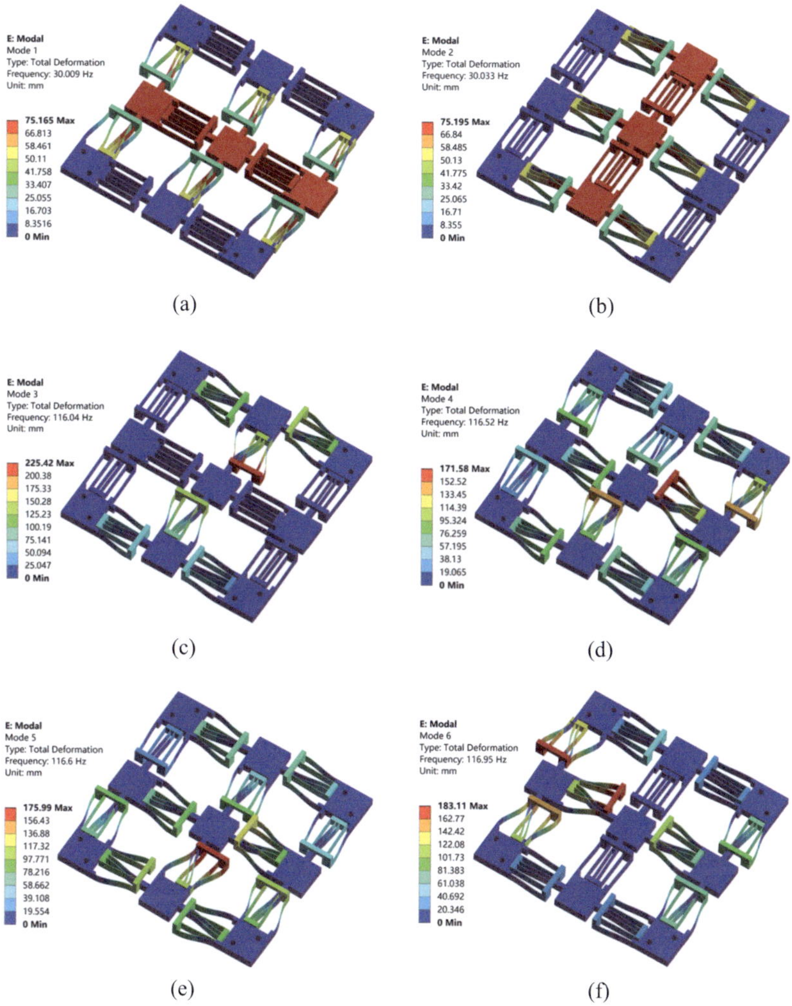

Fig. 9.10 Results of the modal analysis showing the first six mode shapes. **a–b** First two modes: translational motions along the X and Y working axes. **c–f** Higher-order modes: parasitic in-plane rotations and vibrations of the leaf flexures

output stroke reaches centimeter scale (e.g., ±6 mm), the simulation results indicate a minimum lifecycle of 30,492 before fatigue failure. For most application scenarios, such as the microinjection of zebrafish embryos, this fatigue life is acceptable, as motion ranges exceeding 1 cm are typically less frequently utilized. Therefore, a stroke of approximately 1.2 cm is advised to guarantee a sufficient lifecycle for the manipulator. Future work will further investigate improvements to the service life and reliability of the XY manipulator.

Table 9.2 Fatigue life analysis for various output strokes

Motion (mm)	±1	±2	±3	±4	±5	±6	±7	±7.5
Lifecycle	1e8	1e8	1.4894e7	824,220	135,380	30,492	4597	2655

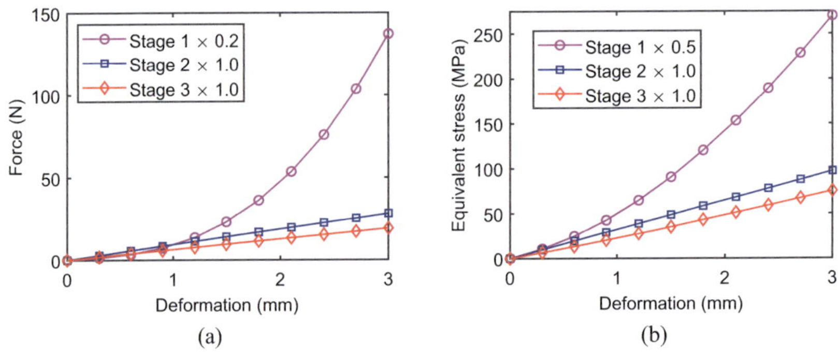

Fig. 9.11 Performance comparison of the three flexure-based mechanisms. **a** Driving stiffness and **b** maximum equivalent von Mises stress for the basic (BPF), double-stage (DSPF), and triple-stage (TSPF) parallelogram designs under identical input displacement

9.4.2.5 Comparative Performance Analysis of Three Stages

The performance of the three aforementioned stages was also simulated using ANSYS software. The mechanisms depicted in Figs. 9.2b, c, and 9.6 are designated as Stage 1, Stage 2, and Stage 3, respectively. For these stages, the key design variables are the leaf flexure's length (l), width (b), and height (h). These parameters are the primary factors influencing the manipulator's deformability, as expressed in the driving stiffness equation.

Under identical displacement actuation, a comparison of the reaction force and equivalent stress for the three mechanisms is shown in Fig. 9.11. Stage 1 exhibits a nonlinear and dramatic increase in both monitored indicators as the deformation increases. In contrast, Stages 2 and 3 demonstrate linear force-displacement relationships and generate lower equivalent stresses.

A further parameter comparison of the three stages is presented in Table 9.3. Although Stage 1 has a small footprint and a high resonant frequency, its nonlinear force-displacement relationship significantly exacerbates the system's control burden. The advantages of Stage 2–namely, its linear stiffness and large travel–justify its increased mechanism size. Stage 3 achieves reduced driving stiffness while maintaining the same planar dimensions as Stage 2. Furthermore, the lower equivalent stress exhibited by Stage 3 indicates its superior potential for large-stroke motion transmission.

Using the performance parameters of Stage 2 as a reference, the stiffness of Stage 3 is reduced by one-third, while its resonant frequency is only reduced by one-

Table 9.3 Performance comparison of the three XY compliant manipulators: the basic (Stage 1), double-stage (Stage 2), and triple-stage (Stage 3) designs

Parameters	Stage 1	Stage 2	Stage 3
Planar area (mm^2)	110 × 110	232 × 232	232 × 232
Input motion (mm)	3	3	3
Force reaction (N)	685.17	28.11	19.26
Stiffness (N/mm)	Nolinear	9.37	6.42
Equivalent stress (MPa)	539.80	97.27	75.14
1st & 2nd frequency (Hz)	≈184	≈37	≈30

fifth. Therefore, Stage 3, composed of triple-stage compound parallelograms, offers distinct advantages in terms of low stiffness, large stroke, and high compactness compared to the first two designs.

9.5 Experimental Testing Results

9.5.1 Prototype Fabrication and Experimental Setup

9.5.1.1 Prototype Fabrication and Test System

Performance testing experiments were conducted to validate the rationality and reliability of the proposed XY compliant manipulator. The manipulator's main component was manufactured via wire electrical discharge machining (EDM), while the connecting elements were fabricated using a milling process. These components were then assembled using bolts and fixed onto the base as a complete unit. The physical device of the fabricated manipulator and the testing system are shown in Fig. 9.12.

The test system comprises the following pivotal equipment:

- Two voice coil motors (VCMs) employed as actuators.
- Two laser triangulation sensors for gathering displacement signals.
- Two force sensors to measure driving forces.
- An industrial-grade embedded real-time controller (NI cRIO-9022).
- Other auxiliary devices (e.g., amplifiers, data acquisition modules).

The major test contents include driving stiffness, working stroke, and resonant frequency.

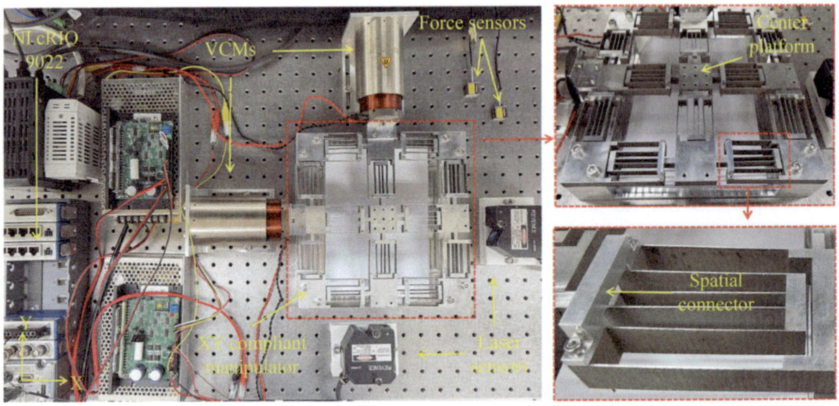

Fig. 9.12 Experimental setup for performance evaluation. The photograph shows the fabricated XY compliant manipulator prototype, voice coil motor (VCM) actuators, laser displacement sensors, and the fixed base within the test frame

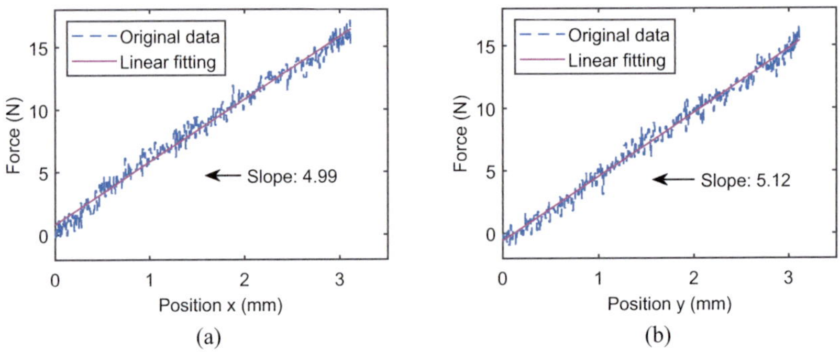

Fig. 9.13 Experimental characterization of driving stiffness. **a** Force-displacement data and linear fit for the X-axis. **b** Force-displacement data and linear fit for the Y-axis. The results confirm the desired linear behavior and axis symmetry

9.5.1.2 Driving Stiffness Test

The first test evaluated whether the manipulator's driving stiffness matched the anticipated design values. The manipulator's input end was propelled by an external platform equipped with a force sensor, while the output end's displacement was monitored by a laser displacement sensor. The force-displacement signals obtained for the two working axes are shown in Fig. 9.13a, b. According to least squares fitting, the stiffness values for the X- and Y-axes are 4.99 N/mm and 5.12 N/mm, respectively, which are consistent with the previous theoretical and simulation estimations. The slightly lower experimental stiffness values can be attributed to manufacturing errors in the compliant hinges, installation errors of the connecting blocks, and calibration errors of the force sensor.

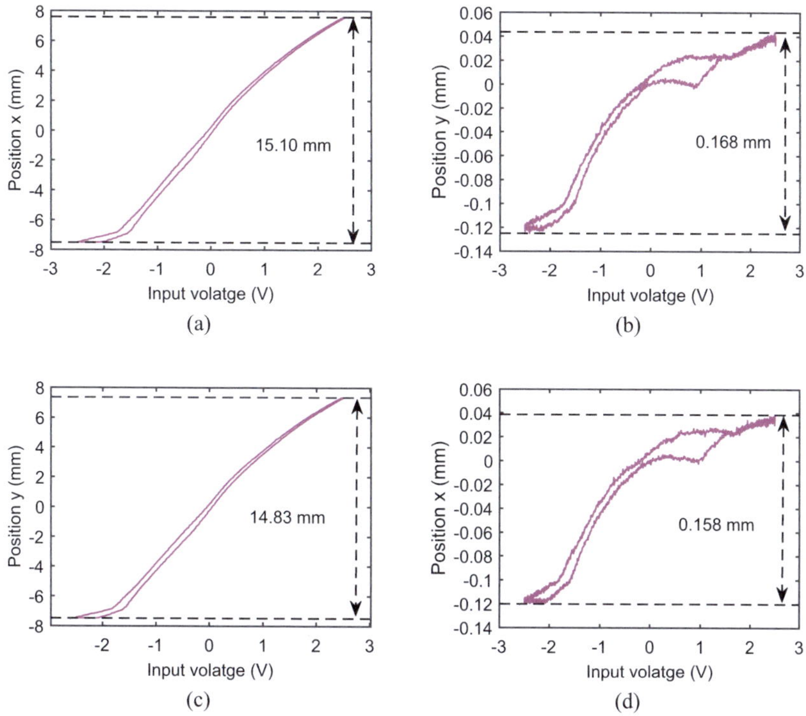

Fig. 9.14 Stroke and cross-coupling performance. **a** X-axis working stroke, **b** X-axis parasitic (Y-direction) displacement, **c** Y-axis working stroke, **d** Y-axis parasitic (X-direction) displacement

9.5.1.3 Working Stroke and Cross-Coupling Test

After mounting the VCMs as actuators on the manipulator, the system's stroke was tested. Two laser displacement sensors were orthogonally positioned at the output end to record the manipulator's stroke in the working direction and the parasitic displacement in the non-working direction, while voltage was supplied to one VCM at a time. The stroke and coupling displacement for the two motion directions are shown in Fig. 9.14.

For the X-axis, the working stroke is 15.10 mm, the range of parasitic displacement is 0.168 mm, and the displacement coupling rate is 1.11%. For the Y-axis, the working stroke is 14.83 mm, the range of parasitic displacement is 0.158 mm, and the displacement coupling rate is 1.06%. These results demonstrate that the manipulator achieves large strokes in both working directions while maintaining low cross-coupling.

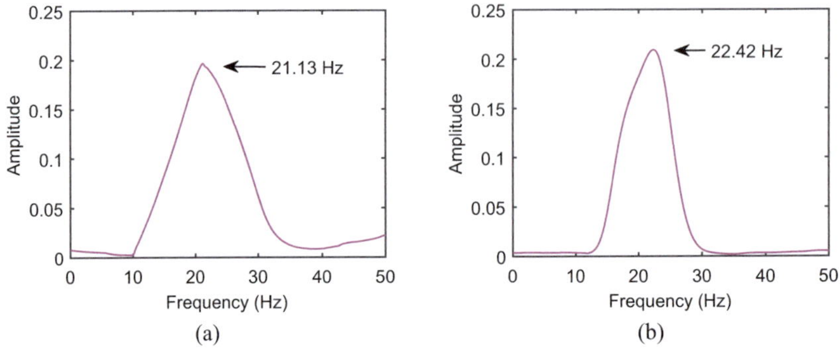

Fig. 9.15 Experimental modal testing results. **a** Frequency response function (FRF) for the X-axis, indicating a resonant frequency of 21.13 Hz. **b** FRF for the Y-axis, indicating a resonant frequency of 22.42 Hz

9.5.1.4 Resonant Frequency Test

To prevent the mechanism from entering resonance during operation, it is crucial to identify its resonant frequencies. The natural frequencies were obtained by applying a swept-sine excitation signal (where the frequency varies with time) to the manipulator. The frequency response results for the two axes are shown in Fig. 9.15. The natural frequencies of the mechanism in the X-axis and Y-axis are 21.13 Hz and 22.42 Hz, respectively.

The measured resonant frequencies are slightly lower than the theoretical and simulation results. This discrepancy can be explained by Eq. (9.10). First, a decrease in the experimental driving stiffness leads to a reduction in K_x and K_y. Second, the coils of the VCMs are attached to the manipulator and contribute to the moving mass, thereby increasing the equivalent masses M_x and M_y. Consequently, the mechanism's resonant frequency decreases. Nonetheless, the measured frequencies remain adequate for quasi-static operations at low frequencies.

9.5.2 Comparison and Discussion

9.5.2.1 Comparison of Theoretical, Simulation, and Experimental Results

The results for driving stiffness, reachable stroke, and resonant frequency obtained from the three assessment approaches are summarized in Table 9.4. The theoretical modeling, FEA simulation, and experimental testing results demonstrate strong mutual verification.

Both the simulation and experimental results confirm that a considerable range of motion is achieved, meeting the desired centimeter-scale travel requirement. The

Table 9.4 Comparison of performance parameters for the proposed XY compliant manipulator and other notable designs from the literature

Specification	Model	Simulation	Testing	[2]	[29]	[36]	[22]	[18]
			X/Y-axis	X/Y-axis	X/Y-axis	X/Y-axis	X/Y-axis	X/Y-axis
Stiffness (N/mm)	5.60	6.42	4.99/5.12	–/–	–/–	47.4 (FEA)	27.5/27.5	50 (FEA)
Stroke (mm)	>10	18	15.10/14.83	10.00/10.00	11.75/11.66	5.19/5.19	1.80/1.78	1.00/1.00
Resonant frequency (Hz)	29.32	30.01	21.13/22.42	18/18	29.3/29.6	48.6 (FEA)	35.62/35.82	55.80/55.99
Size (mm^3)	–	–	$232 \times 232 \times 17$	$255 \times 255 \times 25.4$	$120 \times 120 \times 25$	$220 \times 220 \times 22$	–	$255 \times 255 \times 20$
Volume ratio (%)	–	–	2.45×10^{-4}	6.05×10^{-5}	3.81×10^{-4}	2.53×10^{-5}	–	7.69×10^{-7}

simulation indicates a maximum achievable deformation of 18 mm, which is ultimately constrained by the relative positions of the internal components, as illustrated in Fig. 9.7a. In experimental testing, the mechanism was not driven to its ultimate position to avoid potential damage from rigid impact on the leaf flexures.

Taking the simulation results as the benchmark, the relative errors of the stiffness and natural frequency obtained from the theoretical model are 12.77% and 5.63%, respectively. The experimental results are slightly lower than those from the other two approaches. This discrepancy can be attributed to manufacturing errors in the flexible hinges, assembly errors in the connection bars, and the additional mass contributed by the VCM coils. The performance of the two axes of the proposed XY compliant manipulator is highly consistent, which is beneficial for subsequent motion control.

9.5.2.2 Benchmarking Against State-of-the-Art Designs

The performance of the proposed manipulator is benchmarked against several other typical designs, as shown in Table 9.4. To enable a substantial and fair comparison of the workspace relative to the manipulator's size, a dimensionless index called the volume ratio is proposed. This concept can be referenced from the literature [12], and its expression is given as follows:

$$\text{Volume ratio} = \frac{S_X \cdot S_Y \cdot 1}{\text{Size}} \times 100\% \tag{9.11}$$

where S_X and S_Y are the strokes along the X-axis and Y-axis, the value '1' represents a unit compensation stroke along the Z-axis for dimensional consistency, and "Size" denotes the volume of the manipulator's bounding box (in mm^3). A larger volume ratio indicates that a larger workspace is achieved relative to the mechanism's physical volume.

The volume ratio of the manipulator designed in this work shows obvious advantages over other designs, with the exception of the work presented in [29]. The high volume ratio in [29] is primarily due to its layered design, which effectively utilizes three-dimensional space. Consequently, its vertical dimension (25 mm) is greater than that of this work (17 mm). This insight also points toward a direction for future improvement of our mechanism: a more compact and portable XY manipulator could be developed by integrating the multi-stage compound parallelogram mechanism with a layered design approach.

9.5.2.3 Contributions and Design Implications

Compared to existing works, the manipulator presented in this paper exhibits greater compliance and a longer working stroke. A key contribution is the proposed design concept of using a planar connection at the fixed end and a spatial connection at the moving end. This innovation successfully resolves the application issue of multi-stage parallelograms, where the moving platform is typically concealed. This design strategy enables the moving end to be extended outward, facilitating practical interaction and actuation. It is anticipated that readers can leverage this concept to construct more complex and high-performance compliant manipulators.

9.6 Conclusion

This chapter presented the design, development, and validation of a novel XY compliant manipulator based on a triple-stage parallelogram flexure (TSPF), capable of substantial displacement transmission. In contrast to many existing centimeter-stroke manipulators that rely on hierarchical (multi-layer) designs, the proposed architecture offers the distinct advantages of lower manufacturing cost and reduced vertical space occupancy.

A revised connection scheme–featuring a planar connection on the fixed side and a spatial connection on the moving side–was introduced to overcome the application limitations of multi-stage parallelograms. This design improvement fully utilizes the deformation capacity of the TSPF without compromising its stiffness characteristics. Furthermore, this connection strategy is generalizable and can be extended to other multi-stage parallelogram mechanisms. The use of a multi-stage architecture effectively reduces the mechanism's overall stiffness and distributes stress across the hinges, enabling larger strokes.

Simulation results confirmed that, compared to a manipulator utilizing a double-stage parallelogram flexure (DSPF), the TSPF-based design achieves lower stiffness and a larger stroke without increasing the planar footprint. Experimental testing demonstrated that the fabricated XY compliant manipulator achieves a workspace of $1.51\,\text{cm} \times 1.48\,\text{cm}$, driving stiffness values of 4.99 N/mm and 5.12 N/mm, and first-two-mode resonant frequencies of 21.13 Hz and 22.42 Hz for the two working axes. These results indicate that the performance of the designed manipulator successfully meets the anticipated design requirements.

Future work will focus on developing a nonlinear model to enhance prediction accuracy under large deformations and on further improving the mechanism's long-term reliability and fatigue life.

References

1. Al-Jodah, A., Shirinzadeh, B., Ghafarian, M., Das, T.K., Pinskier, J.: Design, modeling, and control of a large range 3-DOF micropositioning stage. Mech. Mach. Theory **156**, 104159 (2021)
2. Awtar, S., Parmar, G.: Design of a large range XY nanopositioning system. J. Mech. Robot. **5**(2), 021008 (2013)

3. Cui, M., Yang, W., Guan, Y., Zhang, Z.: Fabrication of high precision grating patterns with a compliant nanomanipulator-based femtosecond laser direct writing system. Precis. Eng. **78**, 60–69 (2022)
4. Evke, E.E., Huang, C., Wu, Y.W., Arwashan, M., Lee, B., Forrest, S.R., Shtein, M.: Kirigami-based compliant mechanism for multiaxis optical tracking and energy-harvesting applications. Adv. Eng. Mater. **23**(4), 2001079 (2021)
5. Fan, S., Liu, H., Fan, D.: Design and development of a novel monolithic compliant XY stage with centimeter travel range and high payload capacity. Mech. Sci. **9**(1), 161–176 (2018)
6. Gan, J., Xu, H., Zhang, X., Ding, H.: Design of a compliant adjustable constant-force gripper based on circular beams. Mech. Mach. Theory **173**, 104843 (2022)
7. Gu, Y., Chen, X., Lin, J., Lu, M., Lu, F., Zhang, Z., Yang, H.: Vibration-assisted roll-type polishing system based on compliant micro-motion stage. Micromachines **9**(10), 499 (2018)
8. Hao, G., Yu, J.: Design, modelling and analysis of a completely-decoupled XY compliant parallel manipulator. Mech. Mach. Theory **102**, 179–195 (2016)
9. Herpe, X., Walker, R., Dunnigan, M., Kong, X.: On a simplified nonlinear analytical model for the characterisation and design optimisation of a compliant XY micro-motion stage. Robot. Comput. Integr. Manuf. **49**, 66–76 (2018)
10. Huang, H.B., Sun, D., Mills, J.K., Cheng, S.H.: Robotic cell injection system with position and force control: toward automatic batch biomanipulation. IEEE Trans. Rob. **25**(3), 727–737 (2009)
11. Jadhav, S.B., Dhande, K.K., Deshmukh, S.P.: Design and evaluation of compliant modular XY positioning stage. Aust. J. Mech. Eng. 1–12 (2020)
12. Kim, J.J., Choi, Y.M., Ahn, D., Hwang, B., Gweon, D.G., Jeong, J.: A millimeter-range flexure-based nano-positioning stage using a self-guided displacement amplification mechanism. Mech. Mach. Theory **50**, 109–120 (2012)
13. Lee, C., Stepanick, C.K., Lee, S.K., Tarbutton, J.A.: Cross-coupling effect of large range XY nanopositioning stage fabricated by stereolithography process. Precis. Eng. **46**, 81–87 (2016)
14. Lee, J.W., Li, Y.C., Chen, K.S., Liu, Y.H.: Design and control of a cascaded piezoelectric actuated two-degrees-of-freedom positioning compliant stage. Precis. Eng. **45**, 374–386 (2016)
15. Ling, M., Howell, L.L., Cao, J., Chen, G.: Kinetostatic and dynamic modeling of flexure-based compliant mechanisms: a survey. Appl. Mech. Rev. **72**(3), 030802 (2020)
16. Liu, M., Zhang, X., Fatikow, S.: Design and analysis of a multi-notched flexure hinge for compliant mechanisms. Precis. Eng. **48**, 292–304 (2017)
17. Liu, Y., Zhang, Z.: A large range compliant XY nano-manipulator with active parasitic rotation rejection. Precis. Eng. **72**, 640–652 (2021)
18. Liu, Y., Zhang, Z.: A large range compliant nano-manipulator supporting electron beam lithography. J. Mech. Des. **144**(4), 043303 (2022)
19. Lyu, Z., Wu, Z., Xu, Q.: Design and development of a novel piezoelectrically actuated asymmetrical flexible microgripper. Mech. Mach. Theory **171**, 104736 (2022)
20. Lyu, Z., Xu, Q.: Novel design of a piezoelectrically actuated compliant microgripper with high area-usage efficiency. Precis. Eng. **76**, 1–11 (2022)
21. Roy, N.K., Cullinan, M.A.: Design and characterization of a two-axis, flexure-based nanopositioning stage with 50 mm travel and reduced higher order modes. Precis. Eng. **53**, 236–247 (2018)
22. Shang, J., Tian, Y., Li, Z., Wang, F., Cai, K.: A novel voice coil motor-driven compliant micropositioning stage based on flexure mechanism. Rev. Sci. Instrum. **86**(9), 095001 (2015)
23. Wadikhaye, S.P., Yong, Y.K., Reza Moheimani, S.: A serial-kinematic nanopositioner for high-speed atomic force microscopy. Rev. Sci. Instrum. **85**(10), 105104 (2014)
24. Wang, F., Ma, Z., Gao, W., Zhao, X., Tian, Y., Zhang, D., Liang, C.: Dynamic modeling and control of a novel XY positioning stage for semiconductor packaging. Trans. Inst. Meas. Control. **37**(2), 177–189 (2015)
25. Wang, F., Zhao, X., Huo, Z., Shi, B., Liang, C., Tian, Y., Zhang, D.: A 2-DOF nano-positioning scanner with novel compound decoupling-guiding mechanism. Mech. Mach. Theory **155**, 104066 (2021)

26. Wang, X., Meng, Y., Huang, W.W., Li, L., Zhu, Z., Zhu, L.: Design, modeling, and test of a normal-stressed electromagnetic actuated compliant nano-positioning stage. Mech. Syst. Signal Process. **185**, 109753 (2023)
27. Wu, K., Hao, G.: Design and nonlinear modeling of a novel planar compliant parallelogram mechanism with general tensural-compresural beams. Mech. Mach. Theory **152**, 103950 (2020)
28. Wu, Z., Xu, Q.: Survey on recent designs of compliant micro-/nano-positioning stages. In: Actuators, vol. 7, p. 5. MDPI (2018)
29. Xu, Q.: Design and development of a compact flexure-based XY precision positioning system with centimeter range. IEEE Trans. Industr. Electron. **61**(2), 893–903 (2013)
30. Yang, M., Zhang, C., Huang, X., Chen, S., Yang, G.: A long-stroke nanopositioning stage with annular flexure guides. IEEE/ASME Trans. Mechatron. **27**(3), 1570–1581 (2021)
31. Yang, Y., Wu, G., Wei, Y.: Design, modeling, and control of a monolithic compliant XY-θ microstage using a double-rocker mechanism. Precis. Eng. **71**, 209–231 (2021)
32. Yao, T.F., Duenner, A., Cullinan, M.: In-line metrology of nanoscale features in semiconductor manufacturing systems. Precis. Eng. **47**, 147–157 (2017)
33. Zhang, H., Wu, Z., Xu, Q.: Design of a new XY flexure micropositioning stage with a large hollow platform. In: Actuators, vol. 9, p. 65. MDPI (2020)
34. Zhang, Q., Zhao, J., Shen, X., Xiao, Q., Huang, J., Wang, Y.: Design, modeling, and testing of a novel XY piezo-actuated compliant micro-positioning stage. Micromachines **10**(9), 581 (2019)
35. Zhao, Y., Sun, H., Sha, X., Gu, L., Zhan, Z., Li, W.J.: A review of automated microinjection of zebrafish embryos. Micromachines **10**(1), 7 (2018)
36. Zhu, J., Hao, G., Li, S., Kong, X.: A compact mirror-symmetrical XY compliant parallel manipulator for minimizing parasitic rotations. J. Mech. Des. **144**(7), 073303 (2022)

Chapter 10
Design of a New Compliant Microgripper with Two Working Modes

Abstract This chapter presents a monolithic compliant microgripper with dual piezoelectric-actuation modes that combines the benefits of normally open and normally closed configurations. Designed to overcome the limited range of conventional microgrippers, it offers enhanced adaptability for handling objects of various sizes. The mechanism integrates bridge-type and parallelogram-based displacement amplifiers, which were optimized using a kinematic model and finite element analysis. A prototype was fabricated and experimentally validated, demonstrating a large clamping range of 860.81 μm and a natural frequency of 401 Hz. The results confirm the gripper's suitability for high-speed, precise microassembly tasks.

Keywords Microgripper · Compliant mechanism · Flexible mechanism · Optimization · Normally open gripper · Normally closed gripper

10.1 Introduction

Micro-electro-mechanical systems (MEMS) offer numerous advantages–including small size, low power consumption, and high integration–that are unattainable with traditional electromechanical systems [5]. Advances in microassembly technology have enabled the mass production of MEMS devices at micro/nano scales [4]. As the end-effector of micromanipulation systems, microgrippers are a critical component [12, 13]. Among them, piezoelectric-actuated microgrippers are widely adopted due to their high resolution, fast response, and ease of control [10, 11].

Piezoelectric microgrippers can be classified into two types: normally open and normally closed [7]. Figure 10.1 illustrates their working principles. A microgripper is referred to as *normally open* when its initial jaw gap exceeds the final clamping gap. This type is suitable for manipulating objects smaller than the gripping gap. The object is inserted into the gap, and the jaws close gradually via piezoelectric actuation, eventually securing the target through shape closure. In contrast, the *normally closed* microgripper first expands its jaw gap to accommodate larger objects. Once the object is placed, the actuation voltage is released or reduced, allowing the flexure hinges to recover elastically and grip the object.

Q. Xu, *Piezoelectric-Driven Flexible Robotic Microgrippers*,
https://doi.org/10.1007/978-981-95-4685-5_10

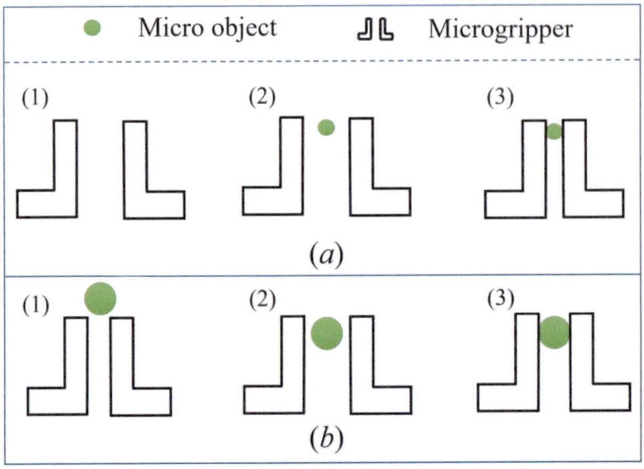

Fig. 10.1 Working principle of **a** normally open and **b** normally closed microgrippers

Numerous designs of both types have been reported. For instance, Zubir et al. developed a compliant normally-open microgripper with a stroke of 100 μm [14]. Wang et al. proposed a monolithic normally open gripper with a displacement amplification ratio of 16, featuring parallel gripping and integrated sensing [9]. Das et al. employed a three-stage amplification mechanism to design a normally-open gripper for micro/nano-manipulation [3]. Currently, normally-closed grippers are less common. Nah et al. presented a piezoelectrically actuated normally-closed compliant gripper, complete with design, fabrication, and testing [6]. Kumar et al. designed a normally-closed gripper based on flexure hinges and conducted parametric studies [8]. Choi et al. developed a micromanipulation system incorporating a normally-closed gripper [2].

Normally-open grippers are suitable for targets smaller than the jaw gap, while normally-closed grippers accommodate slightly larger objects. However, real-world micro-targets often vary in size, limiting the adaptability of both gripper types. A combination of both functionalities could enhance versatility, yet no such hybrid microgripper has been reported in the literature.

This chapter introduces a novel dual-mode compliant microgripper actuated by two piezoelectric actuators. It combines the benefits of both normally open and normally closed operation. As shown in Fig. 10.2, the left side acts as a normally-open gripper, and the right side as a normally-closed gripper. The following sections detail the design, modeling, optimization, simulation, and experimental validation of the gripper. Results confirm that the optimized design exhibits dual-mode operation and excellent gripping performance.

10.2 Mechanism Design and Modeling

10.2.1 Structural Design and Components

Figure 10.2 shows the mechanical design of the proposed microgripper, which consists of a monolithic gripper body, two piezoelectric actuators (PEAs), two gaskets, and two preload screws. The geometric structure combines flexible hinges with rigid links, using rectangular flexure hinges to generate and transmit motion due to their design simplicity, manufacturability, and large deformation capability.

Both the left and right sides of the gripper incorporate a two-stage displacement amplification mechanism, comprising a bridge-type mechanism followed by a levered parallelogram mechanism. The latter ensures jaw parallelism during gripping operations while providing additional displacement amplification.

A notable feature is the use of a convex bridge mechanism on the right side, which generates a pulling force under actuation from PEA 2. This force is transmitted via a connecting beam to the parallelogram mechanism–composed of two parallel lever structures–that amplifies the displacement while minimizing parasitic motion. As a result, the right jaw moves outward, enlarging the gripping gap. In contrast, the left side uses a concave bridge mechanism that produces a pushing force, causing the left jaw to move inward and reduce the gap. This dual-configuration enhances the gripper's adaptability to objects of varying sizes.

Preload is applied to the PEAs by tightening the rear screws against gaskets, which also prevent unwanted rotation and damage. The monolithic design reduces fabrication and assembly errors. The gripper can be mounted onto a precision motion stage via a fixed hole for accurate positioning.

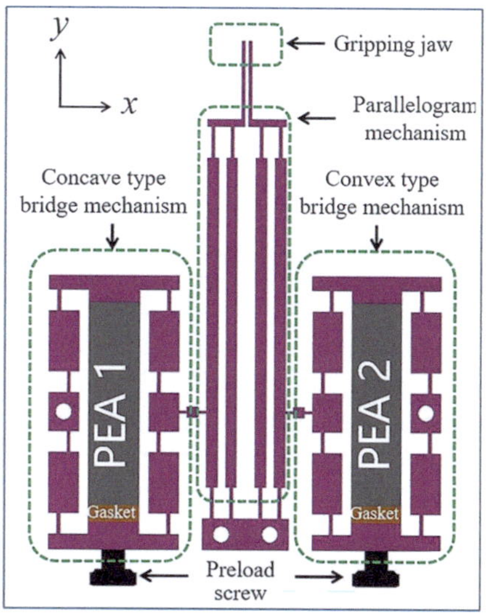

Fig. 10.2 Schematic of the proposed dual-mode compliant microgripper

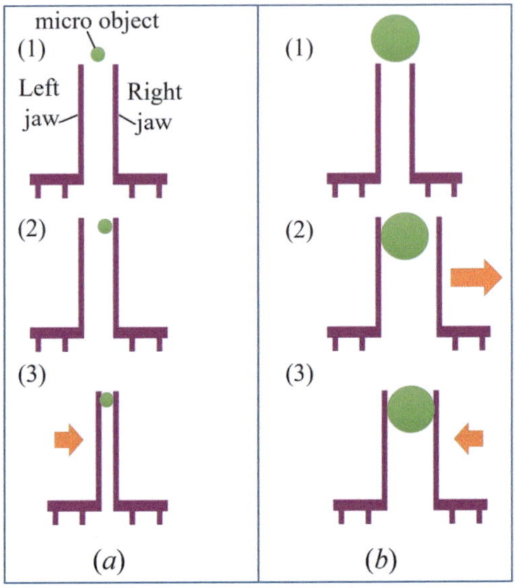

Fig. 10.3 Two working modes of the microgripper: **a** normally open mode and **b** normally closed mode

10.2.2 Working Principles and Operation Modes

The microgripper supports two distinct clamping modes, as illustrated in Fig. 10.3:

- **Normally-open mode (Fig.** 10.3**a)**: When the target is smaller than the initial jaw gap, PEA 1 is activated to close the left jaw and grip the object.
- **Normally-closed mode (Fig.** 10.3**b)**: For objects larger than the gap, PEA 2 is first activated to open the right jaw. After the object is inserted, the voltage is reduced, allowing elastic recovery of the flexures to grip the object.

This dual-mode operation significantly enhances the gripper's versatility in handling micro-objects of different sizes within the same manipulation system.

10.2.3 Kinematic Modeling and Analysis

To analyze the kinematics, a pseudo-rigid-body model (PRBM) was developed for the right-side mechanism (Fig. 10.4a). The displacement amplification ratio R_{amp} is defined as the output displacement x_{out} divided by the input displacement x_{in}.

Using velocity vector analysis (Fig. 10.4b), the relationship between input and output velocities is derived as:

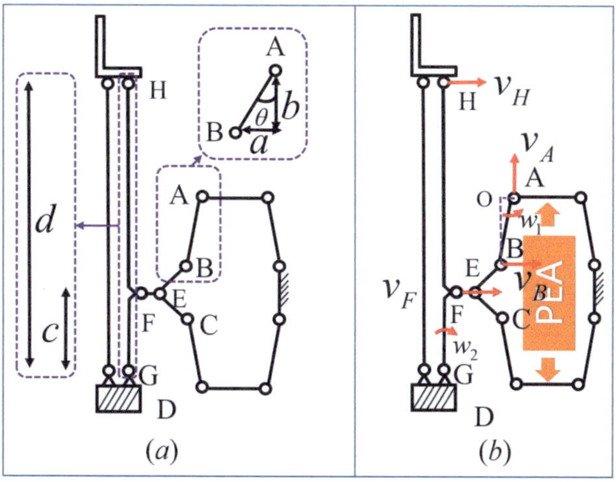

Fig. 10.4 Kinematic modeling: **a** pseudo-rigid-body model and **b** velocity vector diagram

$$v_A = \omega_1 \cdot OA, \tag{10.1}$$

$$v_B = \omega_1 \cdot OB, \tag{10.2}$$

$$v_F = v_B, \tag{10.3}$$

$$v_F = \omega_2 \cdot FG, \tag{10.4}$$

$$v_H = \omega_2 \cdot HG, \tag{10.5}$$

where ω_1 and ω_2 are the angular velocities of links AB and HF, respectively.

The displacement amplification ratio is then approximated as:

$$R_{\text{amp}} = \frac{x_{\text{out}}}{x_{\text{in}}} \approx \frac{v_H}{v_A} = \frac{bd}{2ac}, \tag{10.6}$$

where a, b, c, d are geometric parameters labeled in Fig. 10.4a. This linear relationship indicates that the jaw motion is directly proportional to the actuation displacement, and the amplification can be tuned via these parameters. A similar kinematic analysis applies to the left-side mechanism.

10.3 Optimization of Structure Parameters

10.3.1 Optimization Framework and Methodology

After completing the mechanism design of the gripper, structural optimization was conducted to achieve optimal performance. The optimization aims to identify the best design scheme according to specific objectives under given constraints.

The optimization process, performed in ANSYS commercial software, follows the schematic shown in Fig. 10.5. First, CAD software was used to complete the mechanical modeling of the clamp. Basic simulations were then carried out in the static structural module. Design variables (defined in the geometric model) and objective functions (obtained from static structural analysis) were imported into the parameter set. Essential constraints were quantitatively described in the optimization module. Both direct optimization and response surface-based optimization methods are available; the latter was chosen for this study due to its ability to provide more detailed information.

10.3.2 Definition of Optimization Problem

The end displacements (D_1 and D_2) of the clamping jaws on both sides were defined as the objective functions. Due to the local symmetry of the gripper structure–featuring symmetric bridge and parallelogram mechanisms–five structural parameters were selected as design variables to reduce complexity and eliminate internal stress. These parameters, shown in Fig. 10.6, include:

- Parameters $[b_1, b_2, b_3, b_4]$ related to the bridge mechanism
- Parameter l_1 influencing the magnification of the leverage mechanism.

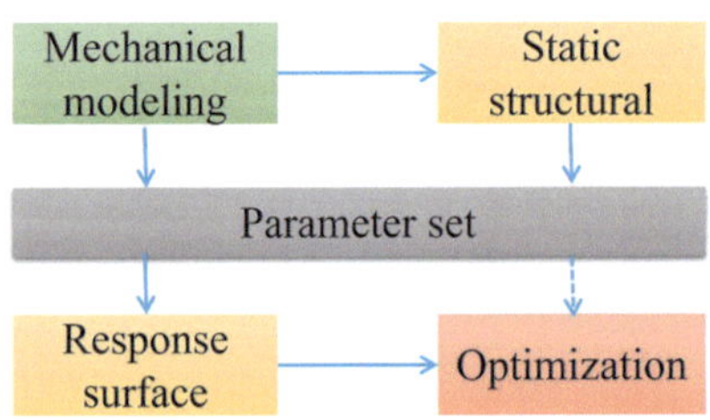

Fig. 10.5 Optimization design workflow for the microgripper

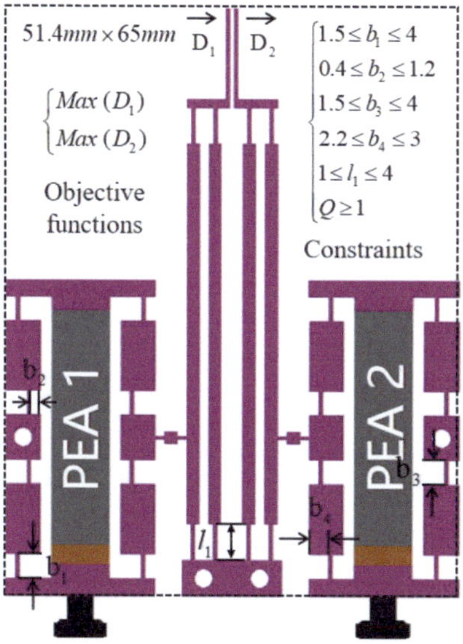

Fig. 10.6 Design variables in the optimization model

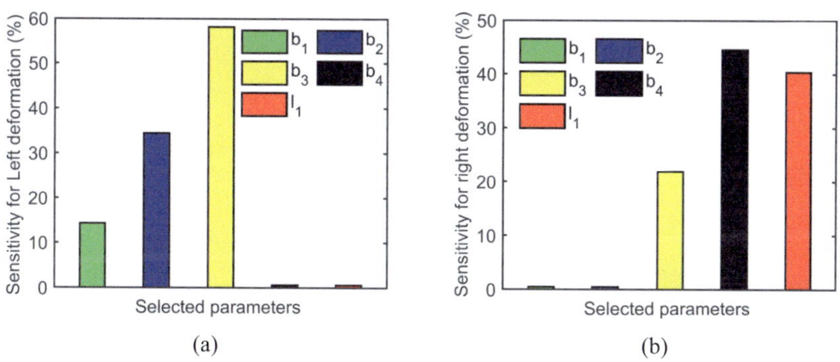

Fig. 10.7 Sensitivity analysis results: **a** left-side deformation and **b** right-side deformation

10.3.3 Sensitivity Analysis

A sensitivity analysis was conducted to evaluate the impact of each design variable on the optimization objectives. The results, shown in Fig. 10.7, indicate that:

- The deformation of the left tip is primarily influenced by parameters $[b_1, b_2, b_3]$
- The deformation of the right gripper is mainly affected by parameters $[b_3, b_4, l_1]$.

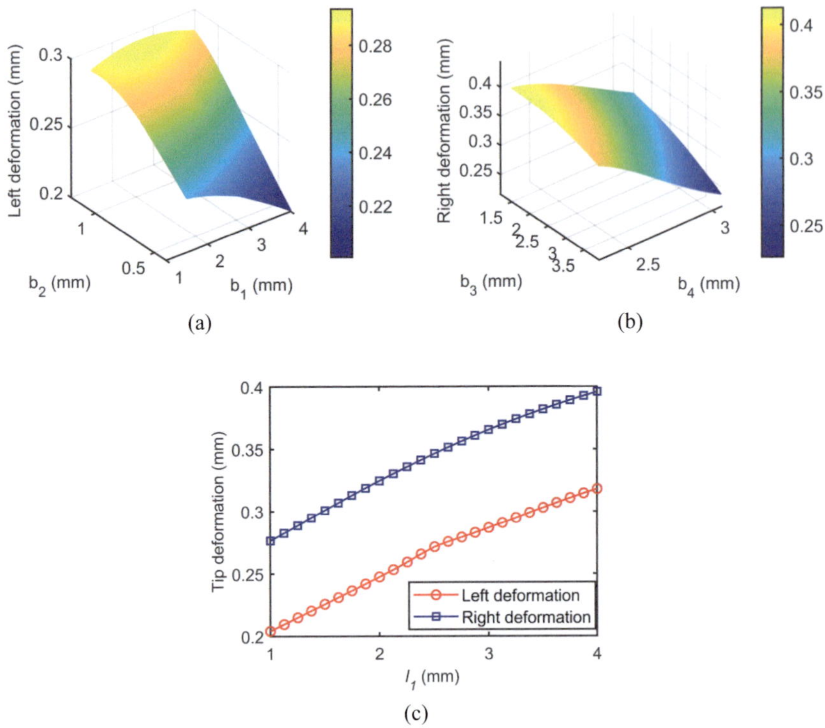

Fig. 10.8 Response surface analysis: **a** b_1 and b_2 versus D_1, **b** b_3 and b_4 versus D_2, and **c** l_1 versus bilateral deformation

10.3.4 Response Surface Analysis

The relationship between design variables and objective functions was further analyzed using response surface methodology. Figure 10.8 shows:

- For the left gripper (Fig. 10.8a): End deformation increases with decreasing b_1 and increasing b_2
- For the right gripper (Fig. 10.8b): Tip displacement decreases with increasing b_3 and b_4
- For parameter l_1 (Fig. 10.8c): An approximately proportional relationship exists with deformation on both sides, though longer flexure hinges may reduce deformation accuracy due to lack of a rotation center.

Table 10.1 Constraints and optimal values of design variables

Design variable	Lower limit	Upper limit	Optimum value
b_1 (mm)	1.5	4	2.6
b_2 (mm)	0.4	1.2	1.1
b_3 (mm)	1.5	4	2.6
b_4 (mm)	2.2	3	2.3
l_1 (mm)	1	4	4

10.3.5 Optimization Algorithm and Results

This multi-objective optimization problem was solved using the Multi-Objective Genetic Algorithm (MOGA), which supports multiple objectives and constraints. Each design variable was constrained within physical limits (Table 10.1), and the safety factor (Q) of the entire mechanism was required to be greater than 1. The optimization model converged after 1689 evaluations, with the final optimal values shown in Table 10.1.

Obviously, it is a multi-objective optimization problem. Hence, the MOGA is used as an optimization algorithm because it supports multiple objectives and constraints. Furthermore, there is a physical constraint space (i.e., constraints) for each design variable. It is important to note that the safety factor (Q) of the whole mechanism should be greater than 1. Finally, the optimization model was converged after 1689 evaluations. Table 10.1 shows the constraints and final values for all the design variables.

10.4 FEA Simulation and Comparison Study

10.4.1 Simulation Setup and Methodology

To verify the reliability of the optimization results, it is necessary to carry out performance simulation based on finite element analysis (FEA). FEA utilizes mathematical approximations to simulate real physical systems to reduce engineering test costs. Here, the simulation environment chosen is commercial ANSYS Workbench software. The PEAs used here were substituted by springs with equivalent stiffness. The preset mounting holes on the gripper are set as fixed support. The simulation study is primarily used to test the motion stroke and natural frequency of the gripper.

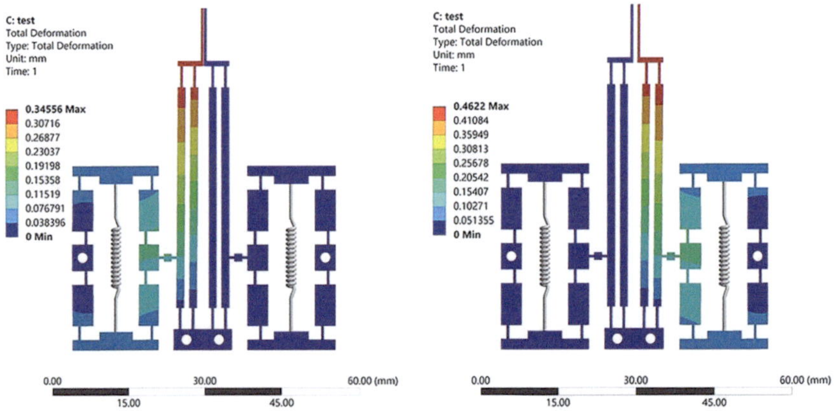

(a) (b)

Fig. 10.9 Simulation results of **a** clamping stroke of the left-side gripper and **b** working stroke of the right-side gripper

10.4.2 Static Performance Analysis

The working stroke of the gripper is simulated in the static structural module, with results shown in Fig. 10.9. The simulation results reveal that the maximum deformation of the gripper occurs at the tip of the clamp. The maximum moving displacement of the left-side gripper is 345.56 μm while the maximum working stroke of the right-side gripper is 462.2 μm. Therefore, the displacement magnification ratios of the left and right side grippers are 11.52 and 15.41, respectively.

The amplification ratios obtained by simulation differ from the theoretical calculation results (16.70 and 19.74). This discrepancy is induced because the in-plane thickness of the rectangular hinge is not fully considered in the theoretical model. In addition, the PRBM is particularly suitable for analyzing large-deformation compliant mechanisms, while the deformation magnitude of the gripper structure is relatively small.

The equivalent stress of the system is mainly distributed in the flexure hinges. The minimum safety factor of the structure is 6.0872, which is much higher than the assigned critical value of 1. The static performance analysis shows that the gripper can work normally without structural damage under PEA actuation.

10.4.3 Dynamic Performance Evaluation

Natural frequency is an important parameter characterizing the dynamic performance of the gripper, simulated using the modal analysis module. Figure 10.10 depicts the first three modes of the gripper and their corresponding natural frequencies.

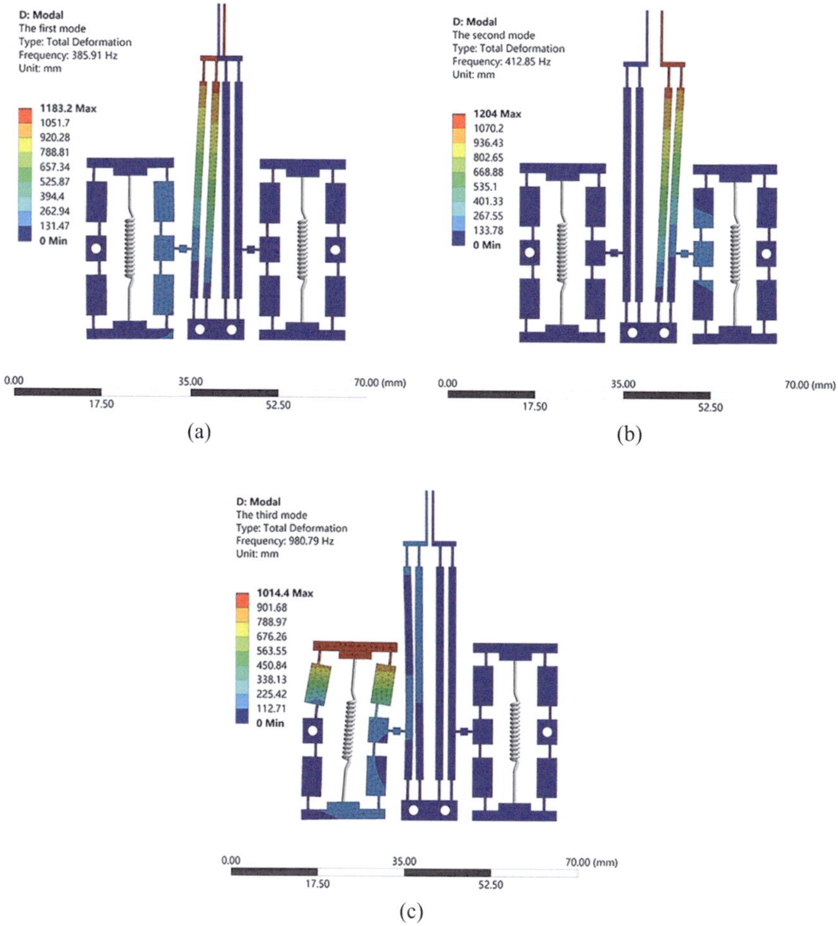

Fig. 10.10 Simulation results of the first-three modal shapes of the microgripper. **a** The 1st natural frequency (385.91 Hz), **b** the 2nd natural frequency (412.85 Hz), and **c** the 3rd natural frequency (980.79 Hz)

The first mode corresponds to the working mode of the left-side gripper, with a natural frequency of 385.91 Hz. Similarly, the natural frequency of the second mode is 412.85 Hz, which represents the working mode of the right-side gripper. The working frequencies of the two sides of the gripper are close to each other, indicating consistent dynamic performance.

The third mode of the gripper is characterized by the left bridge mechanism rotating clockwise around the plane with a frequency of 1014.4 Hz. The significant frequency separation between working modes (first and second) and higher-order modes (third) demonstrates that the designed gripper has good dynamic performance with minimal modal interference.

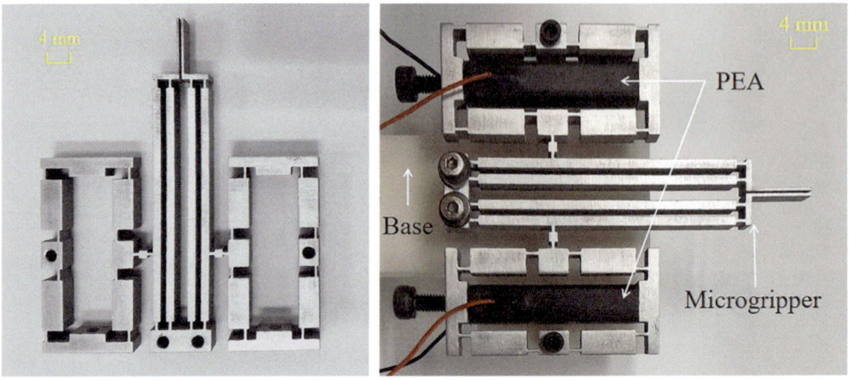

(a) (b)

Fig. 10.11 Photo of the fabricated gripper prototype. **a** Planar microgripper mechanism and **b** assembled microgripper with actuators and mounting components

10.5 Experimental Results

10.5.1 Prototype Fabrication and Assembly

A prototype of the designed microgripper has been fabricated using wire electro-discharge machining (WEDM) technology, as shown in Fig. 10.11a. This processing technology offers advantages of integrated forming and high precision. The overall structural dimensions of the gripper are 51.4 mm × 65 mm × 5 mm. The width of the rectangular flexure hinge is designed to be 0.5 mm due to machining technology limitations. The gap distance between the two jaws is set at 400 μm, which can be adjusted using preload bolts.

The material selected for manufacture is Al-7075 due to its outstanding mechanical properties. The fully assembled clamp is displayed in Fig. 10.11b, showing consistency with the concept design in Fig. 10.1. The gripper is fixed on a base or connector for mounting onto other devices, such as micropositioning stages. The PEA model used is RP150/5 × 5/30, providing a nominal displacement of 30 μm with an output thrust of 900 N.

10.5.2 Experimental Setup and Instrumentation

To examine the capability of the clamp, an experimental apparatus has been constructed. Figure 10.12 displays both the schematic diagram and photograph of the experimental setup. The microgripper is connected to an XYZ displacement platform (model: HTCL25-X, Newport Corp.) through a custom base, providing 3D spatial movement capability.

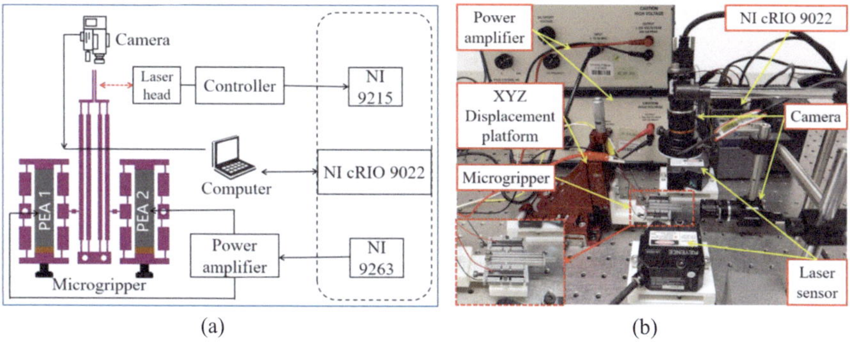

Fig. 10.12 Experimental setup: **a** schematic diagram and **b** physical apparatus

The measurement system includes:

- Two laser displacement sensors (model: LK-H055, Keyence Corp.) with ±10 mm measurement range
- NI 9215 module for signal acquisition transmitted to the master computer
- NI 9263 module and voltage amplifier (model: EPA-104, Piezo Systems, Inc.) for PEA actuation
- Real-time controller (model: NI cRIO 9022, National Instruments Corp.)
- Two industrial cameras (model: MV-CE013-80UM, Hikvision Corp.) for visual monitoring
- Vibration isolation table (model: RS2000, Newport Corp.).

10.5.3 Gripping Performance Characterization

Open-loop experiments were conducted to characterize gripping stroke and natural frequency. Due to the independent operation of left and right grippers, experiments were performed separately for each side.

Figure 10.13 shows the gripping performance under full-scale voltage actuation (0–150 V). The left gripper (Fig. 10.13a–c) achieved a maximum stroke of 342.68 μm, while the right gripper (Fig. 10.13d–f) reached 460.81 μm at 150 V. This results in an overall gripping range from 57.32 to 860.81 μm when considering the 400 μm clearance. These results align well with simulation predictions, with finer adjustments possible via preload bolt tuning.

Both sides exhibited approximately linear displacement-voltage relationships but showed significant hysteresis effects. The maximum hysteresis errors were 83.49 μm (left) and 110.90 μm (right), primarily attributable to the inherent hysteresis of piezoelectric actuators.

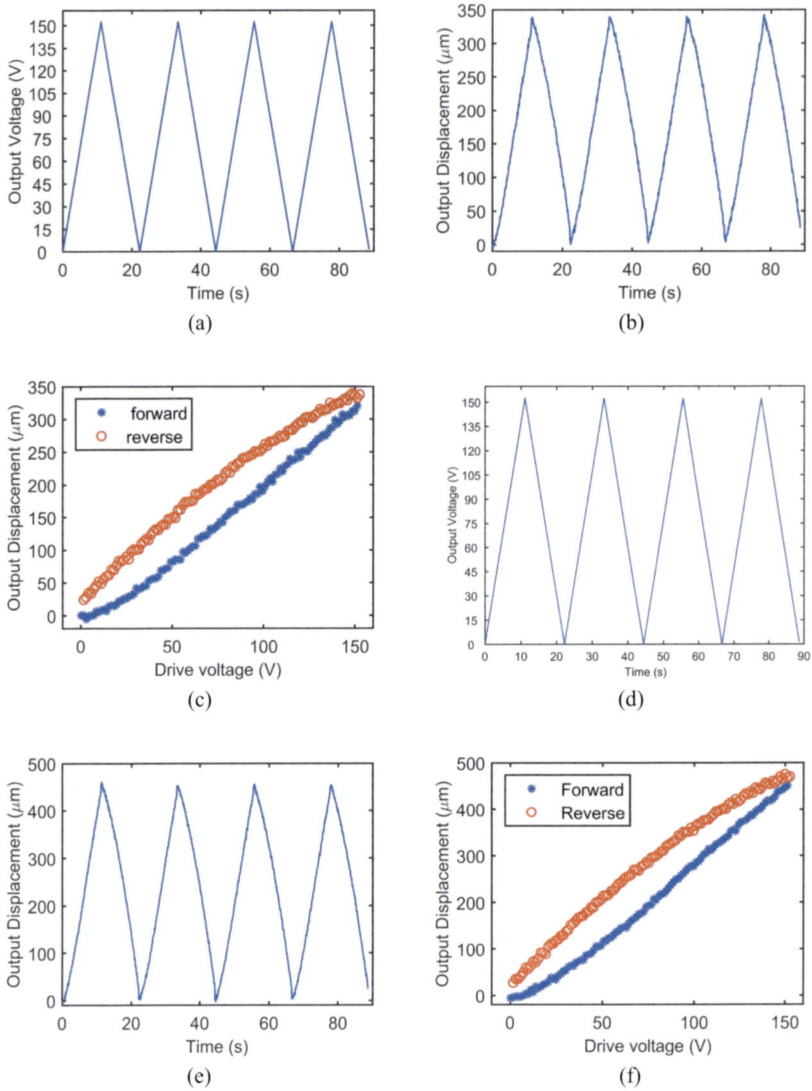

Fig. 10.13 Experimental results: **a** left-side driving voltage, **b** left tip displacement, **c** left-side hysteresis, **d** right-side driving voltage, **e** right tip displacement, and **f** right-side hysteresis

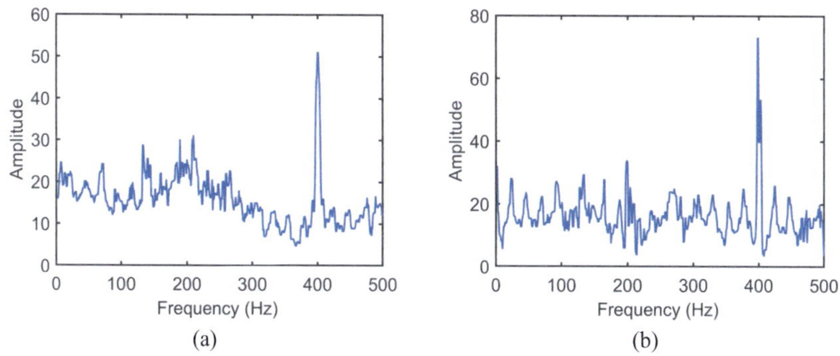

Fig. 10.14 Frequency response spectra: **a** left side and **b** right side

10.5.4 Dynamic Performance Validation

Frequency response tests were conducted using sine waves with 4 V amplitude across 1–500 Hz frequency range. Figure 10.14 shows the experimental spectra for both sides, revealing working mode resonances at 401 Hz (left) and 398 Hz (right). These values show reasonable agreement with FEA simulations (385.91 Hz and 412.85 Hz), with discrepancies arising from added mass of accessories and fabrication tolerances.

10.5.5 Functional Verification

The dual-mode functionality was experimentally validated through grasping tests:

- **Normally-open mode** (Fig. 10.15): A 300 μm diameter micro-wire was successfully grasped using PEA 1 actuation
- **Normally-closed mode** (Fig. 10.16): A 620 μm diameter solder wire (larger than initial clearance) was grasped by first opening with PEA 2, then reducing voltage for elastic recovery.

10.5.6 Performance Comparison

Table 10.2 compares the proposed microgripper with existing PEA-actuated designs. The developed gripper demonstrates unique advantages including:

- Dual operating modes (normally open and normally closed)
- Large clamping stroke (803.49 μm)
- Competitive natural frequency (401 Hz)
- Enhanced adaptability to variable target sizes.

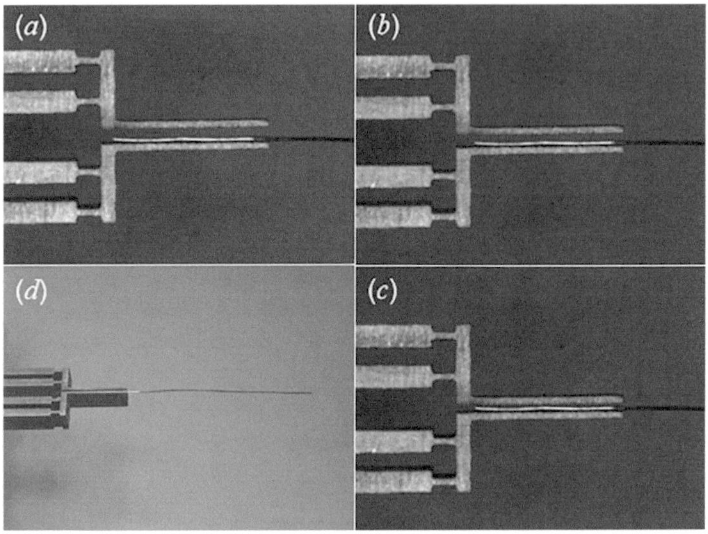

Fig. 10.15 Normally open mode grasping sequence (300 μm wire): **a** approach, **b** grasping, **c** clamped, and **d** lifted

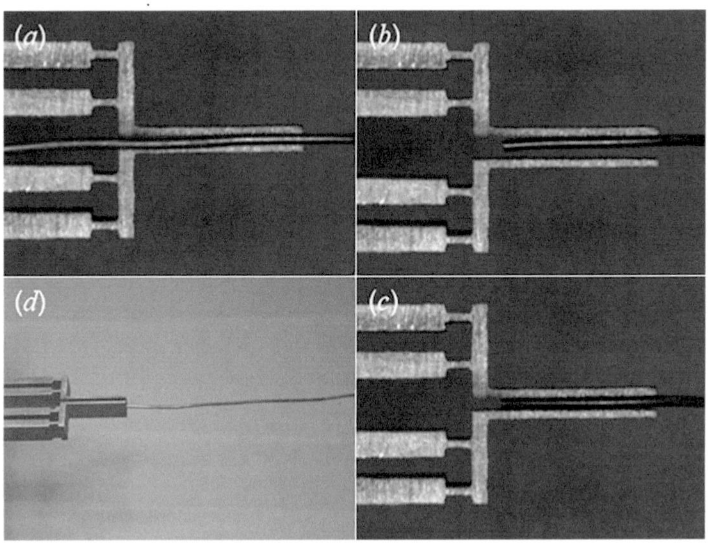

Fig. 10.16 Normally-closed mode grasping sequence (620 μm wire): **a** initial clearance, **b** jaw opening, **c** elastic recovery, and **d** successful grasp

Table 10.2 Performance comparison of PEA-actuated microgrippers

Reference	Working mode	Amplification stages	Gripping stroke (μm)	Natural frequency (Hz)
[14]	Normally open	2	100	–
[9]	Normally open	2	432	–
[3]	Normally open	3	93.52	1044
[1]	Normally closed	2	180	–
[6]	Normally closed	2	170	–
This work	Both modes	2	803.49	401

This dual-mode capability significantly reduces time consumption for gripper adjustment or replacement in applications with variable target sizes.

10.6 Conclusion

This chapter has presented a novel dual-mode compliant microgripper actuated by two piezoelectric actuators, featuring the unique combination of both normally open and normally closed operating principles. The proposed design significantly enhances adaptability to various target sizes, enabling effective manipulation of objects both smaller and slightly larger than the initial gripper gap. This capability allows operators to select the appropriate working mode based on target dimensions, eliminating time-consuming adjustments or gripper replacements.

The mechanical design incorporates a two-stage flexible amplification mechanism on both sides of the gripper. Through pseudo-rigid-body modeling, the kinematic relationships were established and the displacement amplification ratios were derived. Structural optimization using the Multi-Objective Genetic Algorithm (MOGA) identified optimal geometric parameters to maximize working stroke. Comprehensive simulation and experimental validation confirmed the gripper's performance characteristics:

- Left-side working stroke: 342.68 μm
- Right-side working stroke: 460.81 μm
- First two natural frequencies: 401 and 398 Hz (corresponding to operating modes).

Experimental demonstrations successfully verified both operating modes using metal wires of different diameters, confirming the practical functionality of the dual-mode design.

Future work will focus on enhancing the gripper's precision through advanced control strategies. Implementation of displacement and force sensing with closed-loop control will be investigated to improve reliability and manipulation accuracy for microassembly applications.

References

1. Choi, H., Lee, D., Kim, S., Han, C.: The development of a microgripper with a perturbation-based configuration design method. J. Micromech. Microeng. **15**(6), 1327 (2005)
2. Choi, H., Shin, D., Ryuh, Y., Han, C.: Development of a micro manipulator using a microgripper and PZT actuator for microscopic operations. In: Proceedings of the 2011 IEEE International Conference on Robotics and Biomimetics. Karon Beach, Thailand (2011)
3. Das, T., Shirinzadeh, B., Ghafarian, M., Al-Jodah, A., Zhong, Y., Smith, J.: Design, analysis and experimental investigations of a high precision flexure-based microgripper for micro/nano manipulation. Mechatronics **69**, 102396 (2020)
4. Degtiaryov, N., Sukhanov, A., Sukhanov, V., Artemova, A., Firsov, A.: Design and manufacture of miniature power microassemblies using 3d integration technology. In: Proceedings of the 2021 IEEE Conference of Russian Young Researchers in Electrical and Electronic Engineering. St. Petersburg and Moscow Russia (2021)
5. Mishra, M., Dubey, V., Mishra, P., Khan, I.: Mems technology: a review. J. Eng. Res. Rep. **4**(1), 1–24 (2019)
6. Nah, S., Zhong, Z.: A microgripper using piezoelectric actuation for micro-object manipulation. Sens. Actuators, A **133**(1), 218–224 (2007)
7. Nikoobin, A., Niaki, M.: Deriving and analyzing the effective parameters in microgrippers performance. Sci. Iranica **19**(6), 1554–1563 (2012)
8. Raghavendra, M., Kumar, A., Jagdish, B.: Design and analysis of flexure-hinge parameter in microgripper. Int. J. Adv. Manuf. Technol. **49**(9–12), 1185–1193 (2010)
9. Wang, D., Yang, Q., Dong, H.: A monolithic compliant piezoelectric driven microgripper: design, modeling, and testing. IEEE/ASME Trans. Mechatron. **18**(1), 138–147 (2011)
10. Wang, F., Shi, B., Huo, Z., Tian, Y., Zhang, D.: Control and dynamic releasing method of a piezoelectric actuated microgripper. Precis. Eng. **68**, 1–9 (2021)
11. Xu, Q.: Robust impedance control of a compliant microgripper for high-speed position/force regulation. IEEE Trans. Industr. Electron. **62**(2), 1201–1209 (2014)
12. Xu, Q.: Design, fabrication and testing of an mems microgripper with dual-axis force sensor. IEEE Sens. J. **15**(10), 6017–6026 (2015)
13. Yang, S., Xu, Q., Nan, Z.: Design and development of a dual axis force sensing mems microgripper. J. Mech. Robot. Trans. ASME **9**(6), 061011 (2017)
14. Zubir, M., Shirinzadeh, B., Tian, Y.: A new design of piezoelectric driven compliant-based microgripper for micromanipulation. Mech. Mach. Theory **44**(12), 2248–2264 (2009)

Chapter 11
Design of a Piezoelectric-Driven Microgripper With Three Working Modes

Abstract This chapter presents the conceptual design of a reconfigurable piezoelectric-driven microgripper to obtain both excellent static and dynamic performance for complex micromanipulation. For the first time, a flexure-based compliant microgripper is designed to realize three working modes. The analytical model of the microgripper is established based on the beam deformation theory and pseudo-rigid-body modeling approach. The multi-objective optimization is performed based on response surface methodology to determine the structural parameters of the microgripper. A finite element model-based simulation study is implemented to evaluate the gripper's static and dynamic performance. Moreover, a prototype of the microgripper has been fabricated for experimental testing. The results reveal that the microgripper obtains a large displacement amplification ratio of 55.17 in mode I, a high natural frequency of 838.09 Hz in mode II, and a balanced performance in mode III. The micromanipulation capability of the microgripper has been demonstrated by executing gripping-holding-releasing operations of various tiny objects.

Keywords Microgripper · Mechanism design · Piezoelectric actuation · Performance test · Reconfigurable mechanism

11.1 Introduction

Nowadays, with the rapid development of precision engineering, material engineering, and biotechnology, various micromanipulation tasks–including microgripping, micropositioning, and microassembly–are extensively required for different microelectromechanical systems (MEMS) [4, 15, 18, 19]. As a typical end-effector, microgrippers perform ultrahigh-precision gripping-holding-releasing operations, playing an essential role in micromanipulation systems [20, 31]. Recently, MEMS has undergone trends toward greater integration and miniaturization, imposing stricter requirements on the comprehensive performance of microgrippers. Therefore, to satisfy the increasing demands of complex and stringent micromanipulation tasks, it is necessary to develop high-performance microgrippers.

Q. Xu, *Piezoelectric-Driven Flexible Robotic Microgrippers*,
https://doi.org/10.1007/978-981-95-4685-5_11

Based on different actuation principles, various types of microgrippers have been proposed, including electrostatic [3], magnetic [32], and piezoelectric (PZT) microgrippers [28]. In particular, PZT-driven microgrippers have gained extensive attention owing to their rapid response speed and precise positioning resolution [29]. Typical performance indicators for evaluating PZT-driven microgrippers include both static specifications (e.g., displacement amplification ratio (DAR), gripping stroke (GS), and area efficiency (AE)) and dynamic specifications (e.g., first natural frequency (FNF)) [18]. Because piezoelectric actuators usually provide limited output displacement, DAR is considered a primary static performance metric for PZT microgrippers to ensure sufficient gripping stroke for target objects, which also indirectly affects area efficiency. Using various compliant structures, diverse PZT microgrippers with considerable DAR values have been designed. For instance, Yang et al. [34] developed a PZT microgripper with a DAR of 19.3 based on double-rocker and parallelogram mechanisms, achieving a GS of 328.2 μm. Ai et al. [1] employed a Scott-Russell mechanism integrated with a lever mechanism to achieve a DAR of 22.2. Chen et al. [5] presented a microgripper with a two-stage displacement amplifier, realizing a large DAR of 52.6 and a GS of 900 μm. However, the FNF values of these microgrippers were only 117 Hz [34], 70 Hz [1], and 217 Hz [5], respectively, restricting their practical working bandwidth.

As an essential dynamic indicator, FNF directly governs the upper limit of the working bandwidth, with the usable bandwidth typically accounting for approximately 1/5 of the first natural frequency. Several PZT microgrippers with high FNF have been reported. For example, Liang et al. [12] developed an asymmetrical microgripper with an FNF of 531 Hz and a DAR of 13.94. Wang et al. [30] proposed a microgripper with an FNF of 730 Hz and a DAR of 6.38. Das et al. [8] reported a symmetric microgripper with a DAR of 12.5, a GS of 184.04 μm, and an FNF of 854 Hz. Notably, these designs achieved high FNF at the expense of DAR and GS. Figure 11.1 shows the relationship between DAR and FNF for typical PZT microgrippers published in recent years, revealing a trade-off between these two specifica-

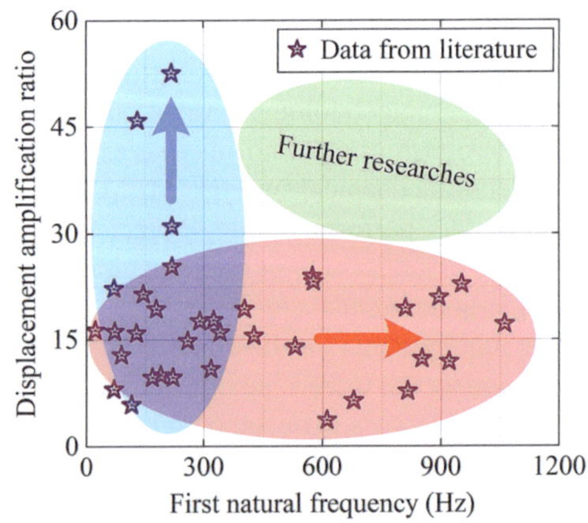

Fig. 11.1 Relationship between displacement amplification ratio and first natural frequency for typical piezoelectric microgrippers

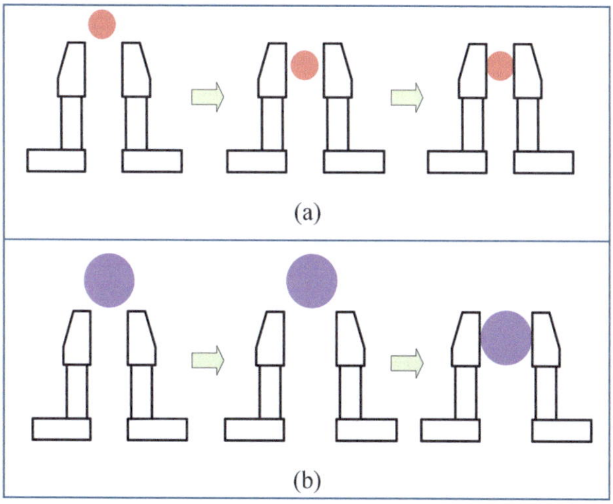

Fig. 11.2 Two types of microgrippers. **a** Normally-open type and **b** normally-closed type

tions. While some designs achieved large displacement amplification (along the blue arrow), others attained high working bandwidth (along the red arrow). Few designs simultaneously achieve both high DAR and high FNF (corresponding to the upper-right region). This limitation restricts the performance of existing microgrippers in complex and variable operating environments. Thus, it is desirable to improve both specifications for practical applications.

Microgrippers can generally be divided into normally-open and normally-closed types. As shown in Fig. 11.2a, in a normally-open microgripper, the actuator drives the jaws inward to grasp target objects smaller than the initial jaw gap [17]. The actuation force must be maintained during grasping, enabling precise control of the gripping force [22]. Numerous normally-open microgrippers have been reported [1, 5, 8, 12, 28–30, 34]. In contrast, as shown in Fig. 11.2b, a normally-closed microgripper can grasp objects larger than the initial jaw gap, offering the advantage of energy efficiency [22]. For example, Long et al. [16], Nah et al. [21], Kumar et al. [23], and Choi et al. [6] have developed various normally-closed microgrippers. A microgripper combining both normally-open and normally-closed functionalities would provide the advantages of both types, enabling a broader gripping range through bidirectional jaw motion. However, such designs are challenging and rarely reported.

Two main challenges exist in microgripper design: first, the difficulty in achieving both excellent static performance (e.g., high DAR, GS, and AE) and outstanding dynamic performance (e.g., high FNF) in a single device; second, the complexity of integrating both normally-open and normally-closed grasping modes into one microgripper. To address these issues, this work proposes a conceptual design for a reconfigurable microgripper. By applying constraints to different selective constraint holes, three working modes are realized in a single microgripper, enabling both high static and dynamic performance as well as both grasping types. Specifically, Mode I provides a large DAR of 55.17, Mode II delivers a high natural frequency of 838.09 Hz, and Mode III offers balanced performance with a DAR of 26.33 and an FNF of 401.58 Hz. Modes I and II operate as normally-open, while Mode III functions as normally-closed. This reconfigurable design achieves superior performance compared to existing microgrippers. The main contribution of this work is the design and development of a reconfigurable microgripper with three working modes for advanced micromanipulation tasks.

11.2 Design and Modeling of the Microgripper

11.2.1 Structural Design of the Reconfigurable Microgripper

Numerous microgrippers based on compliant mechanisms have been designed, with comprehensive atlases [10, 27] providing a wide selection of structures and hinges. Flexure hinges are widely employed in compliant mechanisms [11], with circular flexure hinges (CFHs) and rectangular flexure hinges (RFHs) being particularly common, as illustrated in Fig. 11.3a, b. Given the limited output displacement of piezoelectric actuators (typically tens of micrometers), various displacement amplification mechanisms based on flexure hinges are essential for PZT microgrippers. These include bridge-type displacement amplification mechanisms (BDAMs) and lever-type displacement amplification mechanisms (LDAMs). BDAMs can be categorized into single-output or double-output types, as shown in Fig. 11.3c, d. The double-output BDAM, which generates identical displacement on both sides, is adopted as the first-stage displacement amplifier in the proposed microgripper.

To further amplify the displacement, a novel dual-stage lever displacement amplification mechanism (DLDAM) is designed to enable three distinct working modes. Reconfiguration between modes is achieved by applying constraints to different selective constraint holes. In Mode I of the DLDAM, constraint holes corresponding to two rectangular flexure hinges are fixed, as shown in Fig. 11.3e, resulting in amplified output displacement in the same direction as the input. The finite element method (FEM) simulation result for Mode I is shown in Fig. 11.3f. In Mode II, constraint holes corresponding to two circular flexure hinges are fixed (Fig. 11.3g), producing amplified output displacement in the same direction, as validated by FEM simulation in Fig. 11.3h. In Mode III, constraint holes corresponding to one rectangular and one circular flexure hinge are fixed (Fig. 11.3i), generating a reverse amplified displacement at the output end (Fig. 11.3j).

To minimize parasitic displacement, compliant guide mechanisms are incorporated at both the input and output ends, as shown in Fig. 11.3k, l. A pair of gripping jaws, designed to grasp target objects and integrate force sensors, is illustrated in Fig. 11.3m. The overall structural configuration of the proposed microgripper is

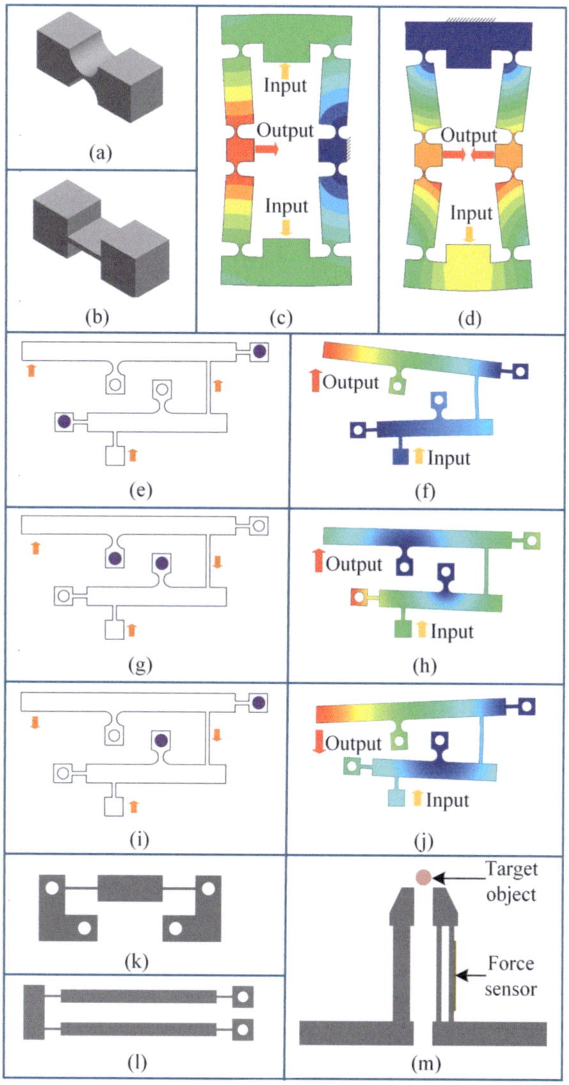

Fig. 11.3 Compliant mechanisms in the microgripper. **a** Circular flexure hinge. **b** Rectangular flexure hinge. **c** Single-output bridge-type displacement amplification mechanism. **d** Double-output bridge-type displacement amplification mechanism. **e** Mode I of DLDAM. **f** FEM simulation of Mode I. **g** Mode II of DLDAM. **h** FEM simulation of Mode II. **i** Mode III of DLDAM. **j** FEM simulation of Mode III. **k** Compliant guide mechanism at input end. **l** Compliant guide mechanism at output end. **m** Gripping jaws

detailed in Fig. 11.4. The microgripper comprises a stack piezoelectric ceramic actuator (SPCA), a double-output BDAM, a DLDAM, two compliant guide mechanisms (CGMs) at the input and output ends, a pair of gripping jaws, nine fixing bolts, eight selective constraint holes, and a preloading bolt. By applying constraints to specific selective constraint holes, the microgripper can operate in different modes, enabling reconfigurable functionality to meet diverse working requirements. Specifically:

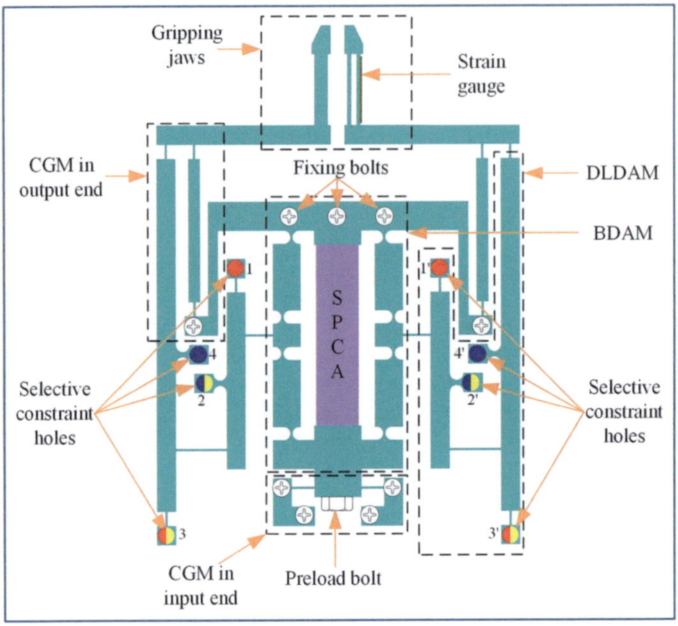

Fig. 11.4 Mechanism of the multi-mode PZT microgripper. CGM: compliant guide mechanism

- *Mode I*: For grasping relatively small objects requiring a large gripping stroke, constraints are applied to holes 1, 3, 1′, and 3′ (marked in red in Fig. 11.4). This configuration maximizes displacement amplification ratio (DAR) and gripping stroke (GS).
- *Mode II*: For high-frequency operations requiring avoidance of resonance, constraints are applied to holes 2, 4, 2′, and 4′ (marked in blue in Fig. 11.4). This mode prioritizes high first natural frequency (FNF) and working bandwidth.
- *Mode III*: For grasping objects larger than the initial jaw gap, constraints are applied to holes 2, 3, 2′, and 3′ (marked in yellow in Fig. 11.4). This mode enables outward expansion of the jaws for normally-closed grasping.

11.2.2 Theoretical Modeling of the Proposed Microgripper

To obtain the theoretical displacement amplification ratio (DAR) of the microgripper in different working modes, we first analyze the displacement amplification of each mechanism component. The theoretical model is established under the following assumptions: elastic deformation occurs primarily at flexure hinges, while other components are treated as rigid bodies; both translational and rotational deformations of flexure hinges remain within the linear range [14].

According to beam deformation theory, the deformation of a flexure hinge can be described using the compliance matrix approach [11, 24]:

$$\Delta = CF = \begin{bmatrix} c_{11} & c_{12} & c_{13} \\ c_{21} & c_{22} & c_{23} \\ c_{31} & c_{32} & c_{33} \end{bmatrix} F$$
$$= \begin{bmatrix} \frac{1}{Eb}\left[\pi\left(\frac{r}{t}\right)^{1/2} - \frac{\pi}{2}\right] & 0 & 0 \\ 0 & \frac{9\pi}{2Eb}\left(\frac{r}{t}\right)^{5/2} + \frac{3\pi}{2Eb}\left(\frac{r}{t}\right)^{3/2} & \frac{9\pi r^{3/2}}{2Ebt^{5/2}} \\ 0 & \frac{9\pi r^{3/2}}{2Ebt^{5/2}} & \frac{9\pi r^{1/2}}{2Ebt^{5/2}} \end{bmatrix} F \tag{11.1}$$

where $\Delta = [\delta_x, \delta_y, \delta_z]^T$ represents the planar deformation of the flexure hinge, C is the compliance matrix, c_{ij} $(i, j = 1, 2, 3)$ are compliance coefficients, $F = [F_x, F_y, M]^T$ is the load vector, E is the elastic modulus, and r, t, and b denote the radius, thickness, and width of the flexure hinge, respectively.

Considering the symmetry of the double-output bridge-type displacement amplification mechanism (BDAM), we analyze one bridge arm (Fig. 11.5a) and its equivalent spring model (Fig. 11.5b). Here, points A and B represent two flexure hinges, with l_1, l_2, and α denoting the horizontal distance, vertical distance, and angle between them. Using Euler-Bernoulli beam theory, the force and moment balance equations are:

$$F_{Ax} = F_{Bx} = \frac{F_{SPCA}}{2} \tag{11.2}$$

$$F_{Ay} = F_{By} = \frac{F_s}{2} \tag{11.3}$$

$$M_A = M_B = \frac{F_{Ax}l_2 - F_{Ay}l_1}{2} = \frac{F_{Bx}l_2 - F_{By}l_1}{2} \tag{11.4}$$

where $F_A = [F_{Ax}, F_{Ay}, M_A]^T$ and $F_B = [F_{Bx}, F_{By}, M_B]^T$ are load vectors at hinges A and B, and F_{SPCA} and F_s are forces at the input and output ends of the BDAM.

Substituting Eqs. (11.2)–(11.4) into Eq. (11.1) yields the deformations:

$$\delta_{Ax} = c_{11}F_{Ax} \tag{11.5}$$

$$\delta_{Bx} = c_{11}F_{Bx} \tag{11.6}$$

$$\delta_{Ay} = c_{22}F_{Ay} + c_{23}M_A \tag{11.7}$$

$$\delta_{By} = c_{22}F_{By} + c_{23}M_B \tag{11.8}$$

$$\delta_{A\theta} = c_{32}F_{Ay} + c_{33}M_A \tag{11.9}$$

$$\delta_{B\theta} = c_{32}F_{By} + c_{33}M_B \tag{11.10}$$

where $\Delta_A = [\delta_{Ax}, \delta_{Ay}, \delta_{A\theta}]^T$ and $\Delta_B = [\delta_{Bx}, \delta_{By}, \delta_{B\theta}]^T$ are deformation vectors.

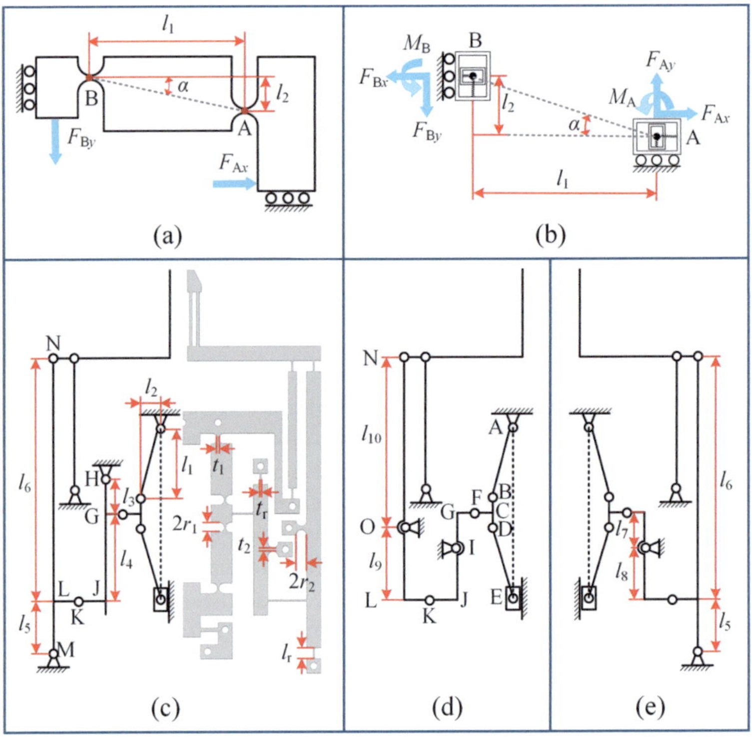

Fig. 11.5 Analytical model of the microgripper. **a** Force analysis of one bridge arm. **b** Equivalent spring model. **c** PRBM of working Mode I. **d** PRBM of working Mode II. **e** PRBM of working Mode III

The total deformation of one bridge is:

$$\Delta_x = \delta_{Ax} + \delta_{Bx} + l_2\delta_{A\theta} \tag{11.11}$$

$$\Delta_y = \delta_{Ay} + \delta_{By} + l_1\delta_{A\theta} \tag{11.12}$$

where Δ_x and Δ_y are horizontal and vertical deformations.

Substituting Eqs. (11.5)–(11.10) into Eqs. (11.11) and (11.12), gives:

$$\Delta_x = \left(2c_{11} + \frac{c_{33}l_2^2}{2}\right)F_{Ax} + \left(c_{32}l_2 - \frac{c_{33}l_1l_2}{2}\right)F_{Ay} \tag{11.13}$$

$$\Delta_y = \left(c_{23}l_2 + \frac{c_{33}l_1l_2}{2}\right)F_{Ax} + \left(2c_{22} + c_{32}l_1 - c_{23}l_1 - \frac{c_{33}l_1^2}{2}\right)F_{Ay} \tag{11.14}$$

The displacement amplification ratio of the BDAM is:

$$
\begin{aligned}
R_{\text{bridge}} &= \frac{2\Delta_y}{2\Delta_x} = \frac{\Delta_y}{\Delta_x} \\
&= \frac{(c_{23}l_2 + \frac{c_{33}l_1l_2}{2})F_{Ax} + (2c_{22} - \frac{c_{33}l_1^2}{2})F_{Ay}}{(2c_{11} + \frac{c_{33}l_2^2}{2})F_{Ax} + (c_{32}l_2 - \frac{c_{33}l_1l_2}{2})F_{Ay}} \\
&= \frac{(2c_{32}l_2 + c_{33}l_1l_2)F_{SPCA} + (4c_{22} - c_{33}l_1^2)F_s}{(4c_{11} + c_{33}l_2^2)F_{SPCA} + (2c_{32}l_2 - c_{33}l_1l_2)F_s}
\end{aligned} \tag{11.15}
$$

Using the pseudo-rigid-body model (PRBM) approach [1, 20, 29, 34], we replace flexure hinges with revolute joints and torsional springs (Fig. 11.5c–e). The DARs of the dual-stage lever displacement amplification mechanism (DLDAM) are:

$$R_{\text{lever I}} = \frac{l_4l_6}{l_3l_5} \tag{11.16}$$

Similarly, by referring to Fig. 11.5d, e, the displacement amplification ratio of DLDAM in modes II and III can be calculated as follows.

$$R_{\text{lever II}} = \frac{l_8l_{10}}{l_7l_9} \tag{11.17}$$

$$R_{\text{lever III}} = \frac{l_8l_6}{l_7l_5} \tag{11.18}$$

The overall DARs for the three working modes are:

$$R_{\text{I}} = R_{\text{bridge}}R_{\text{lever I}} \tag{11.19}$$

$$R_{\text{II}} = R_{\text{bridge}}R_{\text{lever II}} \tag{11.20}$$

$$R_{\text{III}} = R_{\text{bridge}}R_{\text{lever III}} \tag{11.21}$$

For an input displacement S_{in} from the piezoelectric actuator, the gripping strokes are:

$$S_{\text{I}} = R_{\text{I}}S_{\text{in}} \tag{11.22}$$

$$S_{\text{II}} = R_{\text{II}}S_{\text{in}} \tag{11.23}$$

$$S_{\text{III}} = R_{\text{III}}S_{\text{in}} \tag{11.24}$$

This completes the theoretical static modeling. While dynamic modeling of piezoelectric actuators [2, 26] is important for control purposes, this work focuses on structural performance parameters. Equations (11.22)–(11.24) identify key structural parameters for optimization [19], with final optimized values listed in Table 11.1.

Table 11.1 Optimized design parameters

Variable	Symbol	Value (mm)
Horizontal distance of link AB	l_1	0.5
Vertical distance of link AB	l_2	7
Length of link GH	l_3	4
Length of link GJ	l_4	12
Length of link LM	l_5	6.5
Length of link LN	l_6	30.5
Length of link GI	l_7	5
Length of link LO	l_9	10
Thickness of CFH in BDAM	t_1	0.2
Radius of CFH in BDAM	r_1	1.4
Thickness of CFH in LDAM	t_2	0.5
Radius of CFH in LDAM	r_2	1.5
Thickness of RFH in LDAM	t_r	0.2
Radius of RFH in LDAM	l_r	1.5

11.3 Optimization and Simulation Study of the Microgripper

11.3.1 Multi-objective Response Surface Optimization Based on FEM

To enhance the comprehensive performance across different working modes, a multi-objective response surface optimization was conducted using finite element method (FEM). The design variables, listed in Table 11.1 and illustrated in Fig. 11.5, were optimized to maximize gripping stroke (GS) and first natural frequency (FNF) in all three working modes. The optimization problem is formulated as:

$$\text{Maximize:} \begin{cases} S_1 = f_1(X);\ M_1 = f_2(X) \\ S_2 = f_3(X);\ M_2 = f_4(X) \\ S_3 = f_5(X);\ M_3 = f_6(X) \end{cases} \tag{11.25}$$

$$X = [l_1,\ l_2,\ l_3,\ l_4,\ l_5,\ l_6,\ l_7,\ l_9,\ t_1,\ r_1,\ t_2,\ r_2,\ t_r,\ r_r] \tag{11.26}$$

$$\text{Subject to:} \begin{cases} 0.45 \le l_1 \le 0.55;\ 6.3 \le l_2 \le 7.7 \\ 3.6 \le l_3 \le 4.4;\ 11.8 \le l_4 \le 12.2 \\ 5.85 \le l_5 \le 7.15;\ 26.3 \le l_6 \le 29.7 \\ 4.8 \le l_7 \le 5.2;\ 9.8 \le l_9 \le 10.2 \\ 0.18 \le t_1 \le 0.22;\ 1.26 \le r_1 \le 1.54 \\ 0.45 \le t_2 \le 0.55;\ 0.675 \le r_2 \le 0.825 \\ 0.18 \le t_r \le 0.22;\ 1.35 \le l_r \le 1.65 \end{cases} \tag{11.27}$$

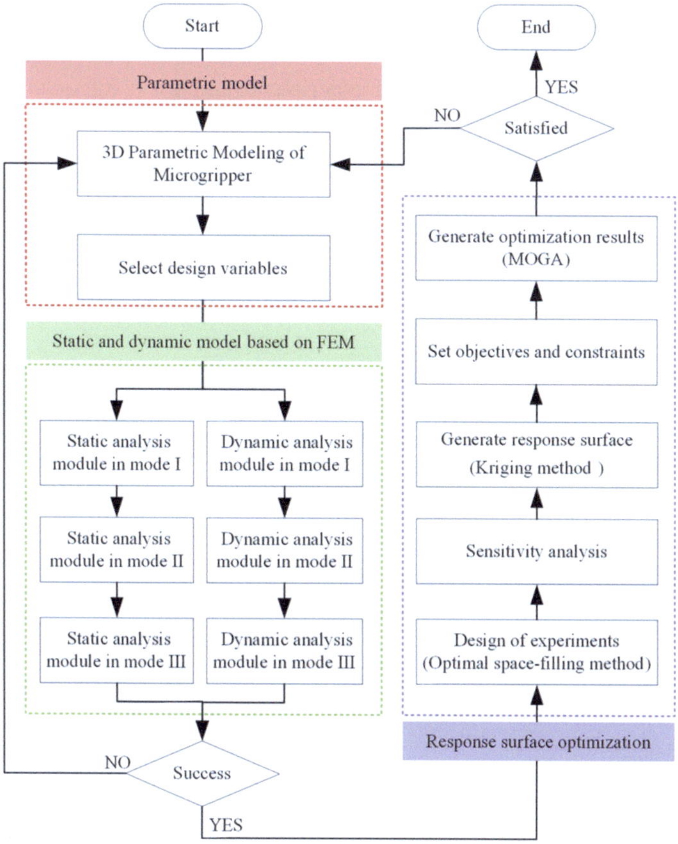

Fig. 11.6 Flowchart of the multi-objective optimization process

The optimization was performed in ANSYS Workbench, following the flowchart in Fig. 11.6. The process involved:

1. 3-D parametric modeling with design variable initialization
2. Static and dynamic FEM analysis for all three working modes
3. Response surface optimization using optimal space-filling design
4. Kriging-based response surface generation
5. Multi-objective genetic algorithm (MOGA) for solution generation.

The optimization process begins with the development of a three-dimensional parametric model of the microgripper within ANSYS software, where all design variables specified in Table 11.1 are defined as parameters to enable automated geometric modifications. Subsequently, finite element method (FEM) analysis modules are established for both static and dynamic evaluations across the three working modes. In the static analysis modules, the gripping stroke (GS) is selected as the optimization objective for each mode, while the first natural frequency (FNF) serves as the objective in the dynamic analysis modules. The response surface optimization phase employs the optimal space-filling design method based on the maximum entropy criterion to generate design of experiment samples. A full quadratic model is utilized to capture both main effects and interaction terms. Response surfaces are constructed using the Kriging method, which provides accurate interpolation of the design space through Gaussian process regression.

11.3.2 Sensitivity Analysis and Response Surface Evaluation

To quantify the influence of each design variable on the six objective functions (S_1–S_3 and M_1–M_3), a comprehensive sensitivity analysis is conducted. The results, presented in Fig. 11.7, reveal significant variations in parameter sensitivities across

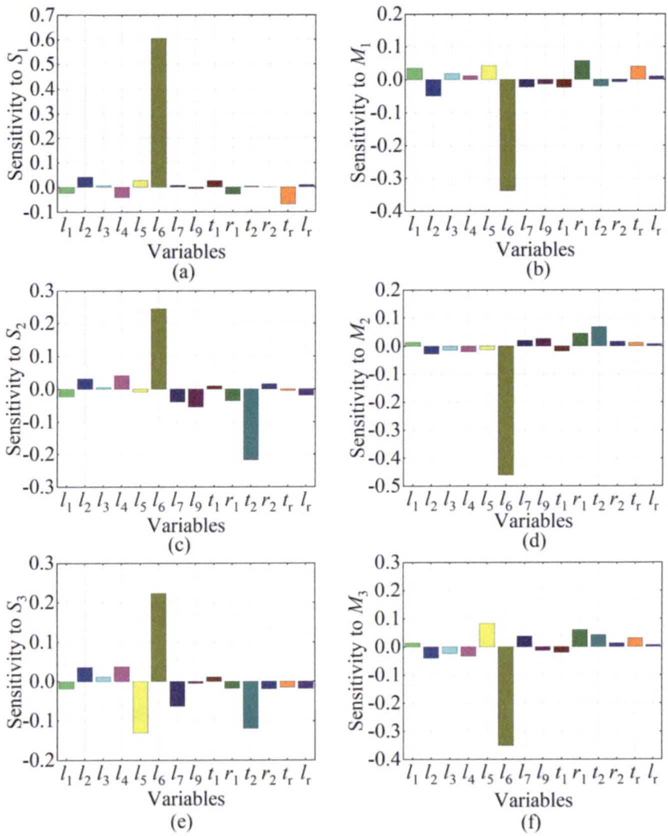

Fig. 11.7 Sensitivity analysis results: **a** S_1, **b** M_1, **c** S_2, **d** M_2, **e** S_3, **f** M_3

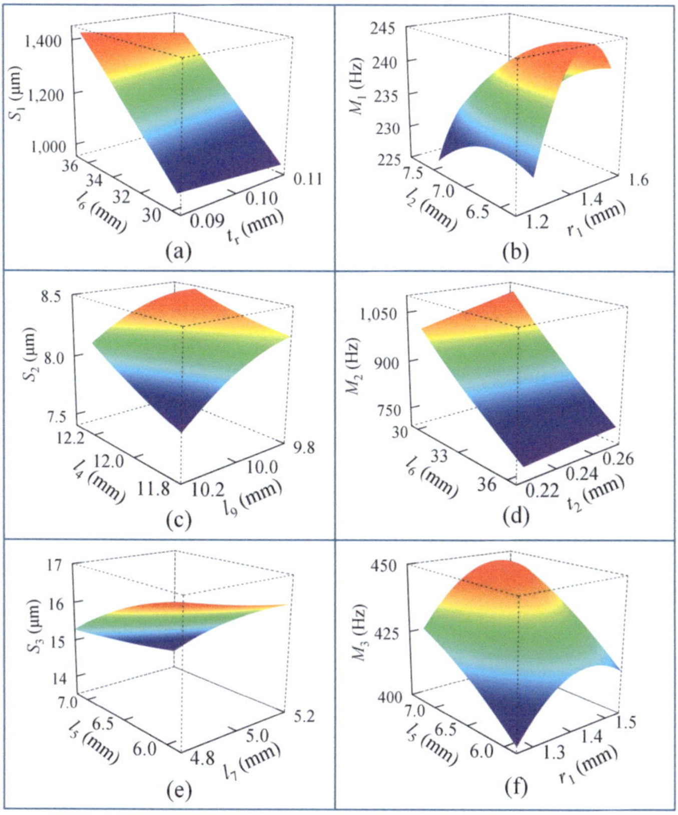

Fig. 11.8 Response surfaces illustrating relationships between design variables and performance objectives: **a** l_6 and t_r versus S_1, **b** l_2 and r_1 versus M_1, **c** l_4 and l_9 versus S_2, **d** l_6 and t_2 versus M_2, **e** l_5 and l_7 versus S_3, **f** l_5 and r_1 versus M_3

different objectives. Notably, the design variable l_6 demonstrates the highest sensitivity for all optimization objectives, indicating its critical role in determining both static and dynamic performance. Furthermore, a consistent trade-off relationship is observed: design variables exhibiting positive correlations with gripping stroke (S) generally show negative correlations with first natural frequency (M), highlighting the inherent performance compromise between displacement amplification and structural stiffness.

To visualize these relationships, three-dimensional response surfaces are generated by plotting pairs of highly sensitive variables against each objective, as shown in Fig. 11.8. These surfaces provide intuitive graphical representations of the complex design space. For instance, Fig. 11.8a demonstrates that S_1 decreases moderately with increasing t_r (negative correlation) but increases rapidly with larger l_6 values (positive correlation). Conversely, Fig. 11.8f reveals a nonlinear relationship where M_3 initially increases then decreases with increasing r_1, suggesting an optimal intermediate value exists for this parameter. These response surfaces enable designers to make informed visual assessments of parameter interactions and identify regions of optimal performance. Following optimization, the final values of all design variables are determined and summarized in Table 11.1.

11.3.3 FEM Simulation Results

To validate the optimized design parameters and evaluate the performance characteristics of the microgripper, comprehensive finite element method (FEM) simulations were conducted using ANSYS software. The microgripper structure was modeled using Aluminum 7075 (Al-7075) alloy with the following material properties: Young's modulus of 71.9 GPa, Poisson's ratio of 0.33, and density of 2810 kg/m^3. The mesh generation employed a path combining algorithm with tetrahedral elements, resulting in a high-quality mesh comprising 48,399 nodes and 29,701 elements.

11.3.3.1 Static Performance Analysis

The static structural analysis module was employed to evaluate the gripping performance under an input displacement of 20 μm applied at the bridge-type displacement amplification mechanism (BDAM). The simulation results, shown in Fig. 11.9a–c, demonstrate the deformed configurations and displacement magnitudes for each working mode:

- *Mode I*: Achieved a total horizontal output displacement of 1218.89 μm, corresponding to a displacement amplification ratio (DAR) of 60.94 (Fig. 11.9a)
- *Mode II*: Produced a total displacement of 324.08 μm with a DAR of 16.2 (Fig. 11.9b)
- *Mode III*: Exhibited external jaw expansion with a displacement of 617.15 μm and DAR of 30.86, enabling grasping of larger objects (Fig. 11.9c).

The area efficiency (AE), calculated based on the overall dimensions (39 mm $\times$ 55.5 mm $=$ 2164.5 mm^2) and maximum displacement, was determined to be 848.25 nm/mm^2.

11.3.3.2 Dynamic Performance Analysis

Modal analysis was conducted to assess the vibrational characteristics and working bandwidth capabilities. The results, presented in Fig. 11.9d–f, reveal the first natural frequencies for each operational mode:

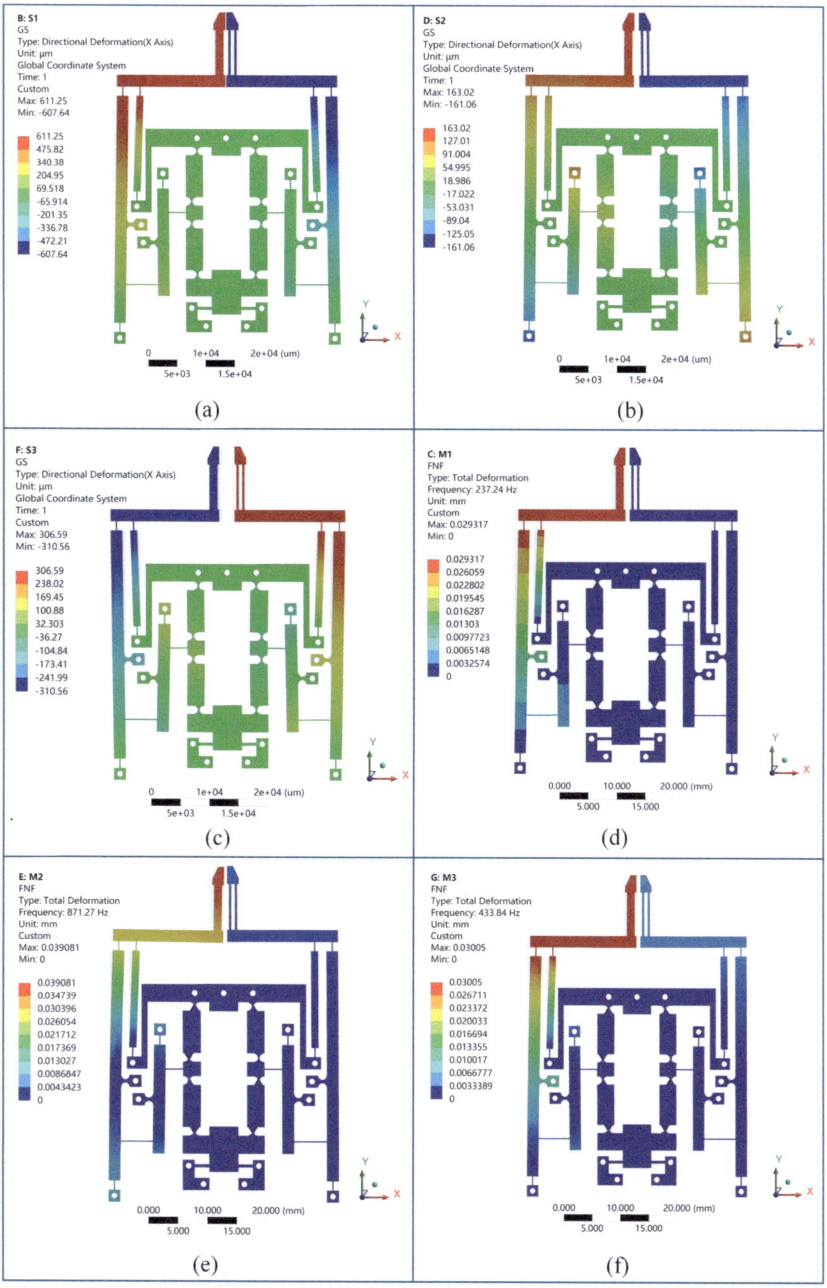

Fig. 11.9 FEM simulation results: **a** gripping stroke in Mode I, **b** gripping stroke in Mode II, **c** gripping stroke in Mode III, **d** first natural frequency in Mode I, **e** first natural frequency in Mode II, **f** first natural frequency in Mode III

Table 11.2 Comprehensive performance summary of the microgripper

Working mode	DAR	GS (μm)	FNF (Hz)	AE (nm/mm^2)
I	60.94	1218.89	237.24	–
II	16.2	324.08	871.27	848.25
III	30.86	617.15	433.84	–

- *Mode I*: First natural frequency of 237.24 Hz (Fig. 11.9d)
- *Mode II*: Enhanced dynamic performance with first natural frequency of 871.27 Hz, suitable for high-bandwidth applications (Fig. 11.9e)
- *Mode III*: First natural frequency of 433.84 Hz (Fig. 11.9f).

The comprehensive performance metrics, summarized in Table 11.2, demonstrate that the optimized microgripper achieves excellent static performance characteristics (high DAR and GS) while maintaining satisfactory dynamic response (FNF) across all three working modes. These results confirm the effectiveness of the multi-objective optimization approach in balancing competing performance requirements for diverse micromanipulation applications.

11.4 Experimental Results

11.4.1 Prototype Fabrication and Experimental Setup

Following comprehensive modeling, optimization, and simulation analyses, experimental studies were conducted to validate the practical performance of the proposed microgripper. The prototype was fabricated using wire electrical discharge machining (WEDM) technology from a solid plate of Aluminum 7075 (Al-7075) alloy, ensuring high precision and dimensional accuracy.

Figure 11.10 illustrates the complete experimental configuration for performance evaluation. The system consists of the following key components:

- *Actuation System*: Stack piezoelectric ceramic actuator (SPCA, model: RP150/5 × 5/20) providing precise displacement input
- *Positioning Platform*: XYZ linear translation stage (model: HTCL25-X) with micrometer resolution for precise positioning
- *Signal Conditioning*: Power amplifier (model: EPA-104) for driving the piezoelectric actuator
- *Displacement Sensing*: Two laser displacement sensors (model: LK-H055) with sub-micrometer resolution
- *Force Measurement*: Calibrated strain gauge force sensor (model: BFH 350-3AA) for gripping force quantification

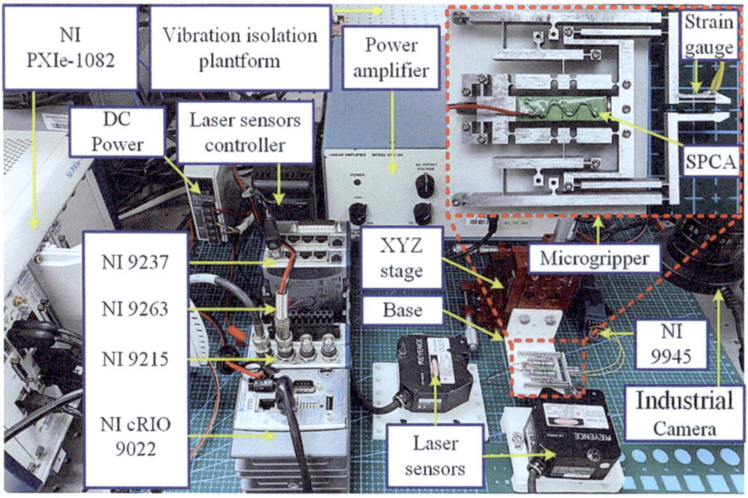

Fig. 11.10 Experimental setup for microgripper performance testing system

- *Visual Monitoring*: Industrial camera (model: MV-CE013-80UM) for real-time process observation
- *Data Acquisition*: NI cRIO-9022 real-time controller with multiple NI compact DAQ modules and NI PXIe-1082 chassis for signal processing.

The signal transmission and data acquisition process follows a systematic workflow:

1. Voltage signals generated by the PXIe-1082 chassis are transmitted through the NI cRIO-9022 controller
2. The analog output module (NI-9263) and power amplifier condition and amplify the signals for SPCA actuation
3. The SPCA converts electrical signals into precise mechanical displacement
4. Laser displacement sensors measure jaw movement and transmit data back through analog input modules (NI-9215)
5. During object contact, the strain-gauge force sensor generates voltage signals processed by bridge input modules (NI-9945 and NI-9237)
6. Force-voltage calibration enables accurate gripping force measurement
7. The industrial camera provides visual documentation of the entire gripping process.

This comprehensive experimental setup enables simultaneous measurement of displacement, force, and visual data, providing a complete characterization of the microgripper's performance across all operational modes.

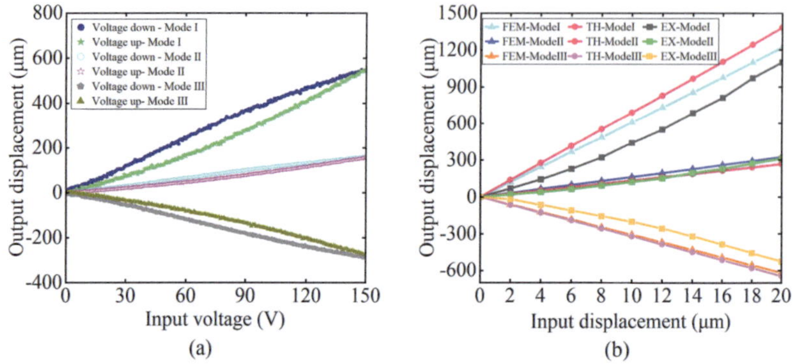

Fig. 11.11 Open-loop experimental results: **a** hysteresis characteristics of output displacement, **b** relationship between output and input displacements across three working modes comparing experimental, theoretical, and FEM results

11.4.2 Open-Loop Performance Test of the Microgripper

The open-loop performance characteristics of the microgripper were evaluated through comprehensive experimental testing. Laser displacement sensors measured the output displacements of the gripping tips across all three working modes while the stack piezoelectric ceramic actuator (SPCA) was driven with forward and reverse voltage signals ranging from 0 to 150 V.

As shown in Fig. 11.11a, the output displacement exhibits significant hysteresis behavior between voltage increasing and decreasing phases, characteristic of piezoelectric actuators. The hysteresis loop demonstrates the nonlinear relationship between applied voltage and resulting displacement. For consistency, inward gripping motion is defined as positive displacement direction, resulting in negative values for Mode III which produces outward jaw expansion.

Figure 11.11b presents the comparative analysis of output-input displacement relationships obtained through three independent methods: experimental measurements, theoretical modeling, and FEM simulations. The results show reasonable agreement across all three working modes. The experimental displacement amplification ratios (DARs) were determined as:

- *Mode I*: 55.17
- *Mode II*: 15.49
- *Mode III*: 26.33.

These experimental values align well with theoretical predictions (68.98, 14.16, and 32.19) and FEM simulations (60.94, 16.15, and 30.86). The observed discrepancies arise from several factors:

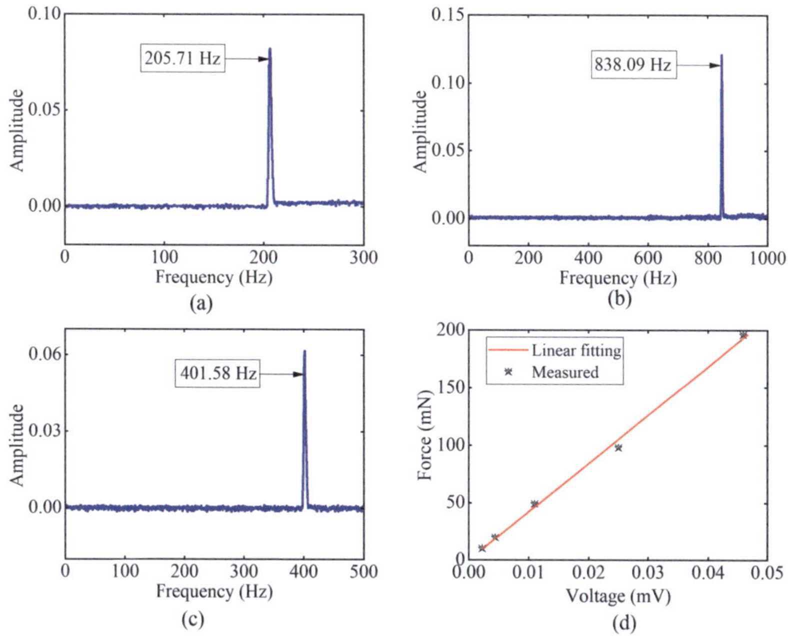

Fig. 11.12 Experimental characterization results: **a** frequency response for Mode I, **b** frequency response for Mode II, **c** frequency response for Mode III, **d** gripping force sensor calibration curve

- Idealized assumptions in the theoretical model regarding material properties and boundary conditions
- Manufacturing tolerances introduced during the wire EDM fabrication process
- Variations in installation methods between different operational modes
- Measurement uncertainties in experimental data acquisition.

Frequency response experiments characterized the dynamic performance, revealing first natural frequencies (FNFs) of:

- *Mode I*: 205.71 Hz (Fig. 11.12a)
- *Mode II*: 838.09 Hz (Fig. 11.12b)
- *Mode III*: 401.58 Hz (Fig. 11.12c).

These experimental FNF values show close agreement with FEM simulation results (237.24, 871.27, and 433.84 Hz), with minor reductions attributed to additional mass from the SPCA and pre-tightening bolt assembly in the experimental setup.

Force sensing calibration was performed using five standard weights applied sequentially to the instrumented jaw. As shown in Fig. 11.12d, the strain-gauge force sensor exhibited excellent linear response with a calibration coefficient of 4201.6 mN/mV, enabling accurate gripping force measurement during manipulation tasks.

11.4.3 Closed-Loop Performance Test of the Microgripper

To evaluate the closed-loop control performance, a proportional-integral-derivative (PID) controller was implemented and tuned using a trial-and-error approach. The optimized control parameters for each working mode were determined as follows:

- *Mode I*: $k_{\text{P-mode-I}} = 55$, $k_{\text{I-mode-I}} = 0.0002$, $k_{\text{D-mode-I}} = 0.00001$
- *Mode II*: $k_{\text{P-mode-II}} = 35$, $k_{\text{I-mode-II}} = 0.0001$, $k_{\text{D-mode-II}} = 0.00002$
- *Mode III*: $k_{\text{P-mode-III}} = 47$, $k_{\text{I-mode-III}} = 0.0002$, $k_{\text{D-mode-III}} = 0.0001$.

The displacement resolution was evaluated using stairway-type reference signals. As shown in Fig. 11.13a, the microgripper effectively tracked the input signals across all working modes. The error analysis in Fig. 11.13b revealed a normal distribution of tracking errors with a standard deviation (σ) of 0.69 μm. With a 95.5% confidence interval (2σ), the motion resolution was determined to be ±1.38 μm.

Further closed-loop performance was assessed using step and sinusoidal reference signals:

- *Step Response*: Signals with amplitudes of 30, 40, and 50 μm were applied to evaluate transient performance and settling characteristics (Fig. 11.13c, e, g)
- *Sinusoidal Response*: Signals with amplitudes of 10, 20, and 30 μm were used to assess dynamic tracking capability and bandwidth (Fig. 11.13d, f, h).

The experimental results demonstrate that the implemented PID control strategy enables effective reference tracking across all three working modes. The microgripper exhibited:

- Precise displacement control with minimal steady-state error
- Stable transient response during step input changes
- Accurate sinusoidal waveform tracking across specified amplitude ranges
- Consistent performance characteristics maintained across different operational modes.

These closed-loop performance characteristics validate the microgripper's capability for precision manipulation tasks requiring accurate displacement control and dynamic response across its reconfigurable working modes.

11.4.4 Gripping Performance Test

The grasping capability of the microgripper was experimentally validated through a series of manipulation tests with various micro-objects, monitored using an industrial camera system. Figure 11.14 demonstrates successful grasping operations across three categories of objects with different size ranges:

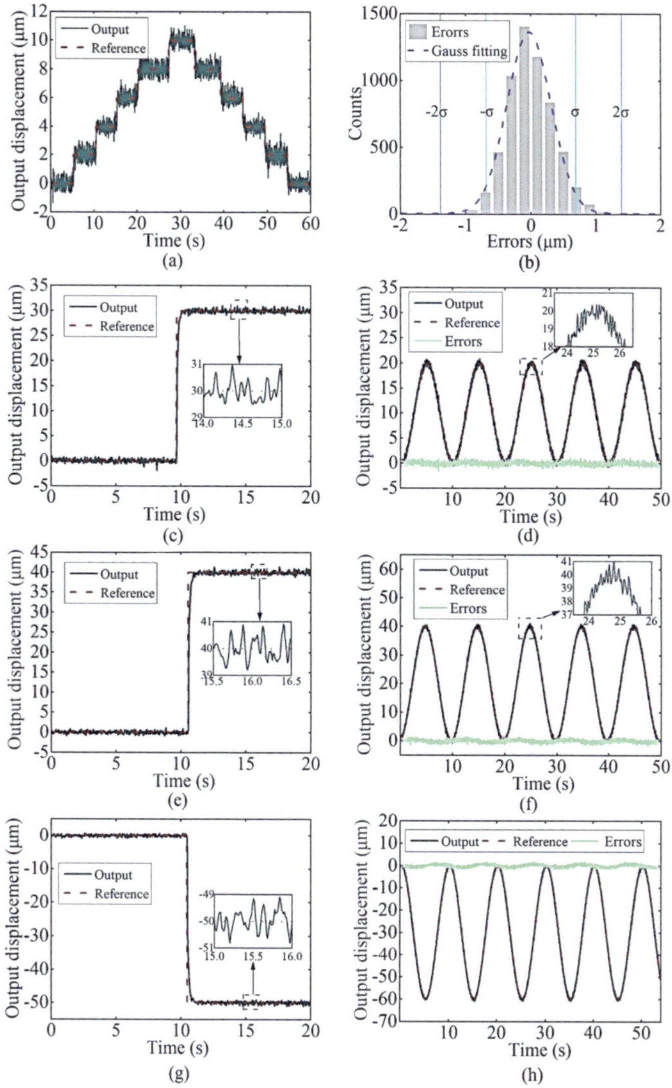

Fig. 11.13 Closed-loop experimental results: **a** Stairway-type signal response, **b** error distribution, **c** step response in Mode I, **d** sinusoidal response in Mode I, **e** step response in Mode II, **f** sinusoidal response in Mode II, **g** step response in Mode III, **h** sinusoidal response in Mode III

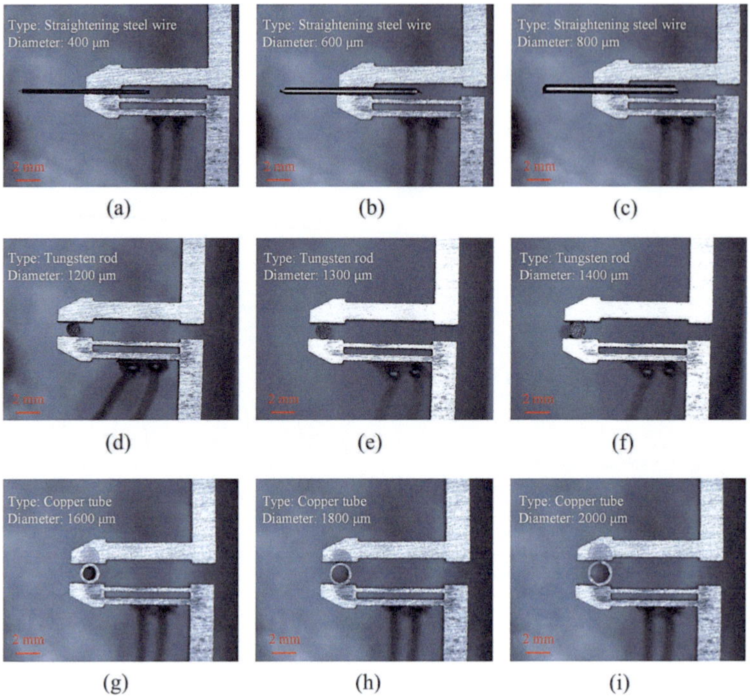

Fig. 11.14 Grasping experimental results: **a** 400 μm steel wire, **b** 600 μm steel wire, **c** 800 μm steel wire, **d** 1200 μm tungsten rod, **e** 1300 μm tungsten rod, **f** 1400 μm tungsten rod, **g** 1600 μm copper tube, **h** 1800 μm copper tube, **i** 2000 μm copper tube

- *Steel Micro-wires*: Diameters of 400 μm, 600 μm, and 800 μm (Fig. 11.14a–c), requiring jaw displacements of 1100 μm, 900 μm, and 700 μm respectively
- *Tungsten Micro-rods*: Diameters of 1200 μm, 1300 μm, and 1400 μm (Fig. 11.14d–f), demonstrating Mode II operation capability
- *Copper Microtubes*: Larger diameters of 1600 μm, 1800 μm, and 2000 μm (Fig. 11.14g–i), utilizing jaw expansion functionality.

These experiments confirm the microgripper's ability to handle objects spanning a wide size range (400–2000 μm) through its reconfigurable working modes, demonstrating exceptional adaptability for diverse micromanipulation tasks.

Figure 11.15 presents detailed displacement and force measurements during a complete gripping-releasing cycle, which can be divided into four distinct operational stages:

1. **Approach Stage**: Input voltage gradually increases, jaws move toward the target without contact. Displacement increases rapidly while gripping force remains constant at baseline level.

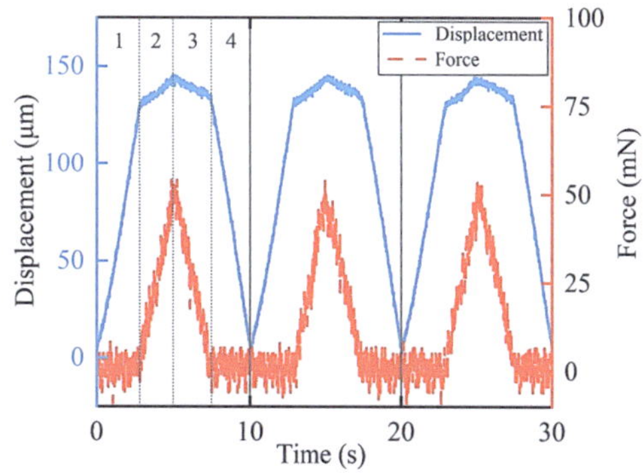

Fig. 11.15 Experimental results of displacement and force characteristics during gripping-releasing operations

2. **Gripping Stage**: Jaws make contact with the target object. Gripping force increases significantly as the jaws apply clamping force, while displacement growth rate decreases due to object resistance.
3. **Releasing Stage**: Input voltage decreases, gripping force reduces to release the object. A holding operation can be inserted between stages 2 and 3 for object transportation or testing.
4. **Retraction Stage**: Jaws separate from the released object, displacement decreases rapidly as the jaws return to their initial position.

This complete gripping-holding-releasing cycle demonstrates the microgripper's precise control capabilities and its suitability for complex micromanipulation sequences requiring accurate force and displacement management. The clear separation between contact and non-contact phases confirms the effectiveness of the integrated force sensing system for detecting object acquisition and release events.

11.4.5 Further Discussion

To demonstrate the superior performance of the proposed microgripper, a comprehensive comparison with state-of-the-art piezoelectric microgrippers is presented in Fig. 11.16. Each data point represents a published design, with circle diameter indicating displacement amplification ratio (DAR) and color representing area efficiency (AE) according to the color map. The position in the coordinate system reflects the gripping stroke (GS) and first natural frequency (FNF) performance metrics.

The proposed microgripper occupies the optimal position in the upper-right quadrant of the plot, demonstrating simultaneous achievement of:

- *Large Displacement Amplification*: Superior DAR compared to existing designs
- *Wide Gripping Stroke*: Extended range of object size accommodation
- *High Natural Frequency*: Enhanced dynamic performance and bandwidth
- *Excellent Area Efficiency*: Compact design with high performance per unit area.

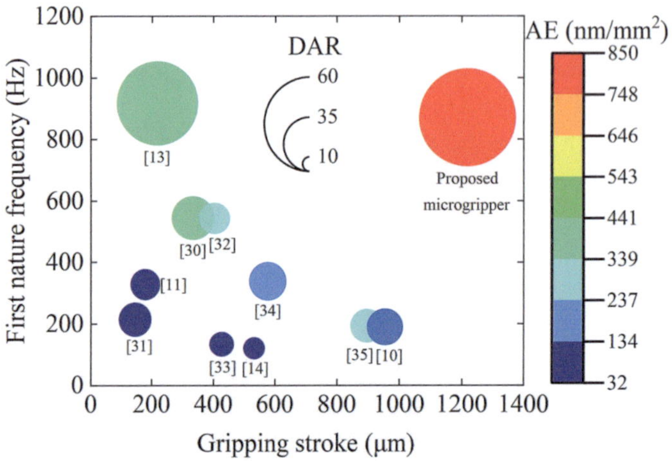

Fig. 11.16 Performance comparison with published PZT microgrippers, showing displacement amplification ratio (circle diameter), area efficiency (color), gripping stroke (x-axis), and first natural frequency (y-axis)

This exceptional comprehensive performance surpasses previously reported microgrippers [5, 7, 9, 12, 13, 19, 25, 29, 33, 34], and is enabled by the novel reconfigurable architecture that allows optimization for different operational requirements through selective constraint application.

The reconfigurability is currently achieved through discrete working modes by applying different fixing schemes. Future work will focus on:

- *Continuous Performance Tuning*: Developing mechanisms for smooth transition between performance characteristics rather than discrete modes
- *Enhanced Dynamic Modeling*: Establishing comprehensive dynamic models of the piezoelectric actuator system for improved control performance
- *Advanced Force Control*: Implementing more sophisticated gripping force control strategies for delicate manipulation tasks
- *Complex Manipulation Capabilities*: Extending functionality to include rotational manipulation, micro-assembly, and force-feedback operations.

These advancements will further expand the microgripper's application scope in precision manufacturing, biomedical engineering, and micro-electromechanical systems assembly, where adaptable manipulation capabilities are essential for handling diverse micro-components and performing complex operational sequences.

11.5 Conclusion

This chapter has presented the comprehensive development of a piezoelectric-driven compliant microgripper with reconfigurable multi-mode capability. The key contributions and achievements can be summarized as follows:

- **Novel Mechanism Design**: A multi-mode dual-stage lever displacement amplification mechanism was developed, enabling three distinct working modes through selective constraint application. This innovative design allows optimization for either large displacement amplification, high natural frequency, or balanced performance depending on operational requirements.
- **Theoretical Modeling**: Comprehensive theoretical models were established using beam deformation theory and pseudo-rigid-body modeling methods, providing accurate prediction of displacement amplification ratios and dynamic characteristics across all working modes.
- **Multi-Objective Optimization**: A systematic optimization framework employing response surface methodology and genetic algorithms successfully determined optimal structural parameters, balancing competing objectives of displacement amplification, gripping stroke, and natural frequency.
- **Experimental Validation**: The manufactured prototype demonstrated exceptional performance metrics:
  - *Mode I*: DAR = 55.17, GS = 1103.41 μm
  - *Mode II*: FNF = 838.09 Hz (high-bandwidth operation)
  - *Mode III*: Balanced performance for intermediate requirements.
- **Versatile Manipulation Capability**: Experimental results confirmed successful grasping of diverse micro-objects ranging from 400 to 2000 μm in diameter, including steel wires, tungsten rods, and copper tubes, demonstrating exceptional adaptability.
- **Superior Comprehensive Performance**: Comparative analysis revealed that the proposed microgripper outperforms existing designs by simultaneously achieving large displacement amplification, wide gripping stroke, high natural frequency, and excellent area efficiency.

The reconfigurable microgripper represents a significant advancement in micromanipulation technology, offering unprecedented flexibility for handling diverse micro-components in applications ranging from microassembly and biomedical manipulation to MEMS fabrication. The successful integration of multiple working modes within a single compact device addresses fundamental trade-offs in microgripper design and opens new possibilities for adaptive micromanipulation systems.

Future work will focus on developing continuous performance tuning mechanisms, enhancing dynamic modeling precision, and implementing advanced control strategies for complex manipulation tasks, further expanding the microgripper's capabilities in emerging precision engineering applications.

References

1. Ai, W., Xu, Q.: New structural design of a compliant gripper based on the Scott-Russell mechanism. Int. J. Adv. Robot. Syst. **11**, 10 (2014)
2. Ballato, A.: Modeling piezoelectric and piezomagnetic devices and structures via equivalent networks. IEEE Trans. Ultrason. Ferroelectr. Freq. Control **48**(5), 1189–1240 (2001)
3. Boudaoud, M., Le Gorrec, Y., Haddab, Y., Lutz, P.: Gain scheduling control of a nonlinear electrostatic microgripper: design by an Eigenstructure assignment with an observer-based structure. IEEE Trans. Control Syst. Technol. **23**(4), 1255–1267 (2015)
4. Cecchi, R., Verotti, M., Capata, R., Dochshanov, A., Broggiato, G.B., Crescenzi, R., et al.: Development of micro-grippers for tissue and cell manipulation with direct morphological comparison. Micromachines **6**(11), 1710–1728 (2015)
5. Chen, F., Gao, Y., Dong, W., Du, Z.: Design and control of a passive compliant piezo-actuated micro-gripper with hybrid flexure hinges. IEEE Trans. Ind. Electron. **68**(11), 11168–11177 (2021)
6. Choi, H.S., Lee, D.C., Kim, S.S., Han, C.S.: The development of a microgripper with a perturbation-based configuration design method. J. Micromech. Microeng. **15**(6), 1327–1333 (2005)
7. Das, T.K., Shirinzadeh, B., Al-Jodah, A., Ghafarian, M., Pinskier, J.: Computational parametric analysis and experimental investigations of a compact flexure-based microgripper. Precis. Eng. **66**, 363–373 (2020)
8. Das, T.K., Shirinzadeh, B., Al-Jodah, A., Ghafarian, M., Pinskier, J.: A novel compliant piezoelectric actuated symmetric microgripper for the parasitic motion compensation. Mechanism Mach. Theory **155**, 18 (2021)
9. Das, T.K., Shirinzadeh, B., Ghafarian, M., Al-Jodah, A., Pinskier, J.: Characterization of a compact piezoelectric actuated microgripper based on double-stair bridge-type mechanism. J. Micro-Bio Robot. **16**(1), 79–92 (2020)
10. Dochshanov, A., Verotti, M., Belfiore, N.: A comprehensive survey on microgrippers design: operational strategy. J. Mech. Des. **139**(7), 070801 (2017)
11. Koseki, Y., Tanikawa, T., Koyachi, N., Arai, T.: Kinematic analysis of a translational 3-D.O.F. micro-parallel mechanism using the matrix method. Adv. Robot. **16**(3), 251–264 (2001)
12. Liang, C., Wang, F., Shi, B., Huo, Z.C., Zhou, K., Tian, Y., Zhang, D.: Design and control of a novel asymmetrical piezoelectric actuated microgripper for micromanipulation. Sens. Actuators A-Phys. **269**, 227–237 (2018)
13. Liang, C., Wang, F., Tian, Y., Zhao, X., Zhang, H., Cui, L., Zhang, D., Ferreira, P.: A novel monolithic piezoelectric actuated flexure-mechanism based wire clamp for microelectronic device packaging. Rev. Sci. Instrum. **86**(4), 10 (2015)
14. Liu, P.B., Yan, P.: A new model analysis approach for bridge-type amplifiers supporting nano-stage design. Mech. Mach. Theory **99**, 176–188 (2016)
15. Llewellyn-Evans, H., Griffiths, C.A., Fahmy, A.: Microgripper design and evaluation for automated μ-wire assembly: a survey. Microsyst. Technol. **26**(6), 1745–1768 (2020)
16. Long, Z., Zhang, J., Liu, Y., Han, C., Li, Y., Li, Z.: Dynamics modeling and residual vibration control of a piezoelectric gripper during wire bonding. IEEE Trans. Compon. Packag. Manuf. Technol. **7**(12), 2045–2056 (2017)
17. Lyu, Z., Xu, Q.: Design and analysis of a new compliant microgripper with two working modes. In: Proceedings of the 2021 IEEE International Conference on Manipulation, Manufacturing and Measurement on the Nanoscale (3M-NANO), pp. 65–70. Xi'an, China (2021)
18. Lyu, Z., Xu, Q.: Recent design and development of piezoelectric-actuated compliant microgrippers: a review. Sens. Actuators A-Phys. **331**, 20 (2021)
19. Lyu, Z., Xu, Q.: Novel design of a piezoelectrically actuated compliant microgripper with high area-usage efficiency. Precis. Eng. **76**, 1–11 (2022)
20. Lyu, Z., Xu, Q., Zhu, L.: Design and development of a new piezoelectric-actuated biaxial compliant microgripper with long strokes. IEEE Trans. Autom. Sci. Eng. **20**(1), 206–217 (2023)

21. Nah, S.K., Zhong, Z.W.: A microgripper using piezoelectric actuation for micro-object manipulation. Sens. Actuators A-Phys. **133**(1), 218–224 (2007)
22. Nikoobin, A., Niaki, M.H.: Deriving and analyzing the effective parameters in microgrippers performance. Sci. Iranica **19**(6), 1554–1563 (2012)
23. Raghavendra, M.R.A., Kumar, A.S., Jagdish, B.N.: Design and analysis of flexure-hinge parameter in microgripper. Int. J. Adv. Manuf. Technol. **49**(9–12), 1185–1193 (2010)
24. Shi, C., Dong, X., Yang, Z.: A microgripper with a large magnification ratio and high structural stiffness based on a flexure-enabled mechanism. IEEE/ASME Trans. Mechatron. **26**(6), 3076–3086 (2021)
25. Sun, X., Chen, W., Tian, Y., Fatikow, S., Zhou, R., Zhang, J., Mikczinski, M.: A novel flexure-based microgripper with double amplification mechanisms for micro/nano manipulation. Rev. Sci. Instrum. **84**(8), 10 (2013)
26. Tian, Y., Shirinzadeh, B., Zhang, D., Alici, G.: Development and dynamic modelling of a flexure-based Scott-Russell mechanism for nano manipulation. Mech. Syst. Signal Process. **23**, 957–978 (2009)
27. Verotti, M., Dochshanov, A., Belfiore, N.: A comprehensive survey on microgrippers design: mechanical structure. J. Mech. Des. **139**(6), 060801 (2017)
28. Wang, D., Yang, Q., Dong, H.: A monolithic compliant piezoelectric-driven microgripper: design, modeling, and testing. IEEE/ASME Trans. Mechatron. **18**(1), 138–147 (2013)
29. Wang, F., Liang, C., Tian, Y., Zhao, X., Zhang, D.: Design of a piezoelectric-actuated microgripper with a three-stage flexure-based amplification. IEEE/ASME Trans. Mechatron. **20**(5), 2205–2213 (2015)
30. Wang, F., Shi, B., Huo, Z., Tian, Y., Zhang, D.: Control and dynamic releasing method of a piezoelectric actuated microgripper. Precis. Eng. **68**, 1–9 (2021)
31. Wang, F., Shi, B., Huo, Z., Tian, Y., Zhang, D.: Design and control of a spatial micromanipulator inspired by deployable structure. IEEE Trans. Ind. Electron. **69**(1), 971–979 (2022)
32. Xie, X., Meng, H., Zhang, H., Sun, L.: Development of a magnetically driven microgripper for piconewton force-controlled microscale manipulation and characterization. IEEE Trans. Ind. Electron. **67**(3), 2065–2075 (2020)
33. Yang, Y., Wei, Y., Lou, J., Tian, G., Zhao, X., Fu, L.: A new piezo-driven microgripper based on the double-rocker mechanism. Smart Mater. Struct. **24**(7), 11 (2015)
34. Yang, Y., Wei, Y., Lou, J., Xie, F., Fu, L.: Development and precision position/force control of a new flexure-based microgripper. J. Micromech. Microeng. **26**(1), 19 (2016)

Chapter 12
Precision Motion Control of a Piezo-Driven Micromanipulator Based on Adaptive Control

Abstract This chapter proposes a reinforcement learning (RL)-based adaptive control design for precision motion control of a two-degree-of-freedom piezoelectric XY nanopositioning system. In this design, an actor-critic structure is developed to eliminate the effects of uncertain nonlinearities and cross-coupling motion between the two working axes. Then, an adaptive parameter adjustment mechanism is designed to optimize the control performance without a priori knowledge of the unknown perturbations. The effectiveness and superiority of the proposed method are verified by performing simulation and experimental studies. The results show that the proposed RL-based adaptive control method provides a better robust performance and smaller tracking error for the nanopositioning system.

Keywords Motion control · Micromanipulator · Nanopositioning · Piezoelectric actuator · Reinforcement learning adaptive control

12.1 Introduction

Nanopositioning devices, capable of providing motion accuracy at the nanometer level, are regarded as ultrahigh-precision motion instruments. These systems have been employed in a variety of engineering applications, including biological micromanipulation and scanning probe microscopy [6, 22, 24]. Owing to their fast response and large blocking force, piezoelectric actuators are frequently adopted to drive nanopositioning stages to generate motion with nanometer resolution [4, 13, 37]. Notably, piezoelectric actuators also introduce hysteresis and creep nonlinearities into the nanopositioning system. To achieve high control performance, the control of piezo-driven nanopositioning systems has garnered persistent attention from both academia and industry in recent years [23, 29, 33]. For instance, the authors in [33] proposed a third-order integral terminal sliding mode control method to guarantee the tracking performance of a piezoelectric-driven nanopositioning system. An active disturbance rejection control strategy was studied in [29] to achieve large stroke and nanometer positioning accuracy. However, the hysteresis inherent in piezoelectric actuators introduces inevitable unknown nonlinearities into nanoposi-

Q. Xu, *Piezoelectric-Driven Flexible Robotic Microgrippers*,
https://doi.org/10.1007/978-981-95-4685-5_12

tioning systems, which can significantly deteriorate performance [34]. Consequently, the development of novel control schemes for piezoelectric nanopositioning stages is highly desirable to improve robustness against hysteresis nonlinearity.

To meet the requirements for micro/nano-manipulation in complex tasks, multiple-degree-of-freedom (multi-DOF) nanopositioning stages have become crucial devices [8, 14, 18, 39], as they provide an adequate workspace and high payload capacity. Fundamental issues such as compliant mechanism design, stabilization, and control of multi-DOF nanopositioning systems have been extensively reported in the literature [10, 11, 19, 21, 27, 31]. For example, the modeling and robust H_∞ control of a 2-DOF piezoelectric cantilevered actuator were studied in [27] to achieve high performance. A mixed H_2/H_∞ control method was proposed in [20] for a 2-DOF nanopositioning system subject to model uncertainties. The development and control methodologies for a 2-DOF micro/nano-positioning stage with out-of-plane payload capacity were considered in [2]. In multi-DOF nanopositioning stages, cross-coupling motion frequently occurs among the working axes, degrading tracking accuracy and increasing controller design complexity [41]. However, few intelligent control strategies have attempted to eliminate hysteresis nonlinearity and cross-coupling motion within a unified framework, which motivates the research presented in this paper.

Reinforcement learning (RL) techniques, which generate optimal actions through interaction with the environment, have attracted considerable attention [7, 12, 26, 40]. Due to their actor-critic structure incorporating adaptive neural networks (NNs), RL-based adaptive control methods can learn unknown system characteristics to achieve high performance [3, 32, 35, 36]. Several successful applications of these methods have been demonstrated [1, 16, 25]. For instance, learning-based control algorithms were proposed for uncertain nonlinear systems with mismatched disturbances in [35, 36]. The trajectory tracking problem for a fully actuated autonomous underwater vehicle was investigated using RL control in [5]. A decentralized tracking optimization control approach was developed for a dual-manipulator coordination system via RL in [38]. In a piezo-driven nanopositioning system, the hysteresis and creep nonlinearities from the piezoelectric actuators inevitably impair control performance. Furthermore, in a multi-DOF nanopositioning stage, cross-coupling motion among the axes also degrades tracking accuracy. Since these issues are difficult to model accurately, the corresponding control problem presents a significant challenge for designing an RL-based adaptive controller. To date, no RL-based adaptive control has been developed for nanopositioning systems in the literature. It remains an open question how to implement RL-based adaptive control to enhance the performance of such systems. Herein, we propose a novel RL-based adaptive controller for a 2-DOF piezo-driven nanopositioning system.

The main contributions of this work are summarized as follows.

1. A reinforcement learning-based adaptive control strategy is proposed for a piezoelectric nanopositioning system. This approach utilizes an actor-critic structure to enhance control performance. In this design, a critic neural network (NN) approximates the long-term performance index of the 2-DOF nanopositioning system,

while an actor NN responds to the critic NN to guarantee tracking accuracy. By continuously monitoring environmental responses, the proposed RL method achieves high-precision control.

2. To eliminate the unknown hysteresis and creep nonlinearities of the piezoelectric actuator and the cross-coupling motion effects within a unified framework, an adaptive compensation mechanism is proposed to compensate for unknown perturbations in real time. As this design requires no a priori knowledge of the hysteresis, creep, or motion cross-coupling, the proposed control method can be easily extended to both single-DOF and multi-DOF nanopositioning systems.

12.2 System Description and Preliminaries

A photograph of the piezo-driven 2-DOF nanopositioning device is shown in Fig. 12.1. With reference to [2, 13, 20, 34], the motion of a piezo-driven nanopositioning system in the XY plane can be generally modeled by the following second-order system:

$$m\ddot{z}(t) + b\dot{z}(t) + kz(t) = pu(t) + d(u, t) + f(z, t) \tag{12.1}$$

where $z(t) = [x(t), y(t)]^T \in \mathbf{R}^2$ denotes the position coordinates; $u(t) \in \mathbf{R}^2$ is the voltage input to the nanopositioning system; $d(u, t) \in \mathbf{R}^2$ represents the hysteresis nonlinearity, creep nonlinearity, and other unmodeled dynamics; $f(z, t) \in \mathbf{R}^2$ is the cross-coupling motion between the x- and y-axes; $m = \text{diag}\{m_x, m_y\} \in \mathbf{R}^{2\times 2}$ denotes the mass matrix; $b = \text{diag}\{b_x, b_y\} \in \mathbf{R}^{2\times 2}$ is the damping coefficient matrix; $k = \text{diag}\{k_x, k_y\} \in \mathbf{R}^{2\times 2}$ represents the stiffness matrix; and $p = \text{diag}\{p_x, p_y\} \in \mathbf{R}^{2\times 2}$ is the piezoelectric coefficient matrix.

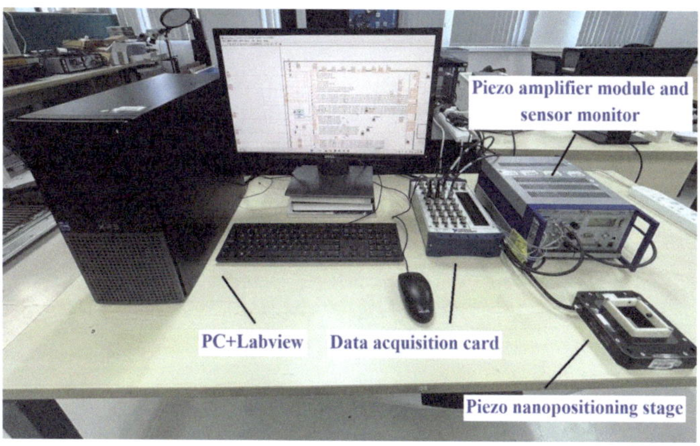

Fig. 12.1 Experimental setup of a piezo-driven XY nanopositioning stage

The original system represented by Eq. (12.1) can be further rewritten as:

$$\ddot{z}(t) + m^{-1}b\dot{z}(t) + m^{-1}kz(t) = m^{-1}pu(t) + \omega(t) \tag{12.2}$$

where $\omega(t) = \left[\omega_x(t), \omega_y(t)\right]^T = m^{-1}(d(u,t) + f(z,t))$. Without loss of generality, it is assumed that there exists an unknown positive constant $\bar{\omega}$ such that $\|\omega(t)\| \leq \bar{\omega}$. In this work, we design an adaptive compensation term to eliminate the effects of hysteresis and creep nonlinearities, as well as cross-coupling motion.

Remark 12.1 According to Refs. [20, 33], the dynamics of nanopositioning systems typically exhibit several resonance peaks, representing high-order characteristics. Consequently, a second-order model cannot fully describe the system dynamics in the high-frequency domain. Nevertheless, the second-order model is widely used in previous works [2, 13, 20, 34] for controller design, as it sufficiently describes the system dynamics up to several hundred Hertz, as reported in [20, 33]. Following [33], the high-order dynamics of the nanopositioning system can be treated as unmodeled dynamics to be handled by the controller. Our experimental results will demonstrate that this control strategy is effective and achieves high tracking performance.

Remark 12.2 Although this work focuses on an RL-based adaptive control method for the 2-DOF nanopositioning system in Eq. (12.1), the proposed method can be extended to a single-DOF or multi-DOF nanopositioning system by adjusting the dimension of the state vector $z(t)$ accordingly. Since no a priori knowledge of the hysteresis, creep in each piezoelectric actuator, or the cross-coupling motion is required, changing the number of motion axes does not affect the controller design process.

Given a desired trajectory $z_d(t) \in \mathbf{R}^2$, the tracking error is defined as:

$$e(t) = z(t) - z_d(t). \tag{12.3}$$

We then define a variable $s(t) = [s_1(t), s_2(t)]^T \in \mathbf{R}^2$ by

$$s(t) = \Lambda e(t) + \dot{e}(t) \tag{12.4}$$

where $\Lambda = \mathrm{diag}\{\Lambda_1, \Lambda_2\}$ is a positive definite matrix.

From Eq. (12.2), it follows that:

$$\begin{aligned} \dot{s}(t) &= \Lambda\dot{e}(t) + \ddot{e}(t) \\ &= (\Lambda - m^{-1}b)\dot{z}(t) - \Lambda\dot{z}_d(t) - m^{-1}kz(t) \\ &\quad + m^{-1}pu(t) + \omega(t) - \ddot{z}_d(t). \end{aligned} \tag{12.5}$$

12.3 Controller Design

In this section, a reinforcement learning (RL) technique employing both a critic and an actor neural network (NN) is utilized to ensure the long-term performance of the nanopositioning system.

12.3.1 Critic NN Design

First, following the reference [9], a long-term performance index $J_1(t)$ at time t is defined as:

$$J_1(t) = \int_t^{\infty} \gamma^{\frac{-\phi+t}{\Delta}} F(s(\phi))d\phi \tag{12.6}$$

where $\gamma \in (0, 1)$ is a discount factor, Δ is a small integration interval, and $F(s(\phi)) = [F_1(s_1(\phi)), F_2(s_2(\phi))]^T \in \mathbf{R}^2$ is given by:

$$F_i(s_i(\phi)) = \begin{cases} 0 & \|s_i(\phi)\| \leq \xi_i \\ 1 & \|s_i(\phi)\| > \xi_i \end{cases}, \quad i = 1, 2 \tag{12.7}$$

where $\phi \in [t - \Delta, t)$ and $\xi_i > 0$ is a given small constant.

Using a technique similar to [9], the long-term performance index at time $t - \Delta$ can be expressed as:

$$\begin{aligned} J_1(t - \Delta) &= \int_{t-\Delta}^{t} \gamma^{\frac{-\phi+t-\Delta}{\Delta}} F(s(\phi))d\phi + \int_t^{\infty} \gamma^{\frac{-\phi+t-\Delta}{\Delta}} F(s(\phi))d\phi \\ &= \gamma^{-1} J_1(t) + \int_{t-\Delta}^{t} \gamma^{\frac{-\phi+t-\Delta}{\Delta}} F(s(\phi))d\phi. \end{aligned} \tag{12.8}$$

Let $b_0 = [b_{01}, b_{02}]^T$ be a vector defined by:

$$b_{0i} = \begin{cases} 0 & \|s_i(\phi)\| \leq \xi_i \\ \frac{\Delta}{\ln \gamma}(\gamma - 1) & \|s_i(\phi)\| > \xi_i \end{cases}, \quad i = 1, 2. \tag{12.9}$$

There exists a positive constant $\bar{b}_0$ such that $\|b_0\| \leq \bar{b}_0$.

By considering Eqs. (12.8) and (12.9), we obtain the following relationship:

$$J_1(t - \Delta) = \gamma^{-1}(J_1(t) + b_0). \tag{12.10}$$

It is notable that the long-term performance index $J_1(t)$ in Eq. (12.6) contains future information that is unavailable at time t. Furthermore, due to the presence of unknown hysteresis, creep nonlinearities, and cross-coupling motion in the nanopositioning system, $J_1(t)$ is difficult to compute directly.

Based on [5], a critic radial basis function (RBF) neural network is employed to approximate $J_1(t)$ as follows:

$$J_1(t) = \theta_1^{*T} \varphi_1(z_1(t)) + \varepsilon_1(t) \tag{12.11}$$

where $\theta_1^* \in \mathbf{R}^{N_1 \times 2}$ is the unknown ideal weight matrix satisfying $\|\theta_1^*\| \le \bar{\theta}_1$ for a positive constant $\bar{\theta}_1$; N_1 is the number of neurons in the hidden layer; $\varphi_1(z_1(t)) \in \mathbf{R}^{N_1}$ is the basis function vector bounded such that $\|\varphi_1(z_1(t))\| \le r_{\varphi_1}$ for a positive constant r_{φ_1}; $\varepsilon_1(t)$ is the approximation error; and $z_1(t)$ is an input vector constructed from the current position, the velocity of the nanopositioning system, and the desired trajectory from t to $t + MT$:

$$z_1(t) = \left[z^T(t),\ \dot{z}^T(t),\ z_d^T(t+T),\ z_d^T(t+2T), ..., z_d^T(t+MT)\right]^T \in \mathbf{R}^{2(M+2)}. \tag{12.12}$$

Here, T is a time interval and M is a prescribed horizon.

Since the ideal weight matrix θ_1^* is unknown, an adaptive method is used to construct its estimate $\hat{\theta}_1(t) \in \mathbf{R}^{N_1 \times 2}$ as follows:

$$\hat{J}_1(t) = \hat{\theta}_1^T(t)\varphi_1(z_1(t)) \tag{12.13}$$

where $\hat{J}_1(t)$ is the estimate of $J_1(t)$. Consequently, at time $t - \Delta$, the estimate is:

$$\hat{J}_1(t-\Delta) = \hat{\theta}_1^T(t)\varphi_1(z_1(t-\Delta)). \tag{12.14}$$

According to Eq. (12.10), an intermediate difference error $e_1(t)$ is defined as:

$$\begin{aligned} e_1(t) &= \hat{J}_1(t) - \gamma \hat{J}_1(t-\Delta) + b_0 \\ &= \hat{\theta}_1^T(t)\Delta_{\varphi_1}(z_1(t)) + b_0 \end{aligned} \tag{12.15}$$

where $\Delta_{\varphi_1}(z_1(t)) = \varphi_1(z_1(t)) - \gamma\varphi_1(z_1(t-\Delta))$.

The adaptive law for updating $\hat{\theta}_1(t)$ is designed as follows:

$$\dot{\hat{\theta}}_1(t) = -c_1\Delta_{\varphi_1}(z_1(t))\left(\hat{\theta}_1^T(t)\Delta_{\varphi_1}(z_1(t)) + b_0\right)^T - c_2c_1\hat{\theta}_1(t) \tag{12.16}$$

where $c_1 = \mathrm{diag}\{c_{11}, c_{12}, ..., c_{1N_1}\} \in \mathbf{R}^{N_1 \times N_1}$ and $c_2 > 0$ are adaptive gains.

12.3.2 Actor NN Design

In this work, an actor NN $J_2(t)$ is employed to respond to the critic NN and is constructed as follows:

$$J_2(t) = \theta_2^{*T} \varphi_2(z_2(t)) + \varepsilon_2(t) \tag{12.17}$$

where $\theta_2^* \in \mathbf{R}^{N_2 \times 2}$ is an unknown ideal weight matrix such that $\|\theta_2^*\| \leq \bar{\theta}_2$ for a positive constant $\bar{\theta}_2$; N_2 is the number of hidden layer neurons; $\varphi_2(z_2(t)) \in \mathbf{R}^{N_2}$ is the basis function vector bounded such that $\|\varphi_2(z_2(t))\| \leq r_{\varphi_2}$ for a positive constant r_{φ_2}; $\varepsilon_2(t)$ is the approximation error; and $z_2(t)$ is the input vector constructed as $z_2(t) = \left[z^T(t),\ \dot{z}^T(t), s^T(t)\right]^T \in \mathbf{R}^6$.

Since the ideal weight matrix θ_2^* is unknown, its estimate $\hat{J}_2(t)$ is given by:

$$\hat{J}_2(t) = \hat{\theta}_2^T(t)\varphi_2(z_2(t)) \tag{12.18}$$

where $\hat{\theta}_2(t) \in \mathbf{R}^{N_2 \times 2}$ is the estimated weight parameter of the actor NN. In this design, the actor NN is designed to minimize the variable $s(t)$, which incorporates the tracking error and the long-term performance index $\hat{J}_1(t)$. Consequently, the adaptive law for $\hat{\theta}_2(t)$ is designed as:

$$\dot{\hat{\theta}}_2(t) = -a_1\varphi_2(z_2(t)) \left(\hat{J}_1(t) + s(t)\right)^T - a_2 a_1 \hat{\theta}_2(t) \tag{12.19}$$

where $a_1 = \mathrm{diag}\{a_{11}, a_{12}, ..., a_{1N_2}\} \in \mathbf{R}^{N_2 \times N_2}$ and $a_2 > 0$ are adaptive gains.

12.3.3 Controller Design

A block diagram of the proposed reinforcement learning-based adaptive control architecture is presented in Fig. 12.2. The critic neural network approximates the long-term tracking performance index. Based on this approximation $\hat{J}_1(t)$, the actor neural network adjusts the control signal. Additionally, an adaptive law for the disturbance estimate $\hat{\omega}(t)$ is designed to compensate for the effects of hysteresis, creep nonlinearities, and cross-coupling motion. This integrated approach ensures robust tracking performance of the nanopositioning system.

The reinforcement learning-based control law $u(t)$ for the nanopositioning system is designed as:

$$\begin{aligned} u(t) = &-p^{-1}mk_u s(t) + p^{-1}b\dot{z}(t) + p^{-1}kz(t) \\ &+ p^{-1}m\ddot{z}_d(t) + u_1(t) + u_2(t) \end{aligned} \tag{12.20}$$

where $k_u \in \mathbf{R}^{2\times 2}$ is a positive definite control gain matrix, and the auxiliary control terms are defined as:

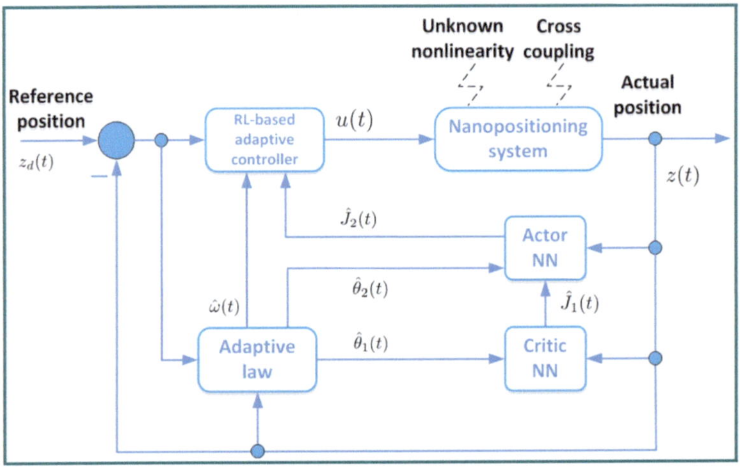

Fig. 12.2 Block diagram of the proposed control scheme

$$\begin{aligned} u_1(t) &= -p^{-1}m\hat{\theta}_2^T(t)\varphi_2(z_2(t)) \\ u_2(t) &= -p^{-1}m\frac{s(t)\|\hat{\omega}(t)\|^2}{\|s(t)\|\|\hat{\omega}(t)\| + \sigma(t)} \end{aligned} \tag{12.21}$$

where $\sigma(t) = \sigma_1 e^{-\sigma_2 t}$ with constants $\sigma_1 > 0$ and $\sigma_2 > 0$. The adaptive law for updating the disturbance estimate $\hat{\omega}(t)$ is given by:

$$\dot{\hat{\omega}}(t) = f_1\|s(t)\| - f_1 f_2\hat{\omega}(t) \tag{12.22}$$

where $f_1 > 0$ and $f_2 > 0$ are adaptive gains. In this design, the parameter $\hat{\omega}(t)$ estimates the upper bound of the disturbance term $\omega(t)$ in Eq. (12.2). The control term $u_2(t)$, constructed using $\hat{\omega}(t)$, actively compensates for the effects of hysteresis, creep nonlinearities, and cross-coupling motion.

Remark 12.3 The adaptive laws in Eqs. (12.16), (12.19), and (12.22) incorporate σ-modification terms to ensure robustness against unknown dynamics and disturbances [9]. Consequently, the persistent excitation (PE) condition is not required to guarantee boundedness of the adaptive parameters [15].

Based on the preceding analysis, the following theorem establishes the stability guarantees for the closed-loop nanopositioning system.

Theorem 12.1 *Consider the nanopositioning system described by Eq.* (12.1) *under the control law Eq.* (12.20) *with adaptive laws in Eqs.* (12.16), (12.19), *and* (12.22). *If the following sufficient conditions are satisfied:*

$$
\begin{aligned}
k_u &> \Lambda + \frac{3}{16} I_2 \\
c_2 &> \frac{1}{2} + 4r_{\varphi_1} r_{\varphi_2} - 2(1-\gamma)^2 r_{\varphi_1}^2 \\
a_2 &> \frac{1}{4} r_{\varphi_1} r_{\varphi_2} + 2 + 16 r_{\varphi_2}^2
\end{aligned} \tag{12.23}
$$

then the error signals of the resulting closed-loop system are semi-globally uniformly and ultimately bounded (SGUUB).

Proof First, we define the parameter estimation errors:

$$
\begin{aligned}
\tilde{\theta}_1(t) &= \hat{\theta}_1(t) - \theta_1^*, \\
\tilde{\theta}_2(t) &= \hat{\theta}_2(t) - \theta_2^*, \\
\tilde{\omega}(t) &= \hat{\omega}(t) - \bar{\omega}.
\end{aligned} \tag{12.24}
$$

Consider the following Lyapunov function candidate:

$$
V(t) = V_1(t) + V_2(t) + V_3(t) + V_4(t) + V_5(t) \tag{12.25}
$$

where

$$
\begin{aligned}
V_1(t) &= \frac{1}{2} e^T(t) \Lambda^2 e(t), \\
V_2(t) &= \frac{1}{2} s^T(t) s(t), \\
V_3(t) &= \frac{1}{2} \mathrm{tr}(\tilde{\theta}_1^T(t) c_1^{-1} \tilde{\theta}_1(t)), \\
V_4(t) &= \frac{1}{2} \mathrm{tr}(\tilde{\theta}_2^T(t) a_1^{-1} \tilde{\theta}_2(t)), \\
V_5(t) &= \frac{1}{2 f_1} \tilde{\omega}^2(t).
\end{aligned} \tag{12.26}
$$

From Eq. (12.4), the time derivative of $V_1(t)$ is:

$$
\begin{aligned}
\dot{V}_1(t) &= e^T(t) \Lambda^2 \dot{e}(t) \\
&= e^T(t) \Lambda^2 s(t) - e^T(t) \Lambda^3 e(t).
\end{aligned} \tag{12.27}
$$

For $V_2(t)$, using Eq. (12.5) yields:

$$
\begin{aligned}
\dot{V}_2(t) &= s^T(t) \dot{s}(t) \\
&= s^T(t) \left[\Lambda \dot{e}(t) - m^{-1} b \dot{z}(t) - m^{-1} k z(t) + m^{-1} p u(t) \right. \\
&\quad \left. + \omega(t) - \ddot{z}_d(t) \right].
\end{aligned} \tag{12.28}
$$

Combining $\dot{V}_1(t)$ and $\dot{V}_2(t)$ gives:

$$\begin{aligned}\dot{V}_1(t)+\dot{V}_2(t)=&-e^T(t)\Lambda^3e(t)+s^T(t)\Lambda s(t)-s^T(t)m^{-1}b\dot{z}(t)-s^T(t)m^{-1}kz(t)\\&+s^T(t)m^{-1}pu(t)+s^T(t)\omega(t)-s^T(t)\ddot{z}_d(t).\end{aligned} \tag{12.29}$$

Substituting the control input $u(t)$ from (12.20):

$$\begin{aligned}\dot{V}_1(t)+\dot{V}_2(t)=&-e^T(t)\Lambda^3e(t)-s^T(t)k_us(t)+s^T(t)\Lambda s(t)+s^T(t)m^{-1}pu_1(t)\\&+s^T(t)m^{-1}pu_2(t)+s^T(t)\omega(t).\end{aligned} \tag{12.30}$$

Since θ_2^{*T} and $\varphi_2(z_2(t))$ are bounded, there exists $r_{\theta_2}>0$ such that $\|\theta_2^{*T}\varphi_2(z_2(t))\|\le r_{\theta_2}$.

From Eq. (12.21):

$$\begin{aligned}s^T(t)m^{-1}pu_1(t)&=-s^T(t)\theta_2^{*T}\varphi_2(z_2(t))-s^T(t)\tilde{\theta}_2^T(t)\varphi_2(z_2(t))\\&\le\|s(t)\|r_{\theta_2}-s^T(t)\tilde{\theta}_2^T(t)\varphi_2(z_2(t)).\end{aligned} \tag{12.31}$$

Applying Young's inequality:

$$\begin{aligned}s^T(t)m^{-1}pu_1(t)&\le\frac{1}{16}\|s(t)\|^2+4r_{\theta_2}^2-s^T(t)\tilde{\theta}_2^T(t)\varphi_2(z_2(t))\\&\le\frac{1}{8}\|s(t)\|^2+4r_{\varphi_2}^2\mathrm{tr}(\tilde{\theta}_2^T(t)\tilde{\theta}_2(t))+\Gamma_1\end{aligned} \tag{12.32}$$

where $\Gamma_1=4r_{\theta_2}^2$.

For the disturbance compensation term:

$$\begin{aligned}&s^T(t)m^{-1}pu_2(t)+s^T(t)\omega(t)\\&\le-\frac{\|s(t)\|^2\|\hat{\omega}(t)\|^2}{\|s(t)\|\|\hat{\omega}(t)\|+\sigma(t)}+\|s(t)\|\bar{\omega}.\end{aligned} \tag{12.33}$$

Note that:

$$\begin{aligned}&-\frac{\|s(t)\|^2\|\hat{\omega}(t)\|^2}{\|s(t)\|\|\hat{\omega}(t)\|+\sigma(t)}+\|s(t)\|\hat{\omega}(t)\\&=\frac{\|s(t)\|\|\hat{\omega}(t)\|\sigma(t)}{\|s(t)\|\|\hat{\omega}(t)\|+\sigma(t)}\le\sigma(t)\end{aligned} \tag{12.34}$$

which implies:

$$s^T(t)m^{-1}pu_2(t)+s^T(t)\omega(t)\le-\|s(t)\|\tilde{\omega}(t)+\sigma(t). \tag{12.35}$$

Considering the adaptive law in Eq. (12.22):

$$\begin{aligned}&\dot{V}_5(t) - \|s(t)\|\tilde{\omega}(t) + \sigma(t)\\&= \frac{1}{f_1}\tilde{\omega}(t)\dot{\hat{\omega}}(t) - \|s(t)\|\tilde{\omega}(t) + \sigma(t)\\&= -f_2\tilde{\omega}(t)\hat{\omega}(t) + \sigma(t).\end{aligned} \tag{12.36}$$

Hence:

$$\begin{aligned}&\dot{V}_5(t) + s^T(t)m^{-1}pu_2(t) + s^T(t)\omega(t)\\&\leq -f_2\tilde{\omega}^2(t) - f_2\tilde{\omega}(t)\bar{\omega} + \sigma(t)\\&\leq -\frac{f_2}{2}\tilde{\omega}^2(t) + \Gamma_2\end{aligned} \tag{12.37}$$

where $\Gamma_2 = \frac{f_2}{2}\bar{\omega}^2 + \sigma_1$.

For $V_3(t)$, using Eq. (12.16):

$$\begin{aligned}\dot{V}_3(t) &= \mathrm{tr}(\tilde{\theta}_1^T(t)c_1^{-1}\dot{\hat{\theta}}_1(t))\\&= -\mathrm{tr}\left(\tilde{\theta}_1^T(t)\Delta_{\varphi_1}(z_1(t))\Delta_{\varphi_1}^T(z_1(t))\hat{\theta}_1(t)\right)\\&\quad - \mathrm{tr}\left(\tilde{\theta}_1^T(t)\Delta_{\varphi_1}(z_1(t))b_0^T\right) - c_2\mathrm{tr}\left(\tilde{\theta}_1^T(t)\hat{\theta}_1(t)\right).\end{aligned} \tag{12.38}$$

From Eq. (12.15), the bound on $\Delta_{\varphi_1}(z_1(t))$ is:

$$(1-\gamma)r_{\varphi_1} \leq \|\Delta_{\varphi_1}(z_1(t))\| \leq (1+\gamma)r_{\varphi_1}. \tag{12.39}$$

Substituting into Eq. (12.38):

$$\begin{aligned}\dot{V}_3(t) \leq &-(1-\gamma)^2r_{\varphi_1}^2\mathrm{tr}(\tilde{\theta}_1^T(t)\tilde{\theta}_1(t)) + (1+\gamma)^2r_{\varphi_1}^2\|\theta_1^*\|_F\|\tilde{\theta}_1(t)\|_F\\&+ \bar{b}_0(1+\gamma)r_{\varphi_1}\|\tilde{\theta}_1(t)\|_F - c_2\mathrm{tr}\left(\tilde{\theta}_1^T(t)\hat{\theta}_1(t)\right).\end{aligned} \tag{12.40}$$

Let $r_\theta = (1+\gamma)^2r_{\varphi_1}^2\|\theta_1^*\|_F + \bar{b}_0(1+\gamma)r_{\varphi_1}$, then:

$$\begin{aligned}\dot{V}_3(t) &\leq -(1-\gamma)^2r_{\varphi_1}^2\mathrm{tr}(\tilde{\theta}_1^T(t)\tilde{\theta}_1(t)) + \frac{1}{4}\mathrm{tr}(\tilde{\theta}_1^T(t)\tilde{\theta}_1(t)) + r_\theta^2\\&\quad - c_2\mathrm{tr}\left(\tilde{\theta}_1^T(t)\tilde{\theta}_1(t)\right) + c_2\mathrm{tr}\left(\tilde{\theta}_1^T(t)\theta_1^*\right)\\&\leq -\left((1-\gamma)^2r_{\varphi_1}^2 + \frac{c_2}{2} - \frac{1}{4}\right)\mathrm{tr}(\tilde{\theta}_1^T(t)\tilde{\theta}_1(t)) + \Gamma_3\end{aligned} \tag{12.41}$$

where $\Gamma_3 = \frac{c_2}{2}\mathrm{tr}(\theta_1^{*T}\theta_1^*) + r_\theta^2$.

For $V_4(t)$, using Eq. (12.19):

$$\begin{aligned}\dot{V}_4(t) &= \mathrm{tr}(\tilde{\theta}_2^T(t)a_1^{-1}\dot{\hat{\theta}}_2(t)) \\ &= -\mathrm{tr}\left(\tilde{\theta}_2^T(t)\varphi_2(z_2(t))(\hat{J}_1^T(t)+s^T(t))\right) - a_2\mathrm{tr}\left(\tilde{\theta}_2^T(t)\hat{\theta}_2(t)\right).\end{aligned} \tag{12.42}$$

Using Eq. (12.13):

$$\begin{aligned}\dot{V}_4(t) \le\ & r_{\varphi_1}r_{\varphi_2}\|\tilde{\theta}_2(t)\|_F\|\tilde{\theta}_1(t)\|_F + r_{\varphi_1}r_{\varphi_2}\|\tilde{\theta}_2(t)\|_F\|\theta_1^*\|_F \\ & + r_{\varphi_2}\|\tilde{\theta}_2(t)\|_F\|s(t)\| - \mathrm{tr}\left(\tilde{\theta}_2^T(t)a_2\hat{\theta}_2(t)\right).\end{aligned} \tag{12.43}$$

Applying Young's inequality:

$$\begin{aligned}\dot{V}_4(t) \le\ & \left(\frac{1}{8}r_{\varphi_1}r_{\varphi_2} + 1 + 4r_{\varphi_2}^2\right)\mathrm{tr}(\tilde{\theta}_2^T(t)\tilde{\theta}_2(t)) + 2r_{\varphi_1}r_{\varphi_2}\mathrm{tr}(\tilde{\theta}_1^T(t)\tilde{\theta}_1(t)) \\ & + \frac{1}{16}\|s(t)\|^2 + \frac{1}{4}r_{\varphi_1}^2r_{\varphi_2}^2\mathrm{tr}(\theta_1^{*T}\theta_1^*) - \frac{a_2}{2}\mathrm{tr}(\tilde{\theta}_2^T(t)\tilde{\theta}_2(t)) + \Gamma_4\end{aligned} \tag{12.44}$$

where $\Gamma_4 = \frac{a_2}{2}\mathrm{tr}(\theta_2^{*T}\theta_2^*)$.

Combining all terms:

$$\begin{aligned}\dot{V} \le\ & -s^T(t)\left(k_u - \Lambda - \frac{3}{16}I_2\right)s(t) - e^T(t)\Lambda^3 e(t) \\ & - \left(\frac{c_2}{2} + (1-\gamma)^2 r_{\varphi_1}^2 - \frac{1}{4} - 2r_{\varphi_1}r_{\varphi_2}\right)\mathrm{tr}(\tilde{\theta}_1^T(t)\tilde{\theta}_1(t)) \\ & - \left(\frac{a_2}{2} - \frac{1}{8}r_{\varphi_1}r_{\varphi_2} - 1 - 4r_{\varphi_2}^2\right)\mathrm{tr}(\tilde{\theta}_2^T(t)\tilde{\theta}_2(t)) - \frac{f_2}{2}\tilde{\omega}^2(t) + \Gamma \\ \le\ & -\alpha V(t) + \Gamma\end{aligned} \tag{12.45}$$

where

$$\begin{aligned}\alpha = \min\Bigg\{&\lambda_{\min}\left\{2(k_u - \Lambda - \frac{3}{16}I_2)\right\},\ \lambda_{\min}\{2\Lambda\},\ f_1 f_2, \\ & \left(c_2 + 2(1-\gamma)^2 r_{\varphi_1}^2 - \frac{1}{2} - 4r_{\varphi_1}r_{\varphi_2}\right)\lambda_{\min}\{c_1\}, \\ & \left(a_2 - \frac{1}{4}r_{\varphi_1}r_{\varphi_2} - 2 - 16r_{\varphi_2}^2\right)\lambda_{\min}\{a_1\}\Bigg\}\end{aligned} \tag{12.46}$$

$$\Gamma = \Gamma_1 + \Gamma_2 + \Gamma_3 + \Gamma_4 \tag{12.47}$$

From Eq. (12.23), $\alpha > 0$, which implies:

$$V(t) \leq \frac{\Gamma}{\alpha} + V(0)e^{-\alpha t}. \tag{12.48}$$

For zero initial conditions:

$$\begin{aligned}
\|s(t)\|^2 &\leq 2\frac{\Gamma}{\alpha}, \quad \|e(t)\|^2 \leq 2\frac{\Gamma}{\alpha}, \\
\|\tilde{\theta}_1(t)\|_F &\leq 2\lambda_{\max}\{c_1\}\frac{\Gamma}{\alpha}, \\
\|\tilde{\theta}_2(t)\|_F &\leq 2\lambda_{\max}\{a_1\}\frac{\Gamma}{\alpha}, \\
\|\tilde{\omega}(t)\| &\leq 2f_1\frac{\Gamma}{\alpha}.
\end{aligned} \tag{12.49}$$

Therefore, all error signals are semi-globally uniformly ultimately bounded. This completes the proof. □

Remark 12.4 In adaptive control methods, the estimation errors $\tilde{\theta}_1(t) = \hat{\theta}_1(t) - \theta_1^*$ and $\tilde{\theta}_2(t) = \hat{\theta}_2(t) - \theta_2^*$ must remain convergent and bounded [9]. To ensure this, we construct the Lyapunov functions $V_3(t)$ and $V_4(t)$ using the adaptive estimation errors $\tilde{\theta}_1(t)$ and $\tilde{\theta}_2(t)$, respectively. This approach is standard in adaptive control frameworks to guarantee overall system stability [1, 9, 30]. Similarly, the Lyapunov function $V_5(t)$ is formulated using the estimation error $\tilde{\omega}(t)$ associated with the upper bound of the perturbation.

Remark 12.5 From Eq. (12.49), it is evident that the error signals in the proposed control design converge to a neighborhood of the origin. Following [30], the size of this neighborhood can be reduced by decreasing the adaptive gains a_2, c_2, and f_2. However, since these gains are sensitive to the dynamic model parameters of the nanopositioning stage, an optimal trade-off between performance and robustness is typically achieved through experimental tuning [1].

12.4 Simulation Results

This section presents a simulation study to demonstrate the effectiveness of the proposed reinforcement learning-based adaptive control method for nanopositioning systems.

12.4.1 Control Performance Demonstration

Following the system identification approach in [34], a swept-sine signal generated via a D/A channel is used to excite the nanopositioning system. The output position is measured in real-time using laser displacement sensors. Frequency response analysis yields the following system parameters for the model in Eq. (12.1) along the x-axis and y-axis:

$$\begin{aligned}\frac{b_x}{m_x} &= \frac{b_y}{m_y} = 300,\\ \frac{k_x}{m_x} &= \frac{k_y}{m_y} = 1.24 \times 10^5,\\ \frac{p_x}{m_x} &= \frac{p_y}{m_y} = 1.05 \times 10^6.\end{aligned} \tag{12.50}$$

Since hysteresis, creep, and cross-coupling effects cannot be precisely modeled, the following perturbation $\omega(t)$ is introduced in the simulation to test the robustness of the proposed controller:

$$\begin{aligned}\omega_x(t) &= 7000 \sin(a_\omega t) + 30000 \cos(y(t)),\\ \omega_y(t) &= 8000 \cos(a_\omega t) + 20000 \sin(x(t)),\end{aligned} \tag{12.51}$$

where $a_\omega = t + 5$. The time-varying frequency a_ω captures the dynamic nature of hysteresis and creep effects, while the terms $30000 \cos(y(t))$ and $20000 \sin(x(t))$ explicitly model cross-coupling between the axes. The desired trajectory is set to $z_d(t) = [2\sin(5t), 2\sin(5t)]^T$ μm for demonstration.

The basis function vectors $\varphi(z_c(t)) = [\varphi_1(z_c(t)), \varphi_2(z_c(t)), ..., \varphi_N(z_c(t))]^T \in \mathbf{R}^N$ for both critic and actor NNs employ Gaussian radial basis functions:

$$\varphi_h(z_c(t)) = \exp\left(\frac{-\|z_c(t) - \mu_h\|^2}{\varpi^2}\right) \tag{12.52}$$

where $\mu_h = [\mu_{h1}, \mu_{h2}, ..., \mu_{hn}]^T \in \mathbf{R}^n$ for $h = 1, 2, ..., N$, and ϖ is the width parameter. We select $N = 5$ neurons with $\varpi = 5\sqrt{2}$, and centers μ_{ih} evenly distributed over $[-2, 2]$. The input vectors $z_c(t)$ are set to $z_1(t)$ for the critic NN and $z_2(t)$ for the actor NN, as defined previously.

Following Theorem 1, the control parameters are selected as: $\Lambda = \text{diag}\{600, 600\}$, $k_u = \text{diag}\{770, 770\}$, $c_2 = 6$, and $a_2 = 20$. Additional parameters are: $\gamma = 0.8$, $c_{1i} = 10, a_{1i} = 2$ (for $i = 1, 2, ..., 5$), $f_1 = 5000$, $f_2 = 0.1, \sigma_1 = 0.001, \sigma_2 = 0.01$, and $\xi_1 = \xi_2 = 0.1$. All weight matrices are initialized to zero.

Figure 12.3 demonstrates the trajectory tracking performance of the proposed controller. The nanopositioning system accurately tracks the desired trajectory despite the significant disturbances described in Eq. (12.51), with the maximum tracking errors of 0.08 μm and 0.06 μm for the x-axis and y-axis, respectively. The corresponding control signals are shown in Fig. 12.4, while Fig. 12.5 displays the con-

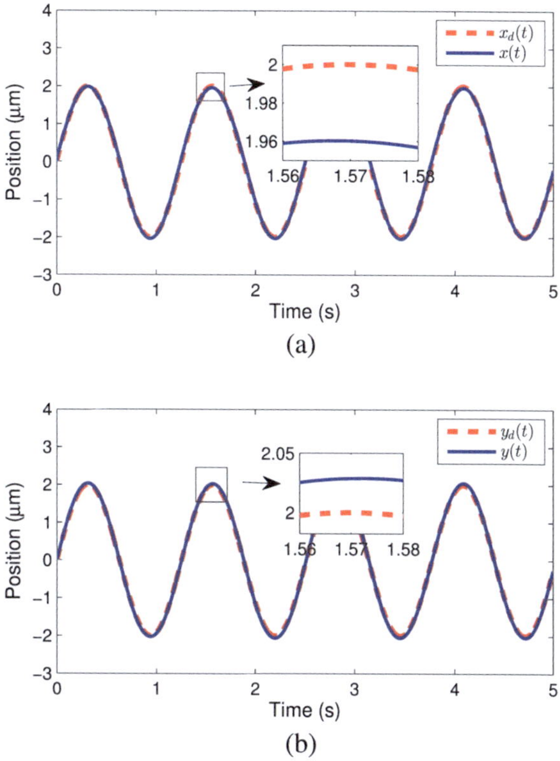

Fig. 12.3 Position tracking performance. **a** x-axis desired trajectory $x_d(t)$ and actual trajectory $x(t)$; **b** y-axis desired trajectory $y_d(t)$ and actual trajectory $y(t)$

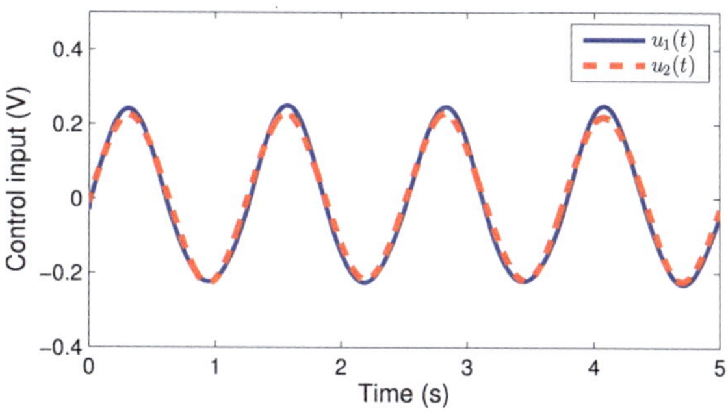

Fig. 12.4 Control input signals generated by the proposed controller

vergence of the adaptive parameters. Figure 12.6 further shows the weight estimates over an extended 20-second period, confirming their boundedness without drift.

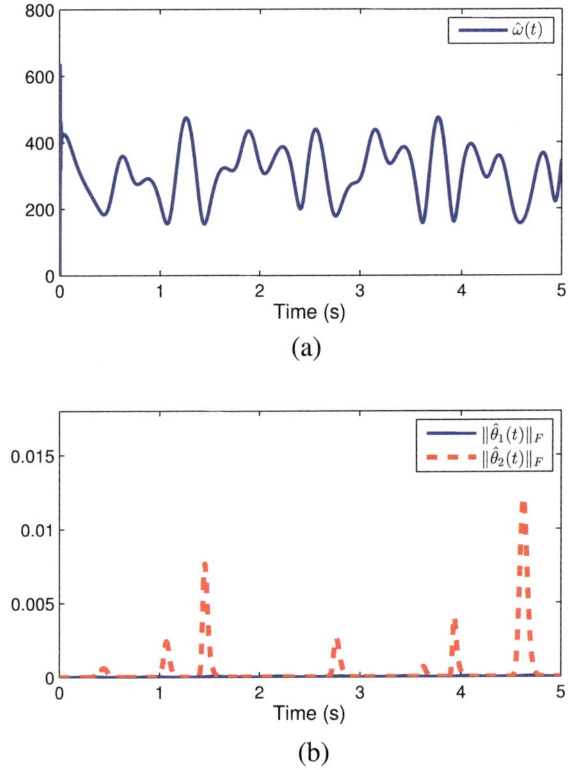

Fig. 12.5 Adaptive parameters evolution. **a** Disturbance bound estimate $\hat{\omega}(t)$; **b** Norms of weight estimates $\|\hat{\theta}_1(t)\|$ and $\|\hat{\theta}_2(t)\|$

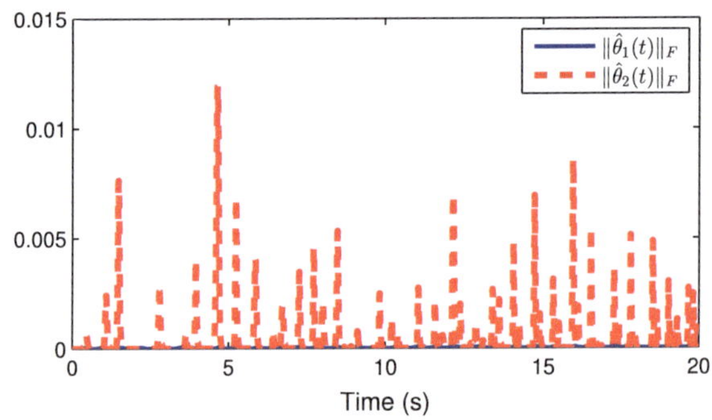

Fig. 12.6 Norms of adaptive parameters $\|\hat{\theta}_1(t)\|$ and $\|\hat{\theta}_2(t)\|$ over 20 s

12.4.2 Comparison Study of Controller Performances

A comparative analysis is conducted to evaluate the performance of the proposed controller against three established methods: adaptive control, PID-based model compensation control, and adaptive sliding mode control. To ensure a fair comparison, all controllers are tested under identical experimental conditions, including initial state values, desired trajectories, and hardware setup.

First, an adaptive controller from [17] is implemented:

$$u(t) = -k_a \frac{s(t)\|\hat{\omega}_d(t)\|^2}{\|s(t)\|\|\hat{\omega}_d(t)\| + \sigma(t)} + p^{-1}b\dot{z}(t) + p^{-1}kz(t) \tag{12.53}$$

where $\dot{\hat{\omega}}_d(t) = d_1\|s(t)\| - d_1 d_2 \hat{\omega}_d(t)$, with gains $k_a = \text{diag}\{0.95 \times 10^{-6}, 0.95 \times 10^{-6}\}$, $d_1 = 10000$, and $d_2 = 0.003$. This method employs an adaptive term $\hat{\omega}_d(t)$ to estimate the disturbance upper bound.

Second, a PID-based model compensation controller from [28] is implemented:

$$u(t) = -k_p e(t) - k_i \int_0^t e(\tau)d\tau - k_d\dot{e}(t) + p^{-1}b\dot{z}(t) + p^{-1}kz(t) \tag{12.54}$$

with gains:

$$\begin{aligned} k_p &= \text{diag}\{0.2, 0.2\} \\ k_i &= \text{diag}\{5 \times 10^{-4}, 5 \times 10^{-4}\} \\ k_d &= \text{diag}\{10^{-3}, 10^{-3}\} \end{aligned}$$

This approach uses model parameters as feedforward terms while PID action compensates for model inaccuracies.

Third, an adaptive sliding mode controller from [21] is implemented:

$$u(t) = p^{-1}m\left(u_a(t) - k_1\dot{e}(t) - k_3 s_e(t) - \beta \text{sign}(s_e(t))\right) \tag{12.55}$$

where

$$\begin{aligned} s_e(t) &= \dot{e}(t) + k_1 e(t) + k_2 \int_0^t e(\tau)d\tau \\ u_a(t) &= [\ddot{x}_d(t) + \phi_x\hat{\theta}_x(t), \ddot{y}_d(t) + \phi_x\hat{\theta}_y(t)]^T \end{aligned}$$

with adaptive laws:

$$\begin{aligned} \dot{\hat{\theta}}_x(t) &= -\Gamma_x \phi_x^T s_{ex}(t) \\ \dot{\hat{\theta}}_y(t) &= -\Gamma_y \phi_y^T s_{ey}(t) \end{aligned}$$

The regressor vectors are defined as:

$$\phi_x = \left[\dot{x}(t), x(t), -\text{sign}(\dot{x}(t)), -\text{sign}(\dot{y}(t)), -1\right] \tag{12.56}$$

$$\phi_y = \left[\dot{y}(t), y(t), -\text{sign}(\dot{y}(t)), -1\right] \tag{12.57}$$

with gains $k_1 = 300$, $k_2 = 0.3$, $k_3 = 1000$, $\beta = 0.1$, and $\Gamma_x = \Gamma_y = 0.1I_5$.

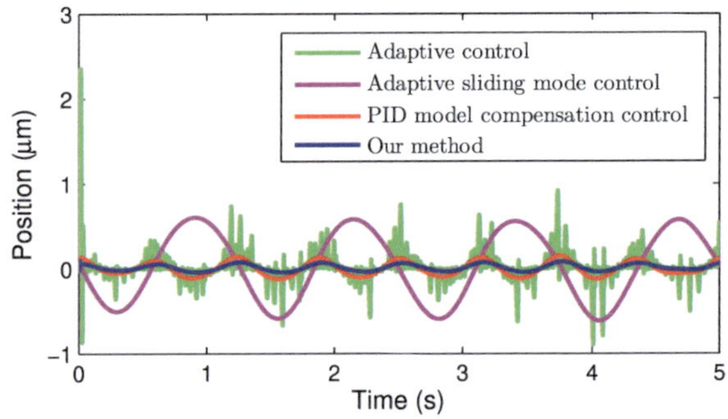

(a)

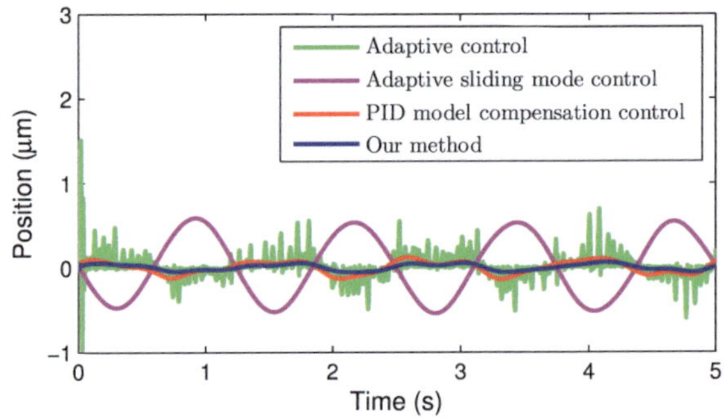

(b)

Fig. 12.7 Tracking error comparison of four control strategies. **a** x-axis errors; **b** y-axis errors

Figure 12.7 presents the tracking performance comparison. The adaptive controller achieves the maximum errors of 0.9 μm (x-axis) and 0.7 μm (y-axis). The adaptive sliding mode controller reduces these to 0.58 μm and 0.52 μm, respectively. The PID-based model compensation controller further improves performance with errors of 0.15 μm (x-axis) and 0.12 μm (y-axis). Notably, the adaptive sliding mode controller's model-free approach struggles to compensate for the large magnitude parameters $m^{-1}b$ and $m^{-1}k$ identified in Eq. (12.50), limiting its performance. The proposed RL-based adaptive control strategy demonstrates superior tracking accuracy under unknown perturbations.

An additional study examines the contribution of the critic NN. We implement a modified version where the actor NN adapts without critic feedback:

$$\dot{\hat{\theta}}_2(t) = -a_2\hat{\theta}_2(t) \tag{12.58}$$

Both methods are initialized with $\hat{\theta}_2(0) = [100, 100, 100, 100, 100]^T$ and use $a_1 = 0.5, a_2 = 2500$. Figure 12.8 shows that the system becomes unstable without the critic NN, while our full critic-actor architecture maintains stable tracking performance.

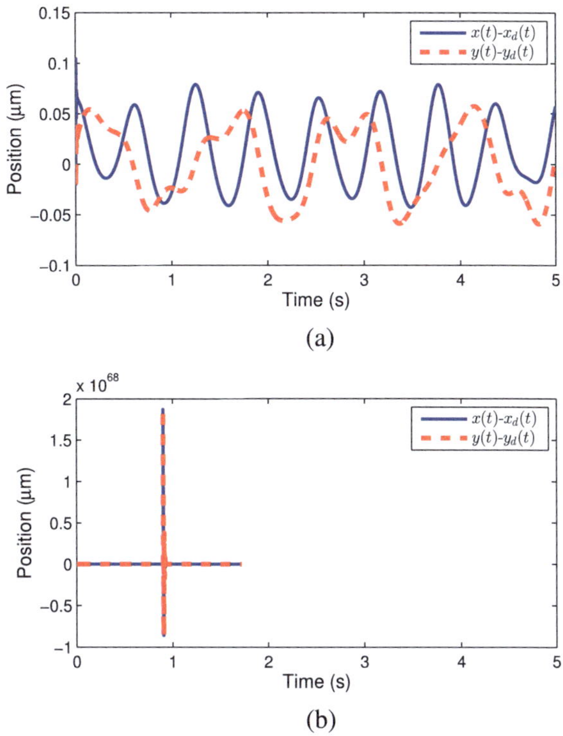

Fig. 12.8 Performance comparison with and without critic NN. **a** Proposed critic-actor controller; **b** actor-only controller using (12.58)

12.5 Experimental Results

This section presents experimental validation of the proposed reinforcement learning-based adaptive control method on a piezoelectric XY nanopositioning stage. The experimental setup, shown in Fig. 12.1, consists of a nanopositioning stage (P-734.2CL, Physik Instrumente), a sensor signal monitor (E-509.C2, Physik Instrumente), and a data acquisition card (PCI-6259, National Instruments). Control input voltages are generated by the 16-bit D/A converter of the PCI-6259 card and amplified by a factor of ten using a high-voltage piezo-amplifier (E503.00, Physik Instrumente). The control algorithm is implemented in LabVIEW software with a sampling rate of 1 kHz. A band-pass filter (0.01–50 Hz) is applied for noise reduction.

Control parameters are identical to those used in the simulation study, with all weight matrices initialized to zero.

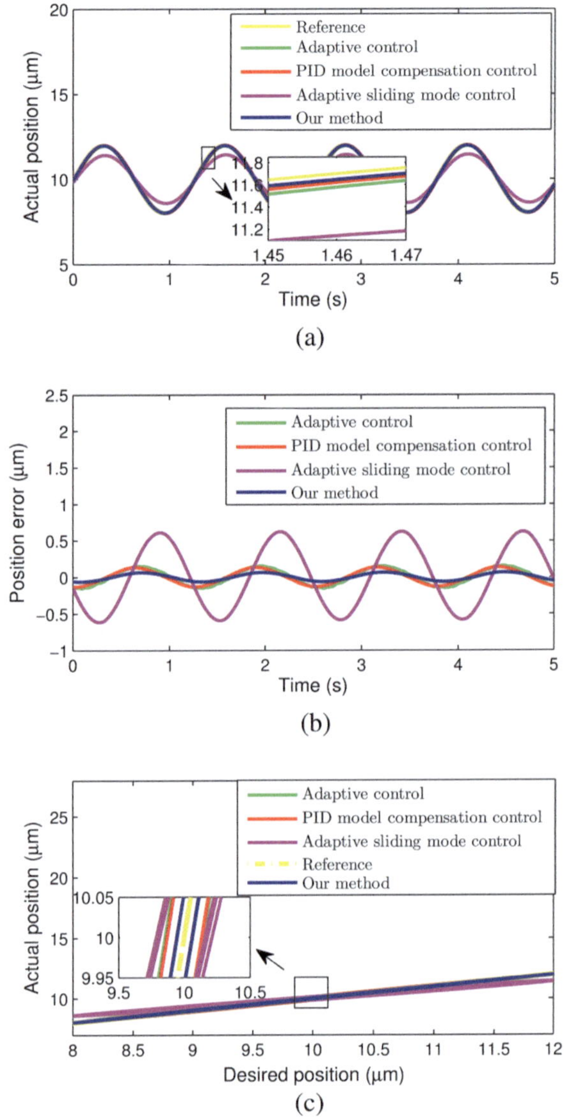

Fig. 12.9 Experimental results for x-axis motion tracking. **a** Position tracking; **b** tracking errors; **c** desired versus actual positions

12.5.1 Robustness Test Against Unknown Nonlinearities

To independently analyze robustness to unknown nonlinearities, each axis is tested separately. For x-axis testing, we set $z(t) = x(t)$ following Remark 12.2, with zero control input applied to the y-axis. Figure 12.9 presents the x-axis tracking results.

Figure 12.9a–b demonstrate that our method achieves the lowest tracking error. The adaptive sliding mode controller [21], which lacks explicit model parameters

$m^{-1}b$ and $m^{-1}k$, produces the largest error. Maximum errors are: 0.064 μm (proposed), 0.135 μm (PID model-compensation), 0.155 μm (adaptive control), and 0.580 μm (adaptive sliding mode). Figure 12.9c shows the desired versus actual position relationship, confirming our method's superior compensation of hysteresis nonlinearity. The performance difference between simulation and experiment arises from unmodeled hysteresis, creep effects, and sensor noise.

Figure 12.10 presents y-axis tracking results. Maximum errors are: 0.066 μm (proposed), 0.137 μm (PID model-compensation), 0.154 μm (adaptive control), and 0.610 μm (adaptive sliding mode). The proposed RL-based adaptive control achieves the best tracking performance, with Fig. 12.10c demonstrating effective compensation of piezoelectric hysteresis. Figure 12.11 shows the smooth, chatter-free control inputs for both axes, confirming the practical viability of the proposed scheme.

We further test the actor-only variant using adaptation law Eq. (12.58) without critic feedback, with identical parameters and initialization as in simulation. Figure 12.12 shows that this configuration leads to instability, while our full method maintains stable performance.

12.5.2 Robustness Test Against Cross-Coupling Motion

We now evaluate the controller's ability to reject cross-coupling effects between the two axes.

Figure 12.13 shows x-axis motion results under y-axis excitation. Figure 12.13b shows y-axis motion at 17.5 μm/s starting at 0.9 s, reaching 8.75 μm before returning to zero at −17.5 μm/s. Despite zero x-axis reference, this motion induces cross-coupling displacement in the x-axis. Open-loop control results in a maximum x-axis displacement of −0.125 μm. The compared controllers produce maximum crosstalk displacements of: −0.090 μm (adaptive control), −0.045 μm (PID model-compensation), −0.036 μm (adaptive sliding mode), and −0.035 μm (proposed). While adaptive sliding mode shows good robustness despite poorer tracking accuracy, our method provides the best overall cross-coupling rejection.

Figure 12.14 shows y-axis results under x-axis excitation. Open-loop control yields a maximum y-axis displacement of −0.038 μm, indicating weaker cross-coupling effects on the y-axis. The controllers produce maximum crosstalk displacements of: −0.027 μm (adaptive control), −0.016 μm (PID model-compensation), −0.015 μm (adaptive sliding mode), and −0.013 μm (proposed). Our method achieves the lowest crosstalk displacement with the fastest transient response.

The proposed control strategy demonstrates effective performance across various trajectories and will be further applied to micro/nano-manipulation systems in future work.

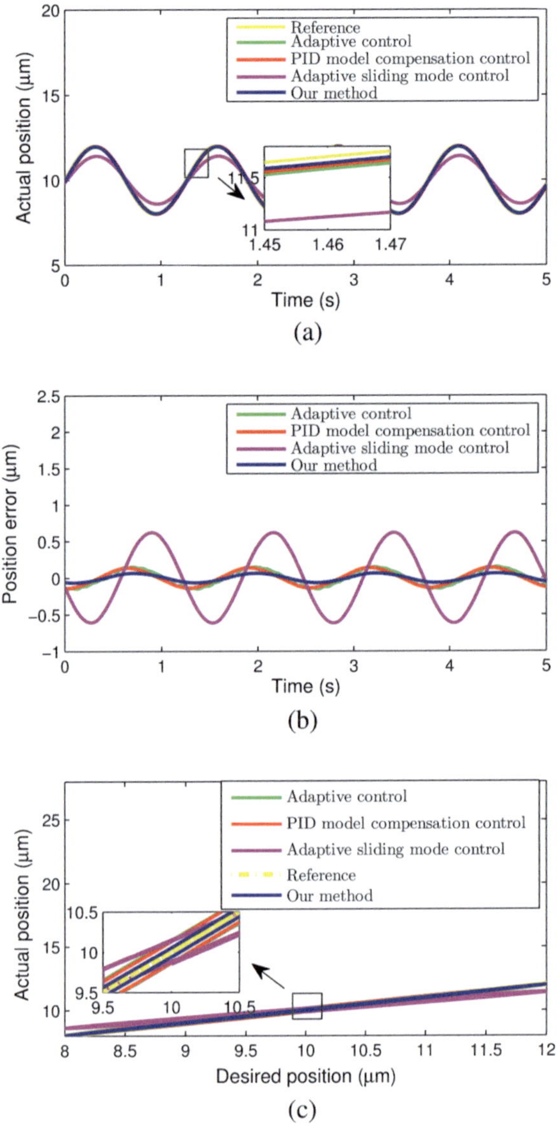

Fig. 12.10 Experimental results for y-axis motion tracking. **a** Position tracking; **b** tracking errors; **c** desired versus actual positions

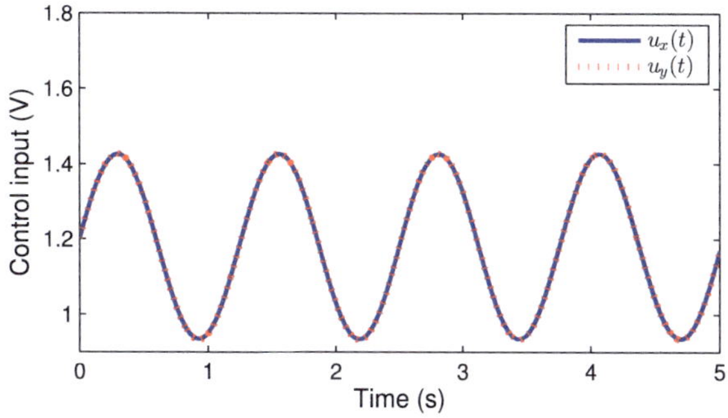

Fig. 12.11 Control inputs generated by the proposed method for both axes

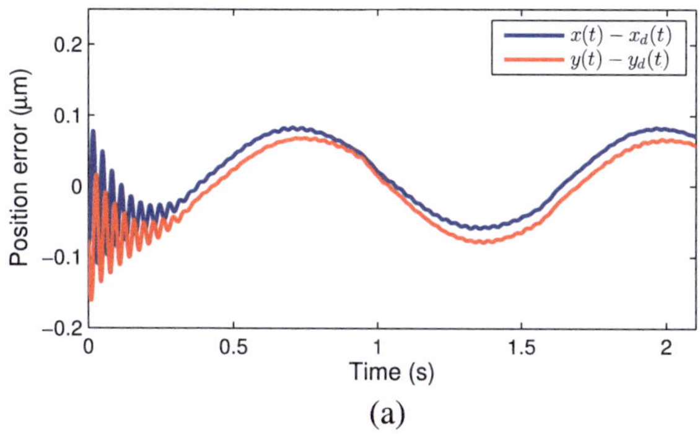

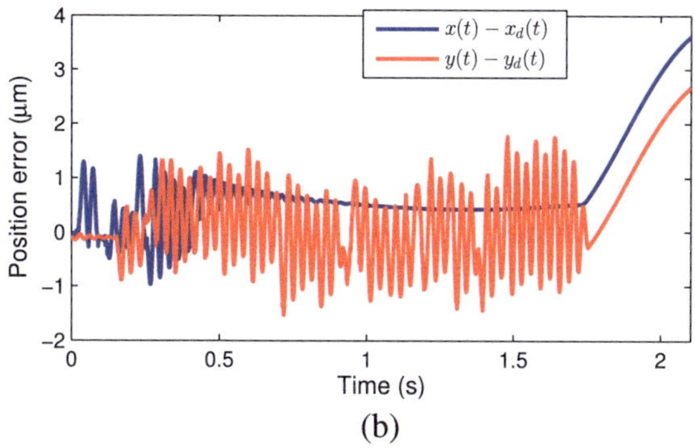

Fig. 12.12 Experimental comparison of two controllers. **a** Proposed critic-actor controller; **b** actor-only controller using (12.58)

Fig. 12.13 Robustness test for x-axis motion under cross-coupling. **a** x-axis position tracking; **b** y-axis input motion

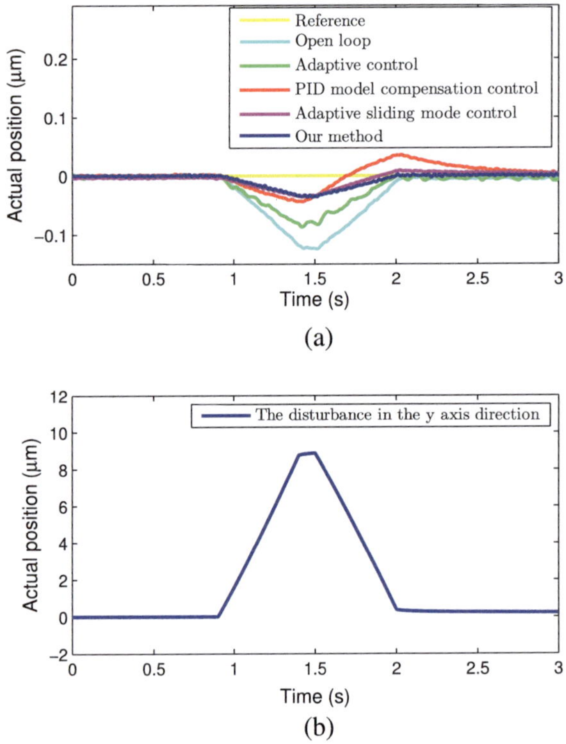

12.6 Conclusion

This chapter presents a novel reinforcement learning-based adaptive control method for piezoelectric nanopositioning systems. The proposed approach employs critic and actor radial basis function neural networks in a dual-network architecture to ensure precise tracking performance. The control strategy effectively compensates for hysteresis and creep nonlinearities inherent in piezoelectric actuators while simultaneously rejecting cross-coupling effects between motion axes, significantly enhancing system robustness.

A key advantage of this method is its adaptive nature: control parameters are updated online without requiring prior knowledge of the hysteresis, creep, or cross-coupling characteristics. The effectiveness of the reinforcement learning-based approach has been rigorously validated through comprehensive simulation and experimental studies on a 2-DOF piezo-driven nanopositioning stage. Results demonstrate superior tracking performance and enhanced robustness compared to existing control methods.

Future work will focus on extending the reinforcement learning framework to address input saturation constraints in nanopositioning systems and applying the proposed control strategy to specific micro/nano-manipulation applications.

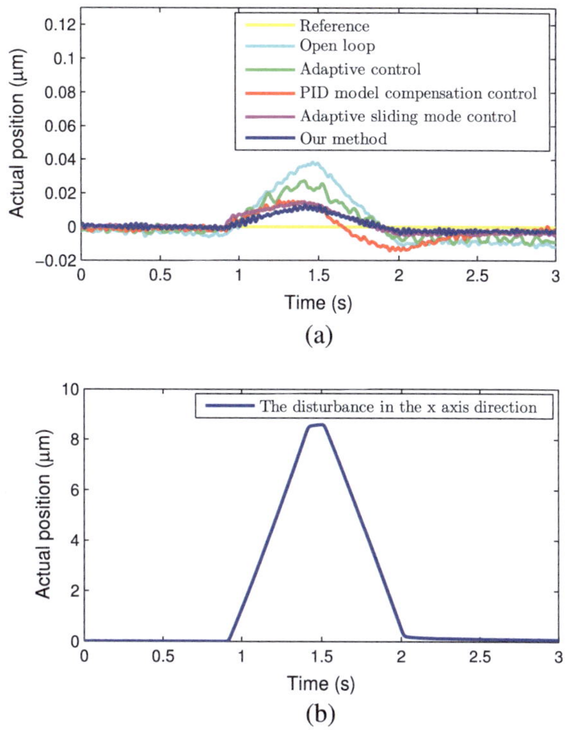

Fig. 12.14 Robustness test for y-axis motion under cross-coupling. **a** y-axis position tracking; **b** x-axis input motion

References

1. Bai, W., Li, T., Tong, S.: NN reinforcement learning adaptive control for a class of nonstrict-feedback discrete-time systems. IEEE Trans. Cybern. **50**(11), 4573–4584 (2020)
2. Cai, K., Tian, Y., Liu, X., Zhang, D., Shang, J., Shirinzadeh, B.: Development and control methodologies for 2-DOF micro/nano positioning stage with high out-of-plane payload capacity. Robot. Comput. Integr. Manuf. **56**, 95–105 (2019)
3. Chen, L., Liu, M., Shi, Y., Zhang, H., Zhao, E.: Adaptive fault estimation for unmanned surface vessels with a neural network observer approach. IEEE Trans. Circuits Syst. I Regul. Pap. **68**(1), 416–425 (2020)
4. Chen, X., Su, C.Y., Li, Z., Yang, F.: Design of implementable adaptive control for micro/nano positioning system driven by piezoelectric actuator. IEEE Trans. Industr. Electron. **63**(10), 6471–6481 (2016)
5. Cui, R., Yang, C., Li, Y., Sharma, S.: Adaptive neural network control of AUVs with control input nonlinearities using reinforcement learning. IEEE Trans. Syst. Man Cybern. Syst. **47**(6), 1019–1029 (2017)
6. Fleming, A.J.: Measuring and predicting resolution in nanopositioning systems. Mechatronics **24**(6), 605–618 (2014)
7. Gao, H., Li, Z., Yu, X., Qiu, J.: Hierarchical multi-objective heuristic for PCB assembly optimization in a beam-head surface mounter. IEEE Trans. Cybern. **52**(7), 6911–6924 (2022)
8. Gu, G.Y., Zhu, L.M., Su, C.Y., Ding, H., Fatikow, S.: Modeling and control of piezo-actuated nanopositioning stages: a survey. IEEE Trans. Autom. Sci. Eng. **13**(1), 313–332 (2014)

9. Guo, X., Yan, W., Cui, R.: Integral reinforcement learning-based adaptive NN control for continuous-time nonlinear MIMO systems with unknown control directions. IEEE Trans. Syst. Man Cybern. Syst. **50**(11), 4068–4077 (2020)
10. Huang, W.W., Hu, C., Zhu, L.M.: Robust repetitive control of nanopositioning stages using the spectrum-selection filter with narrow passbands. IEEE/ASME Trans. Mechatron. **27**(5), 4211–4216 (2022)
11. Kang, S., Lee, M.G., Choi, Y.M.: Six degrees-of-freedom direct-driven nanopositioning stage using crab-leg flexures. IEEE/ASME Trans. Mechatron. **25**(2), 513–525 (2020)
12. Khan, S.G., Herrmann, G., Lewis, F.L., Pipe, T., Melhuish, C.: Reinforcement learning and optimal adaptive control: an overview and implementation examples. Annu. Rev. Control. **36**(1), 42–59 (2012)
13. Kong, L., Li, D., Zou, J., He, W.: Neural networks based learning control for a piezoelectric nanopositioning system. IEEE/ASME Trans. Mechatron. **25**(6), 2904–2914 (2020)
14. Lee, H.J., Woo, S., Park, J., Jeong, J.H., Kim, M., Ryu, J., Gweon, D.G., Choi, Y.M.: Compact compliant parallel XY nano-positioning stage with high dynamic performance, small crosstalk, and small yaw motion. Microsyst. Technol. **24**, 2653–2662 (2018)
15. Lewis, F.L.: Nonlinear network structures for feedback control. Asian J. Control **1**(4), 205–228 (1999)
16. Li, J., Yu, T.: Optimal adaptive control for solid oxide fuel cell with operating constraints via large-scale deep reinforcement learning. Control. Eng. Pract. **117**, 104951 (2021)
17. Liu, M., Ho, D.W., Shi, P.: Adaptive fault-tolerant compensation control for Markovian jump systems with mismatched external disturbance. Automatica **58**, 5–14 (2015)
18. Lyu, Z., Xu, Q.: Design of a new bio-inspired dual-axis compliant micromanipulator with millimeter strokes. IEEE Trans. Rob. **39**(1), 470–484 (2022)
19. Lyu, Z., Xu, Q.: Design and testing of a large-workspace XY compliant manipulator based on triple-stage parallelogram flexure. Mech. Mach. Theory **184**, 105287 (2023)
20. Ma, J., Cheng, Z., Wang, W., Li, X., Zhu, H., Silva, C.W.D., Lee, T.H.: Convex inner approximation for mixed h_2/h_∞ control with application to a 2-DOF flexure-based nanopositioning system. IEEE Trans. Industr. Electron. **69**(2), 1586–1596 (2021)
21. Ming, M., Liang, W., Feng, Z., Ling, J., Mamun, A.A., Xiao, X.: PID-type sliding mode-based adaptive motion control of a 2-DOF piezoelectric ultrasonic motor driven stage. Mechatronics **76**, 102543 (2021)
22. Mitrovic, A., Nagel, W.S., Leang, K.K., Clayton, G.M.: Closed-loop range-based control of dual-stage nanopositioning systems. IEEE/ASME Trans. Mechatron. **26**(3), 1412–1421 (2020)
23. Mohammadzaheri, M., AlQallaf, A.: Nanopositioning systems with piezoelectric actuators, current state and future perspective. Sci. Adv. Mater. **9**(7), 1071–1080 (2017)
24. Motoi, N., Nalbach, M., Ito, S., Thurner, P.J., Schitter, G.: Force-controlled tensile test of collagen fibril by using 2-DOF control system with modeling error compensation. IEEE Open J. Ind. Electron. Soc. **3**, 366–374 (2022)
25. Nguyen, T.T., Nguyen, N.D., Nahavandi, S.: Deep reinforcement learning for multiagent systems: a review of challenges, solutions, and applications. IEEE Trans. Cybern. **50**(9), 3826–3839 (2020)
26. Polydoros, A.S., Nalpantidis, L.: Survey of model-based reinforcement learning: applications on robotics. J. Intell. Rob. Syst. **86**(2), 153–173 (2017)
27. Rakotondrabe, M., Janaideh, M., Bienaim, A., Xu, Q.: Smart Materials-Based Actuators at the Micro/Nano-scale: Characterization, Control and Applications. Springer, New York, Heidelberg, Dordrecht, London (2013)
28. Sarkar, A., Maji, K., Chaudhuri, S., Saha, R., Mookherjee, S., Sanyal, D.: Actuation of an electrohydraulic manipulator with a novel feedforward compensation scheme and PID feedback in servo-proportional valves. Control. Eng. Pract. **135**, 105490 (2023)
29. Tang, H., Li, J., Jia, Y., Gao, J., Li, Y.: Development and testing of a large-stroke nanopositioning stage with linear active disturbance rejection controller. IEEE Trans. Autom. Sci. Eng. **19**(3), 2461–2470 (2021)

30. Tong, S., Wang, T., Li, Y., Chen, B.: A combined backstepping and stochastic small-gain approach to robust adaptive fuzzy output feedback control. IEEE Trans. Fuzzy Syst. **21**(2), 314–327 (2013)
31. Wang, R., Wu, H.: Design and test of a spatial nanopositioner for evaluating the out-of-focus-plane performance of micro-vision. Micromachines **14**(3), 513 (2023)
32. Wang, T., Wang, Y., Yang, X., Yang, J.: Further results on optimal tracking control for nonlinear systems with nonzero equilibrium via adaptive dynamic programming. IEEE Trans. Neural Netw. Learn. Syst. **34**(4), 1900–1910 (2023)
33. Xu, Q.: Continuous integral terminal third-order sliding mode motion control for piezoelectric nanopositioning system. IEEE/ASME Trans. Mechatron. **22**(4), 1828–1838 (2017)
34. Xu, Q.: Adaptive integral terminal third-order finite-time sliding-mode strategy for robust nanopositioning control. IEEE Trans. Industr. Electron. **68**(7), 6161–6170 (2021)
35. Yang, G.: Asymptotic tracking with novel integral robust schemes for mismatched uncertain nonlinear systems. Int. J. Robust Nonlinear Control **33**(3), 1988–2002 (2023)
36. Yang, G., Yao, J., Dong, Z.: Neuroadaptive learning algorithm for constrained nonlinear systems with disturbance rejection. Int. J. Robust Nonlinear Control **32**(10), 6127–6147 (2022)
37. Zhang, C., Huang, X., Yang, M., Chen, S., Yang, G.: Design of a long stroke nanopositioning stage with self-damping actuator and flexure guide. IEEE Trans. Industr. Electron. **69**(10), 10417–10427 (2021)
38. Zhang, K., Zhang, H., Mu, Y., Liu, C.: Decentralized tracking optimization control for partially unknown fuzzy interconnected systems via reinforcement learning method. IEEE Trans. Fuzzy Syst. **29**(4), 917–926 (2021)
39. Zhang, X., Xu, Q.: Design, fabrication and testing of a novel symmetrical 3-DOF large-stroke parallel micro/nano-positioning stage. Robot. Comput. Integr. Manuf. **54**, 162–172 (2018)
40. Zhang, Y.K., Sun, C., Pan, B.: Guidance method of planetary soft landing with GPS model-based reinforcement learning. Flight Control Detect. **4**(5), 34–43 (2021)
41. Zhu, W.L., Zhu, Z., Guo, P., Ju, B.F.: A novel hybrid actuation mechanism based XY nanopositioning stage with totally decoupled kinematics. Mech. Syst. Signal Process. **99**, 747–759 (2018)

Index

Q. Xu, *Piezoelectric-Driven Flexible Robotic Microgrippers*,
https://doi.org/10.1007/978-981-95-4685-5

MIX
Papier aus verantwortungsvollen Quellen
Paper from responsible sources
FSC® C105338

If you have any concerns about our products,
you can contact us on
ProductSafety@springernature.com

In case Publisher is established outside the EU,
the EU authorized representative is:
Springer Nature Customer Service Center GmbH
Europaplatz 3, 69115 Heidelberg, Germany

Printed by Libri Plureos GmbH
in Hamburg, Germany